To Randy & STACY
Regards from the author.

RACE TO THE SEA

The Autobiography of a Marine Biologist

DR. DAYTON LEE ALVERSON, PHD.

iUniverse, Inc.
New York   Bloomington

Race to the Sea
The Autobiography of a Marine Biologist

Copyright © 2008 by Dr. Dayton Lee Alverson, PhD.

iUniverse books may be ordered through booksellers or by contacting:

iUniverse
1663 Liberty Drive
Bloomington, IN 47403
www.iuniverse.com
1-800-Authors (1-800-288-4677)

ISBN: 978-0-595-48680-9 (pbk)
ISBN: 978-0-595-48778-3(cloth)
ISBN: 978-0-595-60776-1 (ebk)

Printed in the United States of America

CONTENTS

PREFACE

I began this book to provide my family with a historical accounting of my life and subsequently decided my story was worth sharing with the public. The book, which spans the early twentieth to the twenty-first century, describes my childhood, experiences in World War II, and life as a marine biologist. I realize that my recollections of these events may have dimmed with the passage of time and thus may not be consistent with the memories of my peers and family.

My experiences in marine science and natural resource management coincided with the great "race to the sea" begun by many of the world's coastal nations following World War II. Ocean exploration was in a state of rapid transition, and modern technology greatly expanded the development of global fisheries. These developments were followed by significant changes in conservation and concepts of sustainable yields. I was active and involved in these events for more than fifty years, from my early belief that the ocean could easily feed the world's hungry (1950s) to my current concern about the consequences of overfishing and the ecological changes caused by fisheries all over the world (1990s–forward).

ACKNOWLEDGEMENTS

I am indebted to a number of individuals who contributed in various ways to the completion of the book. My wife, Ruby Marie, edited the entire manuscript, reminded me of events that I had overlooked, and frequently pushed me back to the computer so I could finish my work. Mr. and Mrs. Clem Tillion of Halibut Cove, Alaska; Mrs. Sharon Parks, of Natural Resource Consultants, Seattle, Washington; Dennis Petersen, Seattle; Rudy Petersen, Seattle; Mrs. Herbjorg Pedersen and Dr. Edward Miles, University of Washington, Seattle were among the many who read and made suggestions for improving the book. Finally, I thank Lisa Kennedy, who undertook the final editing of *Race to the Sea*, and my wife's cousin, Ann McGraft, who retyped the entire book when my computer's hard drive failed.

Chapter I
The Opportunity to Live

Dayton Lee Alverson, I have been told, was born to George D. and Edith M. Alverson at the naval hospital in San Diego on October 7, 1924, smack in the middle of the roaring twenties. I haven't looked at my birth certificate, but it seems to me that the biomass was about eight pounds, a weight that would grow substantially over the years. As my early formative years are well beyond my capacity to recall, the following notes on my early life history are mostly the product of hearsay.

It is not my intent to develop a genealogy of the Alverson family, but a little introductory material will help set the scene for what is to follow. My father, George Dayton—now you know the origin of my first name—was raised with three brothers on a farm just north of Detroit, Michigan. At the age of seventeen, he ran off to join the United States Navy. After completing radio school, he became a Morse code operator and was shipped to San Diego, California. My mother, Edith Margaret Gray, was born in California, but she spent considerable time in Florida as a young girl before returning to her home state. There, she lived with two sisters and a brother on the slopes of Mount Wilson, east of Los Angeles.

From a historical perspective, not a great deal is known about when the Alverson family migrated from Europe to the United States, but we do know that an Alverson stood guard at the Capitol Rotunda when President Lincoln's body was on display. We also know that five

Alversons, including a George and a Frank (my grandfather's name), are listed among veterans of the Revolutionary War. We can't be sure, but it seems reasonable to believe that they formed the roots of our Alverson clan in America.

Beyond Recall

I don't know a great deal about my parents' lives prior to my birth. Family lore states that my father worked on the farm where he was born from the time he was old enough to lead the horses and plow the fields. He was an excellent student in high school, joined the navy at the end of World War I, and subsequently was transferred to California where he met and married my mother.

My mother was quite an athlete in her younger years, a gifted gymnast who could do the "iron cross" when she was in her early teens and an excellent long-distance runner. To this day, she holds the women's record for the Mount Wilson Run, a feat of some importance because she was only nine years old when she ran the race. Before she died, she gave me her gold medal as well as a copy of a testimonial about her that was published in Ripley's "Believe It or Not"—a feature in many of the nation's major newspapers well into the 1950s.

Grandfather Frank Alverson started his early life in Michigan, first as a door-to-door salesman of household goods and later as the owner of a carriage store. He finally became a farmer, which carried him through the latter part of the 1800s and early 1900s. My mother's father had been a professional photographer throughout his life, living in Florida before moving to California at the turn of the century.

Shortly after my birth, my father, a non-commissioned officer in the U.S. Navy, was transferred to the Philippines. My mother, brother (who was fifteen months older), and I were put on a passenger ship, the *President Van Buren*, and shipped off to Manila via Kobe, Japan, and Shanghai, China. We arrived in Kobe on May 1, 1925, and in Shanghai four days later. A rebellion was under way in Shanghai, similar to the Boxer Rebellion that had taken place several years earlier. At any rate, the U.S. Marines were called to protect the *President Van Buren* anchored during her stay off the Bund. The ship's passenger list included Edith Alverson, traveling with son Frank and "infant." I wasn't identified by name but, at the age of seven months, I had gone to China. Not an inkling of this early venture across the Pacific remains in my memory. When we arrived in the Philippines, we were billeted in a small town just to the north of Manila called Los Banjos (which I visited again some seventy-two years later).

While we were in the Philippines, I became allergic to cow's milk, so my parents bought a goat, which became the source of my nutrients for the remainder of my stay. The only other story from our Philippines expedition is that a large boa constrictor tried to make a meal out of me while I was in a crib in the yard, but was thwarted by the local gardener. Some believe that the world would have been a better place had nature taken its course, but I suspect they were all snake lovers. Regrettably, I have no memories of the trip back to San Diego, but I believe that I was about three years old.

The Beginnings of Recall

Sometime in the next year or two, events began to make lasting impressions somewhere in my cerebral cortex. I have vague memories of beginning lifelong friendships with Oscar Olson and Gene Ryan, two youngsters living near our small, white house in east San Diego. Among the first inputs into my hard disk are scrambled images of playing in the empty lot between Oscar's house and mine. The strongest of my early memories, however, is having my tonsils taken out at the navy hospital in San Diego and returning home in an ambulance. I am still puzzled at why this memory remains so vivid. I can remember nothing about going to the hospital or recuperating after the tonsillectomy. At various times in my life I have tried to put these faded memories into focus, but they continue to lack detail.

Several months after we returned from the Philippines, my father took me to visit the battleship *California*, which was anchored, most likely, in San Diego Bay. She was one of the first memories etched deep in my brain. While exploring this magnificent ship, I managed to get lost within a seemingly endless maze of companionways, ladders going between decks, and hundreds of sailors in white hats moving in all directions. Panic set in, followed by high-decibel screams. A friendly sailor took me by the hand and escorted me to the galley, where I was given a bowl of ice cream. This eased my fears of being permanently separated from my family, and my father soon came to my rescue.

I have a strong memory of going to the corner drugstore with my mother, where she bought me a large vanilla ice cream, much to my delight. Unfortunately, as we were returning home, I got angry for some unknown reason, and tossed the cone on the ground. Alas, Mom refused to buy me a new cone.

In undertaking this endeavor to relate my experiences to younger generations, it became evident that my recall of events before the age of four is extremely fuzzy, due partially to the passing of time and partially to rusty memory cells. The combination of these factors has made it impossible for me to differentiate between what I believe to be my own recorded versions of the past and things that were told to me by my family.

A few events from the ages of four and five left rather clear impressions. I remember my mother and father listening carefully to the radio to hear each update of Charles Lindbergh's 1927 trans-Atlantic flight. Somewhat later, my father took my brother and me down to Lindbergh Field, situated on San Diego's waterfront, to look at the *Spirit of Saint Louis*. The plane sat on the ramp just outside the Ryan Aircraft hangar. I looked in awe at the large (in the context of the 1920s), rather peculiar-looking, high-winged, single-engine plane with almost no forward view. We were lucky to see a small passenger plane fly over the city and touch down on the south end of the runway. It was a big day in the life of a youngster.

I also have happy memories of visiting my maternal grandparents, who lived in high mountain areas, rich in opportunities for children. After settling in the Sierra Madre Mountains in the early 1900s, my grandmother and grandfather built a small store on one of the popular walking trails to Mount Wilson. My mother used to take my brother and me on hikes, sometimes on the back of a donkey, up to a little store called the Halfway House and/or the Little Gray Inn. After rest and refreshments, we struggled up the trail to a small clear-flowing stream where we played for a few hours before making the journey home. We enjoyed fishing, hiking, and horseback riding, and listening to the endless stream of people who visited the Little Gray Inn.

My paternal grandparents lived on a farm in Michigan, and I would not meet them until I was six.

On the Road

These carefree formative years soon would be disrupted, as my father's duties with the U.S. Navy were about to forge a new set of adventures that would alter my perspective of the world forever.

In 1930, when I was six, my father was transferred to Tatoosh Island, off the northwest corner of the United States at the entrance to the Strait of Juan de Fuca. About a month before we were scheduled to leave, Father decided that he should introduce his wife and children to his parents in Michigan. He brought home a map and penciled out a route to Emily City, about forty miles north of Detroit. While Frank and I weren't that familiar with American geography, we knew it would be a long trip. Still the vacation that covered almost six thousand miles can be summarized in a few short paragraphs.

We left San Diego at about 5:00 a.m. Frank and I scrambled into the back seat of our 1928 Chevy sedan, and with my father at the controls, we headed east over the mountains. We drove through the dry, hot, desert country, climbed up to Flagstaff, and then down into the high desert of Arizona. The two-lane road took us from just east of San Diego across the arid parts of Arizona and New Mexico. Parts of the road were unpaved and bumpy and, as I remember it, some sections in the desert were made of wooden planks. My brother and I enjoyed the areas where lofty buttes climbed above the desert floor and imagined Indians watching our passage. When we entered New Mexico, we began to see Navajo hogans and occasionally Indians selling baskets and other artifacts along the road.

Texas seemed flat, endless, full of sage and grazing cattle, hot, and, at times, boring (regards to the Texans). There were, however, interesting and fun parts during the trip. It was hot and humid as we entered Arkansas. We grew increasingly uncomfortable and opened the car windows, but the hot outside air did little to improve our attitude. Many of the farm houses were unpainted and run down, but a few green fields stood out. We were moving north along a narrow highway when we saw a large hand-painted sign advertising fresh fruit and melons for sale. We urged our father to stop. An elderly black man with a silver-gray beard came to the car and asked if he could help us. He had some chilled watermelons in an underground storage area. We purchased two medium-sized melons for twenty cents and moved to a picnic table under the shade of a tree to consume the bounty. Perhaps it was because it was extremely hot and we were all parched, but the watermelons were the best that I have ever eaten. They improved the family's mood and to this day, some seventy years later, every time I eat a watermelon, I remember our feast on that hot Arkansas highway.

As we traveled north into Missouri, the country improved in appearance. We drove to a state park in the mountains where water gushed from the sides of rocks to form a small river that could be fished for trout. The farms in the surrounding area looked relatively prosperous compared with those in the western and southern parts of the country. We continued into Illinois, skirting Chicago, driving through a corner of Indiana, and finally up into Michigan to my grandfather's farm near Emily City. The Alverson farm was approximately thirty acres of land that was mostly flat but also slanted down toward some trees and a small pond at one corner of the property. The farm was planted mostly in corn except for a small garden near the house. The home was a two-story structure with the bedrooms upstairs and the kitchen, living room, dining room, and pantry downstairs. A hand pump behind the house supplied the water, and the bathroom was a two-seater outhouse in the backyard.

My brother and I spent a great deal of time playing in and around the barn, watching the farm animals and the hand milking of the five or six cows. We walked the cows to the pasture with my grandfather and rode with him to town on a wagon pulled by a large horse. I suspect we had a wonderful time but I don't recall much about our first trip east. When we left, Frank and I were in big trouble because we had buried a chicken up to its neck in the dirt. I still don't know what possessed us to do it. There was, however, no question about my father's feelings on the matter as he used his belt to whip our behinds. I guess we made a hell of an impression on our grandparents because my relatives remind me about the event every time I return to that part of the country.

We spent the better part of the week at the farm and visiting other Alversons in the region before we headed home. I can't remember the exact route, but we went south and east, swung through Cumberland Gap, and then headed west. On the last day in the desert, the miserable heat that was relieved only by the gallon of lemonade that Mom had made for the trip.

A few days after we arrived in San Diego, we began our journey north, taking a passenger steamer, the *Ruth Alexander*, from San Pedro, California, to Port Angeles, Washington. After the first day out, the weather turned nasty, and a lot of people became ill. The ship smelled of vomit, and the crew was kept busy cleaning up the mess. My youthful strength and vigor, however, refused to be overcome by the pitching and rolling ship. I never missed a meal in the dining room where there seemed to be an endless supply of everything a young child could wish to eat.

In Port Angeles, we boarded a small coast guard vessel and sailed via the Strait of Juan de Fuca to the Makah Indian village at Neah Bay. My parents, brother, and I, as well as our belongings, made up the boat's entire cargo, and I soon realized that we were not on our way to paradise. As we rounded the breakwater off Port Angeles and headed west out of the strait, it started to rain and blow; all I could see was the dark evergreen forest hugging the mountainsides on the north coast of Washington. The small vessel rose and fell sharply in an oncoming sea, and it became increasingly apparent that we were moving steadily away from the life and friendships I had known in San Diego toward a very uncertain future. I began to feel sad and lonely and isolated myself from the family discussion regarding "the great new adventure."

We were on board the coast guard vessel for about six hours before we pulled into Neah Bay, located just inside Cape Flattery at the northwest tip of the United States. At that time, there was no road to Neah Bay. Our belongings were transferred onto a smaller open boat that took us out to the island several miles to the west. The Makah village was shrouded in fog, and a light drizzle was falling. There was nothing there to lift my spirits. We moved out past Wadda Island, and then we saw our future home, which stood like a fortress at the entrance to the strait. Tatoosh Island was perhaps less than a mile across at the top; a rift split the landmass into two sections, which were connected by a small sandy beach that served as the landing area for our small boat. Our belongings were set on the beach, and several men helped us to load them on carts. At the base of a cliff, a powered crane hoisted the luggage to the top of the island. We walked up on a boardwalk that clung to the side of the cliff.

The island was a base for a small naval radio and direction finder station; a large lighthouse guided ships entering the strait en route to ports in Puget Sound and Canada. The top area of the larger of the island's two landmasses was rather flat but had several deep crevasses that dropped down to water-filled caverns formed by the endless pounding of the surf. In addition to the radio station and its two towers and the lighthouse (also the home of the lighthouse keeper and his family), there were a series of small cottages on the southwest tip of the island that housed the men who managed the radio station. The island's children were educated in a one-room schoolhouse.

Tatoosh Island, late 1920's

Lee and Frank Alverson, Tatoosh Island, 1930

Tatoosh Island lighthouse, late 1920's

Much to my surprise, there were eight other children on the island; my brother and I increased the total to ten. As most of the others were about the same age , my fear of doom and isolation rapidly abated. During our first weeks on Tatoosh, my father and mother helped us to investigate the entire island. To our delight, there was a small running stream in which we could play and get filthy. Rocks that ran from the base of the island to the northeast provided an observation area for bird watching, and a series of trail down the side of the cliffs offered access to excellent sport fishing for rockfish, halibut, flounders, and lingcod. The small beach between the two landmasses was a great area for picnicking, beachcombing, and occasionally conversing with salmon trollers who came ashore to share magazines and sea stories with the islanders.

Mail was delivered to the island once a week by a small coast guard boat, weather permitting, and the island had more than enough challenges to keep a young lad busy. I started school—one teacher handled the first through fourth grades—which frequently required walking in the howling winds that constantly battered the island. It was here, at the school house, that Frank and I saw our first silent movie, a Charlie Chaplin comedy. My father often took my brother and me down the cliffs to fish for red snapper, and we had the chance to explore the bird nesting areas that were plentiful on the rocky areas at the base of the island. Although this now probably would be considered politically incorrect, we collected seagull eggs, which were used by my mother to make various baked goods.

I remember very little about our house except that it had a coal stove on which my mother prepared our meals; it also warmed our home. On stormy nights the house creaked, and the wind always seemed to last through the night. The winds and heavy swells that slammed into the base of the island intimidated me. In my mind I visualized the island sinking into the sea, but these fears disappeared when the sun rose or filtered its way through the clouds. On the occasional day when the sun came out and the winds died down, we would scamper over the island searching for mushrooms, which also flourished in the grassy fields surrounding our house, or go down to the small beach to build sand castles and fish with our father. Casting off the rocks, we caught a multitude of red, yellow, brown, and black rockfish (which the locals called sea bass).

We arrived on the island in the early fall and stayed through the winter. Sometime in the early spring of 1931, I awoke to a great excitement in our house. During the night, a steamer inbound for the strait had gone ashore on the rocks just south of the cape. The Coast Guard and natives from Neah Bay had rescued many of the ship's crew, but the steamer *Skagway* remained hard aground. Several days later when the weather improved, Frank and I went with our father by small boat to the grounded vessel, which we spent explored for several hours. Not much was left on board, however, since what the sea hadn't damaged had been stripped and carried away by earlier visitors.

Later in the spring we had the chance to visit the Indian village in Neah Bay, which on a small outboard motor boat was about an hour from the island. I was a little apprehensive, but the trip turned out to be a great adventure. Most of the villagers were involved in fishing, although a few female basket weavers still practiced their trade. There were a number of small fishing docks and a trading store named Washburn's in the bay. My mother and father spent time in the trading post buying baskets and native artifacts and restocking our fishing gear.

The time on Tatoosh passed quickly, and by summer's end my father received orders to relocate to Bremerton, Washington, which represented a big step back into the modern world. We lived beside a golf course within the Bremerton Naval Shipyard that was edged with firs and some hazelnut trees, whose nuts Frank and I collected in the early fall. In Bremerton my parents gave us our first pet, a cat that we named Andy. He was a smart critter that we trained to do a number of tricks, including shake hands, roll over, sit up, and jump over our outstretched arms. He also would follow us into the woods or to search for golf balls in the grass on the golf course.

The horse was rapidly disappearing as a mode of common transportation, and a network of roads was connecting the country from east to west and south to north. Seeking to become fashionable, my parents purchased a 1930 Chevy, and we began to explore the surrounding country. We traveled south to Shelton, north to Port Ludlow and around the bay to Port Orchard. We stopped many times to look at streams filled with spawning salmon, but that was more than sixty years ago and how the world would change.

During our short stay in Bremerton, I met a boy my age who lived on the other side of the golf course, and we became friends. One day, however, we argued over some now-forgotten, stupid issue. This led to a skirmish, and my friend hurried home, crying. I suddenly realized that I had lost my only friend. Tears came to my eyes and I wanted, in the worst way, to recapture the friendship.

Sometime in the spring of 1931, we were on the move again, as my father was attached to a ship stationed in San Pedro. We packed up our Chevy and headed south—cat and all. The road south wound endlessly through the heavily forested areas and past Tacoma. We crossed the mountains of southern Oregon in heavy rain, with mud and water covering much of the road. After two days of drudgery, we made it to Redding, California, and stopped at a small motor court. It was evening but the sun was out, and my brother and I went for a walk. Our cat must have followed us, but we were unaware of its presence. The next morning Andy was nowhere to be found and, after a lengthy search, our first pet was ultimately left to forage for himself. I hope that he was adopted by a friendly family and not hunted down by a hungry coyote.

When we arrived in Southern California, there was a great deal of talk about the Depression and the fact that thousands of people were out of work. A new politician named Roosevelt suggested that the government should initiate programs to assist the unemployed and also repeal Prohibition. We moved into a small house in a town near San Pedro called Wilmington. Our home was just the street from a large pepper farm that seemed to stretch endlessly toward a grouping of small trees. A large wooden barn near the farmhouse appeared ready to collapse, but it served as a great gathering place for the neighborhood kids, including the farmer's children. Some of the mothers, including my own, expressed concern that the barn wasn't safe, but to little avail. We all joked that the next Los Angeles earthquake would cause the barn to fall to the ground in a heap.

During our stay in Wilmington, I made it to the second grade. It was there that I first came into contact with the type of bully that is present in all schools. Somehow I managed to annoy one of the third graders, and I became a "marked boy." This aggressive lad promised to beat the stuffing out of me after school, so for the next week or so I stuck close to my older brother as a practical solution to my personal national defense problem.

My memories of my life to age seven are a series of snapshots that form a not-too-cohesive collage of my path on earth in the early part of the twentieth century. I have a much better recall of the events that followed our time in Wilmington.

We had hardly settled into our new neighborhood when my father, who had been promoted to chief petty officer, was transferred to the Hawaiian Islands. He left first via a naval transport, and we were to follow on a private cruise ship. I had never been overly enthusiastic about hopping around the country like a nervous frog, but the assignment to Hawaii seemed to indicate a positive turn.

By the early months of 1932 our family had formed rather close bonds with some of my mother's relatives, my aunt and uncle along with my two cousins, Katherine Jo and Florence May. They drove us to Terminal Island, where we boarded the Matson liner *Malolo*. We took our luggage to our stateroom, which had a nice porthole view. We went topside, and shortly thereafter visitors were called to leave the vessel. A band was playing, and streamers were flying from the side of the ship; the ship's horn let off a couple of loud blasts, and my brother Frank and I thought we had been elevated to the level of kings. My mother, brother, and I stood on the deck waving to our relatives until they faded from sight.

The ship seemed gigantic compared to the coastal steamer we had taken to Washington and the small coast guard cutter that had hauled us out to Tatoosh Island. It had several elegant dining rooms, entertainment lounges, and, much to our surprise, a swimming pool located on a lower deck. Several decks had open sitting and walking areas around the periphery of the main cabin, and on in the open area around the stern of the vessel, we could play a variety of games. Inside the main cabin, large stairwells ran between decks. Halfway up one stairwell, a group of slot machines successfully baited many of the passengers; it was always a crowded area.

After a day on board we quickly learned the daily routine and the expected protocol. My brother and I were quite taken with the pomp and ceremony surrounding meal times. A young man with a small xylophone or dinner bell would walk around the open decks and inner companionways, playing a brief tune and announcing that dinner, lunch, or breakfast was being served. My mother would make us dress up for all the evening meals, but we didn't mind because the great diversity and quality of food was worth the effort. Meats, salads, soups of all types, as well as fruits, vegetables, breads, and an endless supply of desserts were ours to order. The waiters dressed in tuxes and referred to my brother and me as "Master Frank" and "Master Lee." This enhanced our self-esteem and probably increased the tip Mom left the waiters.

On the second day out, Frank and I went to watch other passengers playing the slots. It was mid-morning, and there were only a few couples at the machines. A tall, good-looking man and a beautiful young lady were at the quarter slot machine, which paid big money in those days. After watching for about twenty minutes and seeing the machine spit out coins in response to three plums and later three oranges, a lady standing next to us suggested that we ask the man playing for his autograph. Since neither my brother nor I recognized the guy, we asked the obvious question. The lady looked at us like we were uncultured little urchins (which we were) and said that he was an upcoming movie star, Fred MacMurray. The name didn't help much because we had seen only one or two movies at that point.

Nevertheless, I scampered down to our stateroom, found my autograph book—which every "in" kid had in those days—and ran back to the micro casino. When Fred MacMurray had had his fill of the one-armed bandits, I requested his autograph and he politely provided it. He was the first celebrity to sign my book and, as it turned out, the only one. He had just been married, and he and his wife were on their honeymoon. I have kept the book for more than sixty-five years, and Fred's name is still there, along with those of a number of school friends, cousins, and my mother. We saw the MacMurrays off and on throughout the cruise, but it was only some years later, after he became a big star, that I recognized that the signature was a conversation piece.

The next day my mother took Frank and me to the game area, where passengers were gambling on a horse race. The gamblers moved small, wooden, colored horses along an oval track painted on the deck in response to a series of dice throws. Mom allowed us to bet on one game each. I went with the red horse and, after a tough race, it won, earning me $27, an absolute fortune for a young lad during the Depression years. I used the money wisely, spending it on candy, cheap souvenirs, and playing the slots when no one was around. Thus, my great fortune was depleted well before we reached Honolulu.

CHAPTER II
ADVENTURES IN PARADISE

We arrived in Honolulu five days after leaving San Pedro. It was a beautiful warm sunny day, the band was playing at the Aloha Tower, streamers were flying from the side of the ship, and young Hawaiian lads were diving for coins passengers threw from the decks of the *Malolo*. We thought we had arrived in paradise. Our time in Honolulu was very short, however, as we were taken immediately to a small, seagoing tug called the *Royal T. Frank*. There, the four of us crammed into a small stateroom. I had always thought that Honolulu and Hawaii were synonymous, but I was about to learn a little geography.

We sailed from Honolulu Harbor just before sundown and headed south past Diamond Head and Koko Head, dormant volcanoes of Oahu. The *Royal T. Frank* was about 110 feet in length with very little freeboard and none of the amenities of the *Malolo*. The weather rapidly deteriorated as we turned into the channel between Molokai and Oahu. Within an hour we were in heavy weather, and the boat was heaving, bucking, and being tossed about like a cork. Blue water broke over the bow and swamped the deck. We were told to stay in our cabins. As we turned south along the west side of Molokai, although the weather abated somewhat, it was still very rough, and my brother vomited continuously. My father was able to escort me to the galley where I ate lavishly, the only passenger who wanted to take part in the evening meal.

The next morning the tug was riding easier, running before a gentle swell. My brother and mother remained in the comfort of their beds. Looking through the porthole, I could see the sugarcane fields that dominated the coastline and the occasional sugarcane mill above the sharp cliffs that made up Hawaii's Hamakua Coast. My father took me down to breakfast. From the galley, he pointed out the swells breaking over a distant breakwater—which, he explained, was the outer protection of Hilo Harbor—and added that we would soon arrive at our destination on the Big Island, Hawaii. When we entered the bay, I saw a ring of coconut palms circling the waterfront and sugarcane fields above the small town. A sugarcane mill belched steam on one side of Hilo, which was nestled close to the harbor and seemed to run up the hillside behind the bay. The protruding Hamakua Coast on the north and a long arching breakwater to the south formed the harbor. Our voyage to paradise came to an end as the boat was tied to one of the piers. I'm sure my mother and brother did not enjoy the trip from Honolulu, and even I was pleased to be on terra firma again.

We soon settled into a small two-bedroom home located several hundred feet from where the breakwater made landfall. The area was rich in lush tropical vegetation, including coconut palms, mango, banyan and banana trees, guava bushes, and vines. Two large radio towers, several hundred feet in height, dominated the landscape. Between them stood a building with radio equipment that was operated by about half a dozen Navy radiomen, who lived in several small homes scattered around the grounds. Adjacent to and just to the east of the base, a small semi-enclosed bay had formed between the breakwater and the main shipping docks. A grove of evergreens mixed with coconut palms grew between the shoreline and our house. Frank and I would spend the next three and a half years exploring and establishing friendships in this enchanted wonderland—the bay, breakwater, docks, and tropical flora around our new home.

Settling In

We had just moved into our Hilo home when we learned that a devastating earthquake had hit the Los Angeles and Long Beach areas. The radio reports suggested that our former home just east of San Pedro had been hit hard, and we wondered how our friends had fared. My father found out from our former neighbors that most of the homes on the block had toppled from their foundations, and serious cracks had developed on the sides of many of the stucco homes, including our old house. The old wooden barn that had been our favorite play site survived the major shake-up without any damage.

A crew of about five men, several of whom had children about our age, manned the Hilo navy base at the time we arrived—a wonderful opportunity, we thought, to make a few new friends. We had hardly been in Hilo a month, however, when the navy shipped out everyone except my father, who was to stay and maintain and operate the station on his own. Overnight, our new friends were gone. Like everyone else in the early years of the Depression, the military was forced to tighten its belt.

Our house was hardly a stone's toss from Hilo Bay and the beach, so my mother began to teach us how to swim. The closest shoreline was, for the most part, a steep rocky beach that fell off quickly into deep water. However, near the base of the breakwater was a small sandy beach where my mother allowed us to play and swim. We were soon joined by a number of local Hawaiian children who all swam like Johnny Weissmuller, making us feel very inadequate. This made us strive even harder to become self-reliant in the water. Once we could manage a few strokes without touching bottom, we began to make short sorties into the deeper water and back to the beach. After several days we gained confidence and I, at least, assumed that there was little left to learn.

To show off my swimming prowess, I extended my offshore swims farther and farther. Everything went fine until I returned to the shallow water and put my feet down on what I thought was the sand below. I had misgauged the depth and the bottom was about a foot-and-a-half below my extended toes. Panic set in, and I began to flounder. My brother and our friends tried to rescue me, but they couldn't get me to a solid footing. My mother watched for a few seconds, saw that I was getting into increasing trouble, and then waded out fully clothed, grabbed me by the hair and pulled me into shallow water.

Needless to say, my area of play at the beach was restricted for the next week or so, but within a short time Frank and I became accomplished swimmers. Soon we were swimming across the bay, a distance of two hundred yards; then out to the "pile of rocks" near the breakwater, about a quarter of a mile from shore; and then almost anywhere in the inner bay area. At that point, my mother decided that we could swim without parental oversight, and we were allowed to do whatever our Hawaiian friends chose to do in the water. Of course, our parents didn't know that this included swimming on the offshore side of the breakwater, where the surf pounded the rocks from which we dove into the water. Corals and spiny sea urchins were plentiful on the seabed, and we occasionally saw sharks swimming farther offshore.

Our move to Hilo required my brother and I to return to ground zero regarding friends, something to which we had become accustomed. Although we had been to Asia when we were very young, we had always lived in predominantly white communities and had never really had close contact with any other racial group. As far as I can remember, my parents never discussed any aspect of racial issues in either a positive or negative sense. If we had parentally ingrained racial biases, I was unaware of them.

The small naval enclave at the end of the Hilo breakwater was next to a Hawaiian land-grant area. As such, all of the residents were required to be of Hawaiian lineage, and thus the law determined the racial makeup of our neighbors. Frank and I soon learned that Hilo's residents included Hawaiians, Japanese, Chinese, Portuguese, Filipinos, other Asians, a few northern Europeans, and mixes of all the above. The islands were touted as a harmonious melting pot of the world's races, but cultural and socio-economic differences both encouraged racial bonding and created barriers between the dominant ethnic groups. Fortunately those barriers were often less confining for children than adults. My brother and I quickly became known throughout the area as the two "haoles" (white kids or foreigners) who lived near the breakwater. To some, this likely was a parochial putdown, but to others it was just a way to identify the new kids on the block.

Our first friends were Buddy Terrel and Dede Puehow. Buddy was half-Hawaiian and half-Irish while Dede was a full-blooded Hawaiian. Buddy was a year or so older than Frank, and Dede was about a year younger. Both had joined our swimming lessons although they were completely at home in the water. They also helped us learn the local pidgin English and useful Hawaiian words and phrases—both clean and dirty. Later, a boy named Abraham, whose last name I have forgotten, joined our group of friends. Our first conversion to local custom was to abandon our shoes and, with except for occasional visits to my parents' friends and attending Sunday school, we did not put them back on until we left the islands.

Our introduction into the local Hawaiian community was enhanced by the fact that my father directed a group of Hawaiian men from the Works Progress Administration (WPA), which had been established by the Roosevelt administration to assist unemployed people during the Depression. The only major employer in Hilo during that time was the sugarcane industry, but it could accommodate only a fraction of the population. Many of the men working for my father lived in the Hawaiian land-grant community, and as our parents began to get acquainted with the workers, Frank and I were introduced to their families and children. The door was opened even further when, after several months, the WPA workers hosted a luau to celebrate—either our arrival or something else.

A date was set, and preparations began. The party was to take place on the navy station near the beach, where there was a large banyan tree. Buddy and Dede told Frank and me what to expect. The luau included purchasing and preparing the pig, hanging it in a dark room so it wouldn't be soured by the moon, digging the pit, heating rocks that would be placed inside the pig's gut cavity, and burying of the pig for cooking. These descriptions raised the whole family's expectations, and we eagerly looked forward to our first Hawaiian feast.

Around midmorning on the day of the luau, four or five of the Hawaiian WPA crew arrived at the party site. They began digging the pit, collecting the fire wood, setting up benches, cutting tea leaves to wrap various foods that were to be cooked along with the pig, and making other general preparations. At the same time, another group was at the local shoreline with cast nets, working to obtain a supply of fish. Frank, Buddy, Dede, and I watched the ongoing activities with great anticipation. Sometime before noon the Hawaiian wives and girlfriends showed up, laden with flowers, leis, and decorations for the tables. The food included tropical fruits of every kind, size, and color, including coconuts, mangos, guavas, bananas, pineapples, and other fruits whose names I've forgotten, as well as poi, breadfruit, *opihi* (limpets or snails), and other traditional luau items.

When the fire pit was ready, the Hawaiian crew picked up heated rocks with their bare hands and placed them in the pig's gut before it was lowered into the pit. Sweet potatoes, fish, fruit, and other goodies were placed on a bed of leaves, and some of the food was wrapped in tea leaves and put into the pit. Then everything was covered with wet gunnysacks, tea leaves and soil. By this time the crowd of Hawaiians, navy personnel, and their respective families had arrived. Dede introduced us to her older brother, Buster, and younger brother, David, while Buddy acquainted us with Benjamin and Chicken Little. It's funny, but I never found out Chicken Little's real first or family name. He and Benjamin were a mixture of Hawaiian and who knows what. All that matters is that they were our playmates and friends during our time in Hilo.

The luau was soon in full swing. Guitars, Hawaiian music, and local *okolehao* (booze) encouraged dancing and dismantled the racial barriers that might have inhibited social interaction. My brother and I joined the Hawaiian youths in a swim at the beach, collected some *opihi* off the breakwater, and then returned to the festivities. In our absence, a fight had broken out between some of the young Hawaiians, but others quickly broke it up, and the party went on well into the night. The luau helped broaden Frank and my friendships with the surrounding community and, most likely, my parents' relationships with our Hawaiian neighbors.

Buddy, who was perhaps nine when we arrived, quickly became our closest friend and would accompany us on all of our adventures. He was an excellent swimmer and spear fisher and had learned to throw a cast net early in life. He also could strip a coconut in a matter of seconds. His home was just across the street from ours on the main highway to town. It was a two-story wooden house with the kitchen and main living quarters downstairs. Banana and guava trees and his mother's orchid plants surrounded the living area, which was open and unscreened on one side. Their dog, cat, and chickens had free access to the lower floor, all which added to the charm of the place.

When we entered public school in Hilo, we were disappointed to find out that neither Buddy, who was enrolled in the local Catholic school, nor any of our other Hawaiian friends would be with us. The schools were divided into pidgin and standard English, and we attended the Hilo English Standard School. Our Hawaiian friends were at the Hilo Union School, which taught in pidgin English and was across the street from the private school. Frank and I thought that if we could learn pidgin English quickly we could transfer to Union, but my mother was against the idea. Most of the white children who lived in the area attended the English Standard School, but there were also a number of Hawaiian, Japanese, Chinese, Filipinos, and multiracial children in attendance.

The Hilo school had several local cultural traditions that were new to us. Even third-graders were required to take afternoon naps. We brought straw mats to school and, at the appropriate time, rolled them out on the floor and closed our eyes. I hated naptime because I could not keep my eyes shut and squirmed around a lot, and thus was continually in trouble with my teacher. Another local touch was that all students were taught *Song of the Islands* in Hawaiian. To this day, I can remember one or two verses and immediately recognize the music when I hear it. Finally, across the street from the Union School was a large outdoor faucet and sink, where the students brushed their teeth every morning—hygiene was the first class of the day.

Exploring Our Surroundings

Once our parents considered us good swimmers and allowed us to roam the local neighborhood, they greatly enriched our lives by acquiring a dog and a small outrigger canoe.

We arrived home from school one day to find a male puppy, half-German shepherd and half-English sheepdog. We named him Kamehameha, after the great Hawaiian warrior who had conquered most of the Hawaiian Islands and united them under his leadership. The dog certainly turned out to be a king in our eyes. Kamehameha grew quickly, was smart, learned a bag full of tricks, and seemed to know instinctively that he was our guardian. He seldom barked but would let out a vicious, deep growl if anyone other than a family member came to the front or back porch.

Kamehameha followed us everywhere. One day on the beach at the base of the breakwater, a young Japanese boy picked up a rock, threw it at Kamehameha, and hit him on the flank. The dog ran up the breakwater in a couple of leaps and nipped the boy on the butt, sending him scampering. Kamehameha, however, did not try to chase him, apparently satisfied that he had done what was necessary. He was a very friendly dog and enjoyed playing with all of our friends, but he would not tolerate fighting between my brother and me or others. He was our constant companion.

The outrigger canoe our parents gave us was about twelve feet long, neatly painted with a yellow hull and blue capping along the top. It gave us access to all of Hilo Bay, significantly extending our water world, but its handling required skill and experience that we did not possess. On our first try, Frank and I went around in circles while Buddy stood on the shore and tried to give us instructions. After several minutes of frustration, we finally got the message: the aft rower had to use his paddle to sweep the water around the stern to offset the outrigger's tendency to turn the canoe in one direction.

By this time we had added Kamehameha and the canoe to the family, Frank and I had scouted the surrounding area, and our parents had established our operational limits. We could use the boat within the small bay below and adjacent to our house between the breakwater and the harbor piers. Our parents also allowed us to swim and fish in the same area and off the breakwater and to follow the coastal lava outcropping south of the breakwater for about a half a mile to an area of protected ponds called Kulapie. On the south side of the ponds was a forest of banyan and other vine-like trees— an excellent place for playing Tarzan. East of the land-grant homes, heavy tropical trees and vines lined a small landing field that now constitutes the main Hilo airport. We also were permitted to go as far as a small Japanese country store, about a mile up the road to town.

These boundaries defined an area that our parents didn't know nearly as well as we did. We saw those limits as general guidelines and broke them whenever the call of adventure overwhelmed our concerns about safety. In fact, our trespasses became more frequent as we grew older and became more familiar with the area and confident in our abilities.

Boating

Feeling more comfortable with the outrigger canoe, Frank and I began to explore the inner bay, from the breakwater to Hilo Bay. Halfway up the bay toward the dock roads, we found a spring where the temperature was always a good ten degrees cooler than the surrounding waters; it was a great place to swim on hot days. Our friends told us that a local mullet population inhabited the spring, so we made plans to go to the Japanese store to supply ourselves with the necessary fishing paraphernalia.

The next day we managed to talk our folks out of fifteen cents. Yes, fifteen cents was enough to buy fishing gear and bait. With Buddy, who had six cents, we headed down the road to the Japanese store, which was about twenty-minutes away. It had a little bit of everything: clothes, sandals, tools, groceries, fresh fruit, and plenty of penny candy. Along one side of the store there was a great variety of bamboo poles, ranging from about six to twelve feet in length, and inside, hooks, lures, and fresh bait. Buddy explained that most of the fish that we would find had small mouths and were known as "steal baits." Following his advice, we chose nine-foot poles, light line, small hooks, and a bag of shrimp for bait. Twenty-one cents was more than adequate to purchase enough gear for each of us, as well as a few pieces of penny candy.

We headed back to the bay, assembled our gear, and set out in the outrigger, edging along the inner bay along the breakwater and trying to hook several inquisitive but elusive tropical butterfly fish. A pile of rocks on the breakwater became one of our favorite fishing spots. It formed an excellent habitat for the reef fish as they could quickly retreat to the protected recesses between and under the rocks. Of all sizes and colors, they swam out, carefully examining our bait, flashing sideways, and frequently neatly cleaning our hooks. Still, by evening we had more than a dozen four- to six-inch reef fish, which we cleaned and took home for dinner.

The first day of fishing only whet our appetites. We wanted bigger fish and a more substantial challenge. The following morning we acquired a spear from our friends and set out again with our poles, bait, and the wooden-handled spear, which had a steel head and barb. When we took the outrigger out for a fishing trip or other adventure, we usually took one of our Hawaiian friends or our dog along. Buddy joined us on this trip, and we worked the breakwater, moving slowly seaward. We found some larger fish, including several species of sea bass and parrotfish, but these would not stay within spearing range and we could not lure them to our baited lines. Nevertheless, we kept working our way out to the breakwater edge, well beyond our parents' established limits. With goggles on and a sling spear fashioned from a piece of an inner tube attached to a short, hollow stem of bamboo, Buddy caught a fifteen-inch parrotfish. Frank and I were basically skunked; we realized that we didn't have the proper gear or diving skills to take advantage of nature's tropical fish garden.

Buddy said he could do better with his sling spear working further inshore, so we dropped him off on the rocks. Then Frank and I began to paddle slowly toward the inner bay, which was more than a quarter of a mile inland. After five or ten minutes, my brother spotted a giant manta ray cruising on the surface about 200 feet ahead of us. Frank had more nerve than I did and was sure he could take it with our spear. As we drew close, it became obvious that we faced a monster whose wings were more than half the length of our canoe. I began to panic and plead with my brother not to spear it, fearing that it would somehow sink or overturn the small craft. He was calmer, however, and prepared to launch the spear into the manta's back.

As we approached the ray, it turned and headed directly toward the canoe, apparently unaware of the "haole" fishermen who sought to end its life. It moved steadily on course, and I could see it was going to pass directly under our boat. I once again told my brother that it was too big and could sink us, but my brother launched the spear and harpooned the manta in the thick of its back. The ray, now between our canoe and the outrigger, took a single leap, cleared the outrigger, and then headed toward the bottom of the bay. We had secured the spear to a steel ring on the canoe with approximately fifteen feet of line, which pulled taught as the manta began to drag down the outrigger. Soon the outrigger was several feet under water, and the canoe was keeling over sharply. Frank and I quickly shifted our weight to the opposite side of the canoe in an attempt to right the boat, and Frank cut the line with a dull knife. The line broke, and we regained control. We returned home with no fish, but with a lot more experience.

With the exception of the occasional cargo ship that brought supplies or the freighters that hauled out crude sugar molasses, Hilo was not a busy harbor. Nevertheless, at times adventurous mariners would sail into Hilo Bay, and ships from navy base at Pearl Harbor would drop anchor in the outer bay. Most of the sailing craft would seek the protection of the inner bay and anchor behind the wharf, and we could easily approach them in the outrigger. Frank and I would load up the outrigger with coconuts (shucked by our Hawaiian friends), bananas, guavas, mangoes, and the occasional pineapple, all of which could be gathered in the surrounding area. We then would make our way out to the visiting craft and negotiate a sale. More often than not, the mariners would buy the entire lot. A major sale would net us twenty-five to thirty-five cents, a great sum of money, which we used to purchase more fishing gear and bait for ourselves and the breakwater gang. It was, however, a very uncertain business as visiting craft came only infrequently, and the visitors generally stayed only a week or so before they moved on to more exciting ports.

As we grew more acquainted with our local surroundings, we strayed even farther from the boundaries set by our parents. One day we decided to go to the end of the breakwater. Unable to find any of our Hawaiian friends, we placed Kamehameha in the middle of the outrigger and headed out. The breakwater extended well over a mile out to sea, and the trip was longer than we had anticipated. It was a relatively calm day, although there was a fairly large swell and the waves occasionally broke over the outer ring of the breakwater.

When we reached the end of the breakwater, we saw that the sea had moderate to large swells but no breakers. The front face of the waves steeply sharpened as they passed the breakwater, and we felt like we could ride them for about 200 yards. After watching for sometime, we concluded that it was safe and paddled a short distance into the opening between the north shore and the breakwater. We paused for a minute before deciding to take a chance and ride a wave. On the first try we ran down the face of the swell, 100 to 200 yards into the bay—we were elated! With this success, we decided that we had invented a new sport and quickly headed back out to catch another wave. After rowing out a good distance, we saw a wave coming in, larger and steeper than the first swell. We got a running start and began to slide rapidly down the wave front. The boat gained tremendous momentum but then abruptly turned sideways, and in a split second flipped over, outrigger and all. Frank, the dog, and I were all thrown into the sea as the wave crashed into the bay.

We could have made it to the breakwater, which was about 150 yards away, but we had to rescue the boat. We swam back to the canoe and after a short struggle, managed to flip the outrigger over and right the boat. It was swamped and, although we had a bailing can, it would have taken a lot of effort and luck to clear and reenter the canoe without flooding it again. While Kamehameha swam in circles around the boat, we managed to bail out enough water so that the canoe had a little freeboard, and then we pushed it slowly toward the breakwater. The swells soon moved us into calmer waters, and we eventually got the boat to the lee side of the breakwater. There, we grounded the outrigger on a submerged rock, bailed it out, and got back in. The salvage work took about three-quarters of an hour, and the dog swam patiently with us until we neared the inner end of the breakwater. Then he swam off and climbed onto the rocks. When the canoe was ready, we climbed back in and made our way to the safety of the inner bay. We never mentioned our mishap to our folks, but we also never tried to surf with an outrigger canoe at the mouth of Hilo Bay.

We usually kept close tabs on the conditions in the outer bay, and when the weather was good, we rounded the freight docks and worked our way along the shoreline to Coconut Island in the southern half of greater Hilo Harbor. We were attracted to this part of the bay because it had several good fishing locations, where we occasionally caught good-sized jacks, mullet, and sea bass. Our favorite spot was offshore of the slough that went up to the fish auction and market. We usually stayed close to the shore, where we could easily take care of the canoe and ourselves if we were turned over or swamped. Sometimes, however, when the waves were very calm, we ventured well out into the outer bay.

Whenever navy ships visited the harbor, Buddy, Frank, and I filled our canoe with local fruits and set out to see if the U.S. Navy wanted to buy local products. Once there were perhaps four or five vessels in the bay, including several destroyers and one large supply ship. The weather was excellent as we moved out of the inner bay. After we passed the docks, we headed for the supply vessel, perhaps a half to three-quarters of a mile offshore. (Distances always seem greater to a child, so it might have been closer than my memory suggests.) En route, the wind picked up and, within minutes, a howler was blowing through the bay. We were caught in a heavy tropical squall midway between the inner bay and the supply ship.

With three people in the outrigger, there wasn't a lot of freeboard, and as the sea grew rougher, we began to take on water. We were not particularly scared, but we were unsure whether we should head back or try to make it to the supply vessel. The thought of making a sale overrode our concern for safety. We quickened our paddling pace and hoped for the best. Within moments, a heavy deluge joined the wind, and we began to take on even more water. When we were about one hundred yards from the ship, it became evident that we wouldn't make it without swamping. Twenty yards to go, we were waist deep in water, and our payload of fruit was lost. Most of the fruit floated away, and we made no attempt to recover it as we were into busy pushing the canoe toward the ship's gangway.

We managed to get the canoe next to the gangway, and several members of the crew pulled us—and our boat—onto it. Only after they had rescued us did they ask such embarrassing questions as, "What the hell are you doing out there in a ten-foot canoe?" and "Where are your brains?" We had nothing to say in our defense, but our treatment improved significantly after my brother told them that our father was in the navy and in charge of the local radio station. We were taken to the ship's galley, where we were given donuts and ice cream. After about a half-hour, they gave us a tour of the vessel and then ushered us back to our canoe. As we went down the ladder, we were surprised to see that the ferocious storm had dissipated as quickly as it had begun.

We had lost the opportunity to make a sale, but we were quite pleased with the outcome. We rode back to the inner bay discussing our great adventure, one that we decided not to mention to our parents. However, navy communications were working well, and the ship's radio personnel told our folks about the whole episode. Our parents put the canoe out-of-bounds for a week, and Frank and I were subsequently more careful when we ventured outside the inner bay. The visit to the navy ship was not the last meeting with our military friends stationed at Pearl Harbor. Around 1935, the first flight of multiple military craft from the mainland to the Hawaiian Islands took place. It was quite an event, and there was a great deal of local press coverage and discussions on radio about the flight. The station at Hilo was one of the major communication links with the PBYs (flying boats) during the flight, and my father was excited to be involved. When the group flight arrived safely at Pearl Harbor, we were all relieved.

Several months later, a dozen or more PBYs from Pearl Harbor landed in Hilo Bay. Frank and I jumped into the canoe and headed out to the flying boats, now anchored just off the docks in the bay. As we arrived at the nearest one, a crewmember warned us to be careful because the plane could easily puncture. Once alongside, we were invited on board to look over the plane. The crew bought some of our fruit, and later one of them rowed ashore with us. We visited two other planes, sold more fruit, and were treated like royalty. The crew of the first PBY visited with our parents on shore and promised to dip the wings of the plane when they left. The next day my mother, father, Frank, and I, along with several friends, were on the breakwater to watch their departure. As the planes flew over, one dipped its wings several times, and I felt a certain national pride that we were a part of the U.S. Navy.

I have no idea who suggested my parents buy us an outrigger canoe, but it was the most wonderful present for two young "haoles" living on Hilo Bay. It was a never-ending source of adventure, a fishing platform, a friend maker with the local breakwater gang, and a means for chasing childhood dreams.

The Breakwater

The Hilo breakwater, whose land originated perhaps 100 to 150 yards from our home, emerged from a low-lying section of the coast formed from remnant lava flows among small coral outcroppings. At low tide, a series of small ponds were formed on the inshore end of the breakwater, while on the other side there was a sandy beach. The breakwater ran for more than a mile in a northerly direction toward the Hamakua Coast, curving slightly shoreward about a half-mile out, helping to form and protect the inner bay.

Construction of the breakwater involved the careful fitting together of large, irregularly shaped chunks of rocks. Ranging from two to as much as five feet on the flatter sides, the rocks were laid and fitted to make a barrier about sixteen to eighteen feet above sea level (depending on the tide) and twenty feet across the top. The lower part was considerably broader than the top portion of the breakwater, which thus formed a trapezoid shape. Although the sides were relatively steep, a person could easily climb them, and the breakwater was a constant lure to those who lived in its vicinity. With its unique fauna and flora, it attracted youngsters, teenagers, and the elderly, who came there to try their luck with a pole and line or to cast a net.

The outer slope was pounded continually by the swells and, at times, breaking surf. The force and energy of incoming and reflecting waves made it difficult for the smaller reef fish to find protection and a suitable habitat on the seaward side. Nevertheless, both small and large reef fish could be seen twenty to forty feet seaward, and offshore sharks were sometimes spotted cruising along the outer end of the breakwater. A major attraction of the seaward slope of the breakwater was its healthy population of limpet or "Chinese hats." The larger of the limpets, *opihi*, were considered a delicacy by local Hawaiian and Asian inhabitants. Reaching a size of about two inches across, they populated the rocks between the low and high tide marks, and even higher as long as the incoming waves had adequately wet the surfaces.

The outer slope of the breakwater plunged directly into the sea except at the large turn. There at its base, the swells, surf, and currents worked together to form a small beach about eighty feet in length. It was composed of coarse sand and pebbles and sloped relatively steeply into the deeper offshore waters. The sand continued for about fifty feet offshore before the bottom gave way to lava rocks and coral mixed with sand, where a dense population of spiny black sea urchins occupied the bottom. At the outer edge of the sandy beach, the water was about ten feet, and on a calm day the surf would break about forty to fifty feet from shore. The bay side of the breakwater did not have a limpet population, but hosted a diverse range of sea life along its entire length. The submerged rocks formed a gigantic tropical aquarium filled with colorful wrasses, parrotfish, puffers, damsels, butterfly, squirrel, pipefish, etc. Although the surf did not pound the inner wall of the breakwater, waves would break over the top and tumble down on the bay side when the surf was high.

Both sides of the breakwater served as fishing platforms, each with different opportunities and risks. To fish the bayside, we needed an eight- to ten-foot bamboo pole with a single line, a leader, tiny hook, and a bag of bait, usually shrimp. We surveyed the types and quantities of fish and then made our way down to a comfortable rock and began to fish. Catching from six to a dozen fish was relatively easy, and trying to hook the great variety of reef fish consumed a large part of our days; when fishing got boring, we would take off our shirts and go swimming. We always cleaned the catch and took it to our parents, although we would sometimes fillet some of the fish and eat them raw with our Hawaiian friends. These eating habits were not, of course, discussed with our father and mother.

We also engaged in night fishing for squirrelfish, which we could never catch during the day. The red-and-white squirrelfish were attracted to lights, so we would head out with a kerosene lantern. Within a half hour, we generally had all the fish we could eat. The catch, however, was always confined to a single species and most of the take was small in size, from four to eight inches long.

Opihi hunting was always a challenge because they inhabited the seaward side of the breakwater. When it was calm, we could spot our prey, dash down the face of the breakwater, pry off the limpets, and make it back up the rocks to the top before a swell crashed on top of us. Things became a little dicey, however, if the waves were large or the limpets were at or below the water's edge. In those situations, hunting the limpets was almost always a group effort. This led to a contest to see who could venture down the farthest and who could get the most limpets, guaranteeing that all would get splashed and, once in a great while, washed out to sea. Once in the ocean, it required patience to get back without being cut or bruised, but as long as the swell was not large, it was seldom dangerous. One day several of us had to swim along the breakwater for several hundred yards until we could surf into the sandy beach.

We frequently used the breakwater beach for body surfing, although it wasn't suited to that purpose, because it was too small and too steep, the waves broke close to the shore, and the ride was short and turbulent. If we rode too far to the side, we could end up on the rocks. But it was easily accessible, so we surfed there often and took our share of bruises.

On most days, the swells pounding the seaward side were small and seldom crashed over the breakwater, except at its outer end. On the infrequent stormy days, large swells charged the breakwater like attacking soldiers, submerging portions in a mass of blue-gray water, which then cascaded down the inner side into Hilo Bay. This presented us with a special challenge: how far out into the breakwater could we get before we were drenched or washed into the bay? When the spill over the top was light to moderate, but only then, we would move out to the breakwater, carefully watching the sea. As a large wave approached, we would scamper down the inner side to the bay edge. When it broke over the top, most of its energy was dissipated against the outer wall or its porous upper surface, but the overflow still drenched us. We then raced back to the top of the breakwater and moved farther out until the next threatening wave arrived, repeating the process until we became too wary of testing the heavier seas. At that point, we dove into the bay and swam back to the beach.

Only once during our stay at Hilo did we suffer the wrath of the sea. It was after a hurricane; the swells were so large that great sections of the breakwater were constantly overrun by heavy surf. The waves crashing over the seawall and into the bay created their own swells, making small boating uncomfortable. We and most of our friends sat for several hours watching nature's show, expecting that portions of the breakwater would be obliterated. By the next morning, the storm had abated and the bay was calm. Frank and I went to inspect the damage, but the breakwater had stood the test; it remained intact, protecting the inner harbor. Some of the marine fauna, however, had not fared as well. The small beach and tide pools at the land end of the breakwater were covered with snail shells of all sorts, and we collected several hundred small and large cowries as well a variety of conches. There was such an abundance of shells that we returned home to get a large cardboard box to carry them.

It was always exciting to stand at the outer end of the breakwater and peer into Hilo Harbor, north toward Hamakua Coast, and southeast at the endless ocean expanse. Walking there was never easy. The top surface was uneven, and we had to watch carefully so that our feet did not slip into cracks between rocks. Even on calm days, the occasional wave would break over the top, making it necessary to keep an eye on both the walking surface and the incoming swells. When we saw that a wave was going to break over the top, we would seek shelter against the inner side of the breakwater and, at the most, get cooled off with the spray. When we reached the outer end of the breakwater, we would sit and watch the incoming waves, vessels in transit, local sport fishermen testing their skills, and the burning sugarcane fields on the Hamakua slopes. After sitting there twenty minutes or so, we would make our way home, a walk that was always much longer and somehow lonelier than the one that took us there.

Perhaps the most exciting and colorful sight we witnessed was an unbelievable run of what I now believe were surgeon fish coming around the end of the breakwater into the inner bay. Reddish brown in color, they ranged from two to ten inches in length and were compressed like butterfly fish. A friend had a Hawaiian name for them that I have long forgotten. All I know is that they swam in such dense schools that they blocked out the bottom of the bay, moving like a gigantic snake steadily toward the beach. Hundreds of people went to the base of the breakwater, where they took the milling fish with dip nets, small beach seines, hand spears, and bamboo sling spears. The parade of tropical reef fish continued for about three days, and Buddy, Dede, Frank and I took our share home for fish fries. There were various explanations of this phenomenon, but our Hawaiian friends said that runs of this species occurred at various points around the island about once every seven years. We were lucky that the surgeon fish chose the breakwater for their family reunion during our stay in Hilo.

The Neighborhood

Although the sea took up a great deal of our time, there was always much to do in the diverse terrain surrounding our house. In front and behind the house the land was composed of broken lava with clusters of tropical vegetation. Toward the beach and the slopes of Mauna Loa, there was lush vegetation that included a tropical forest. The navy radio station occupied a piece of land whose lava outcroppings had been cleared and leveled by a Hawaiian WPA crew. They covered the rubble with a thin layer of topsoil and planted a lawn, which stretched up to the road to the center of Hilo. The rest of the government property was left in its natural state.

As the grass grew, Frank, our friends, and I played sports like football. Unfortunately, we did not possess a football or the funds to purchase one, so we used a dried, unhusked coconut instead. This may seem a little dangerous, but it was relatively light, and our passes were seldom more than ten to fifteen feet. One day, about six of us were playing when a car slowed down and came to a stop. A middle-aged man got out of the car, watched us for a few minutes, then asked, "Would you lads like a new leather football?" We unanimously answered, "Yes," and he invited us to his sports store located on the main street in town. When the store opened the next morning, Buddy, Frank, and I were there waiting. As it turned out, the man who offered us the ball was the storeowner. We were grateful for his kindness and proud to be the owners of a new Wilson football, which we used throughout the fall until we tired of the game.

When spring arrived, we wanted to try our hand at baseball. It dawned on us that we were in the same situation as when we started playing football; we did not have the essential equipment. We remembered the kindness of the storeowner with the football. Thus, Buddy, Dede, my brother, and I organized a game of baseball using a stick and a ball of string and played along the road around the time we expected the owner to pass. The staged show worked, and we were rewarded with a new bat and softball. We were delighted with our gifts, although we always felt that we had gone too far. It wasn't quite the same as the unexpected donation of the football.

Sometime during the period when the WPA was making improvements to the radio station grounds, my father decided that the government property next to the road to Hilo would look better if it were edged with a row of coconut trees. He purchased about twenty small coconut palms and told Frank, Buddy, and me to plant them. At first we eagerly dug holes for the young trees and hauled them from the house. But the work became tedious and tiring, and it took us the better part of three days to complete the task. About two feet tall when we set them in the ground, the trees grew fast. By the time we left Hilo, they were more than twelve feet tall. The trees were still there in the 1990s and must have been close to thirty feet high.

Just south of Kulapie, banyan and other trees with lots of branches, hanging vines, and roots presented great opportunities for climbing and swinging, which attracted many of the youths from the land-grant area. The gathering place was near the center of this mini-jungle where a large banyan tree was crowded by adjacent trees with intertwined limbs. We could move from tree to tree in a circle of eight to twelve trees without touching ground. We played a variety of games, but tree tag was the most popular and drew the largest crowd. At times, as many as ten to fifteen kids would swarm through the trees like a family of chimps, moving up and down the vines and swinging from one branch to another. Occasionally, someone would fall and get banged up or have to lie on his back with the air knocked out of his lungs, but no one was ever seriously hurt.

Then there was "Bang Maukie" (Bang, You're Dead), the local version of cowboys and Indians, which we learned to play soon after we arrived in Hilo. The main radio station made an excellent staging ground for Bang Maukie because it was built off the ground on a series of wooden support columns that we could hide behind. One hand with the thumb up, the first finger extended, and the other fingers pulled back into the palm formed the lethal weapon, and the bullet was fired by saying "Bang Maukie" to anyone you saw. If you pulled, pointed your gun, and got the words out of your mouth before your adversary could, you won. Of course, we seldom agreed on who was first, so everyone continued to play, shouting "bang, you're dead" repeatedly. None of us owned toy guns, holsters, or cowboy outfits, but the game worked just as well without them.

About a hundred feet behind our house stood a dead tree whose truck was filled with hundreds of holes, home to a large family of bumblebees. Their large size and contrasting colors of black and yellow made them magnificent creatures, but we always gave them a wide berth. One day, however, when Buddy, Frank, and I apparently were looking for mischief, we decided to throw a few rocks at the tree trunk and watch the bees' reaction. After the first rock hit, a dozen or so bumblebees emerged from their hole and circled about ten or twelve feet from the tree, but did not approach us. One hit, no stings. This encouraged us to move in closer, and we hit the trunk of the tree with a two-by-four. This enraged the bees, and they came out in a swarm, surrounding us in a cloud. For some reason they didn't sting any of us, and after a few minutes returned to their holes in the trunk.

Sting or no sting, we decided to refrain from further attacks on the bee tree. We discussed this adventure with our other friends; some said that the bees did not possess stingers, and others claimed the opposite but said that their sting was not serious. The question was never really answered to our satisfaction, but we never returned to test any of the hypotheses, and the bees remained secure in their home.

After several years in Hilo, Buddy and his family had to move to the island of Kauai. This came as a terrible shock to Frank and me. We felt that a member of our family was being taken away from us. Buddy departed two days later, really putting a hole in our breakwater gang. Buster Puehow, who was several years older, had other interests, and that left us with his sister Dede and Abraham. The world suddenly became lonelier, fishing less attractive, and boating boring. Fortunately, Buddy returned after several months and helped reconstruct our world.

Not long after Buddy returned, Buster came across an enormous octopus while he was spear fishing near the breakwater. He waded out onto the adjacent reef and thrust his spear into the beast's head. The octopus, however, was not about to give in and struggled for almost ten minutes. Finally, Buster was able to take control, and he dragged the octopus on shore. Its legs were over ten feet in length. Buster became quite the celebrity and had his picture in the local Hilo newspaper. Those of us who either saw the struggle or merely heard of Buster's adventure felt that we were part of the story.

Discovering the Island

From time to time my parents would plan a trip, sometimes with family friends. There was a myriad of interesting places to visit on the island. The largest island in the chain, Hawaii is probably the most diverse in terms of climatic and ecological zones. The Hilo side is noted for its heavy rainfall while much of the western side of the island is dry and semi-arid. Two main volcanic mountains, Mauna Kea and Mauna Loa, which form the core of the island, rise to over 13,000 feet and at times are capped in snow. A striking shift in the island's fauna and flora can occur in relatively short distances up the slopes of the mountains or around the island. Lush tropical foliage and desert country are in close proximity, separated by several miles of mountains and influenced by the prevailing trade winds.

In the early 1930s, Hilo was the chief town on the island and the hub of the intense sugarcane industry. About twenty mills stretched from south of Hilo, along the Hamakua Coast, to the island's northern tip. The scattered mills supported a number of small villages along the coast. On the Kona side, which now supports the island's thriving tourist industry, a small village called Kialua was known for its Kona Inn. There were no golf courses, majestic hotels, tourist tours, opulent apartments and condos, or commercial airport. It was just another small, quiet community on the dry side of the island. Nevertheless, the climate and the interesting sites that attracted both local islanders and the occasional visitor formed the basis of the significant tourist industry that emerged in the post-World War II era.

My folks were particularly fond of the drive north along the Hamakua Coast. The road passed through miles of sugarcane fields and wound down into gorges and valleys crossed by small rivers and streams that originated on the slopes of Mauna Kea. The valleys were filled with lush tropical vegetation, including flowering trees with white, yellow, and red blossoms, and there were some houses near the ocean. About an hour north of Hilo, a small lava peninsula jutted into the surf, which constantly pounded the northwest side of the island. Called Laupahoehoe, the point was the site of a small village comprised of a couple of stores, a schoolhouse, and several homes. People who worked in the cane fields on the mountain slopes above the village occupied the peninsula. Frank and I loved to descend the winding road down to the peninsula's outer point. If the swells were small, we were allowed to swim off the rocks. Even better, we were always treated to ice cream before we left the village.

Shortly after World War II, a tidal wave crashed across the Laupahoehoe Peninsula, destroying the town's buildings. All of the children and teachers, who were inside the school, were lost. The parents survived because they were working in the cane fields well above the town. The village was never rebuilt, and today the only reminder of it is a monument to the lost children and schoolteachers. The same wave destroyed the breakwater protecting Hilo Harbor, flooded the bay frontage, and drowned several other people.

From Laupahoehoe, our drive would continue north along the coast, passing through a number of villages until we eventually came to Waipio Valley. A rock-strewn road dropped steeply down into the valley. We frequently walked down the road, but I never really enjoyed the trip because I was preoccupied with the thought of having to climb back up to our parked car. Furthermore, the valley gave me a feeling of loneliness, perhaps because it seemed so isolated from the rest of the world. Once there, however, it was fun to walk along the riverbank where local children fished for several species of fresh-water fish, farmers worked the large taro patches, and one or two stores provided food and supplies to visiting tourists.

We also liked to drive up the mountain behind Hilo and go over the pass between Mauna Kea and Mauna Loa. It was a short ride to the pass, and from there the road descended steeply down the leeward side of the island. It wasn't far down to Waimea, where we always stopped for a bite to eat. This was cowboy country and near the famous Parker Ranch, which raised cattle. From Waimea, the road continued down through lava outcroppings and sparse clumps of cactus until we came to the Kawaihae region of the coast. From the beach at Kawaihae, the Parker Ranch cattle were taken to be loaded on a freighter, which stood several hundred yards offshore. The heads of the cattle were lashed to the outriggers of canoes, usually several on a side, and the canoes were paddled to the waiting freighter. Straps or harnesses were secured under the cattle, which were then hoisted onboard and transported to neighboring islands.

A few miles down the coast road from the Kawaihae landing, there was an excellent camping site near an extensive swimming area protected by an offshore reef. We went there once to camp with some friends of my parents who had a daughter our age. They were of Scottish descent, and their girl was blessed with golden red hair, a beautiful face, and a great personality. Frank and I both immediately fell in love, and we had a great time swimming and chatting with her. But, alas, things fell apart rapidly when we were heard using some foul language. I don't remember the specific words uttered, only that we were placed on restriction and our friendship with the Scottish lass terminated.

We made several trips around the island, first heading south through the volcano country and then north along the east side. On the way, we usually stopped and camped on the beach at Honaunau, where we enjoyed the surf and outdoor cooking. Once we spent the night on the beach during a period of high meteor intensity. As I looked up at the sky, it seemed that the stars were continually raining down on Mother Earth. I watched nature's fireworks well into the night, thinking that by morning there would be few stars remaining in heaven. The next day we drove on to Kailua-Kona, investigating the coffee plantations on the slopes of the mountain and other tourist sights. When we were in that area, we would have lunch at Kailua and then cross the barren lava flows that stretched along the coast to Kawaihae. Then we would head up the mountain and back over the pass to Hilo. Although the Big Island had a tremendous variety of places to visit, my first love remained Hilo. Despite its reputation for rain, it had a charm and beauty of its own.

The Search for Pele

Soon after we had arrived in Hilo, our young Hawaiian friends told us about Pele, the Hawaiian god of fire, who lives in the depths of the volcanoes. Over the next several years, we were told of Pele's powers and the signs in the sky that signaled her potential appearance. These stories were not told as amusing anecdotes, but with conviction and sincerity. Buddy, in fact, was convinced that he had seen the fire god over Kilauea; although he was raised Catholic and was a committed Christian, he still mixed his religion with a sprinkling of Hawaiian beliefs.

Kilauea Crater, located on the slopes of Mauna Loa in the vast Halemaumau (House of Everlasting Fire) Crater, was approximately thirty-five miles from Hilo. The crater and a string of older fire pits that ran down the slopes toward the coast were major attractions at the Volcano National Park. Kilauea was well known for its frequent eruptions. My mother and father had taken Frank, Buddy, and me to the park several times, but Pele had been quiet since we moved to the island. Buddy assured us, however, that it was only a matter of time before "she" would make her presence known. Of course, people on the Big Island had lived with eruptions for centuries.

One day, Buddy told us that a particular cloud formation was a sign that Pele would soon arrive. We attempted to obtain a more specific prediction, but Buddy refused to divulge any details, saying only that an eruption would happen soon. Frank and I assumed "soon" would be that afternoon or the next day. We told the story to our parents, who said that these stories were just local myths and that there was no reason to believe them. Pele was just a pagan god whom the old Hawaiians had believed in for centuries, but then Buddy didn't seem to be an old Hawaiian.

When the following day passed and the crater remained silent, we began to dismiss the story and went to bed with no thoughts of Pele. That night about 2 a.m., we were awakened by a phone call. After a short discussion on the phone, my father told us to get dressed and that the volcano was undergoing a spectacular eruption. He went over to the radio station and sent a message to the naval radio station in Honolulu, and within minutes we were in the car and on our way to Kilauea. Eruptions often occurred in the crater, and the Park Service had established a viewing station on one side of the fire pit.

We could see a red glow in the sky even from Hilo, and as we moved closer to the park entrance, the glow became progressively brighter. We took the road that circled around the rim of Halemaumau and made our way down to the far side of the fire pit. As we approached the crater, we heard a great rumbling and saw a bright orange-red light above the pit. From the viewing station we could see lava cascading from the far side of Kilauea down to the bottom of the crater, which at that time was about 800 feet deep. The lava gushing from the crater's side, about 400 feet from the top of the fire pit, crashed down to the bottom in a gigantic red fall. The bottom of the pit was filled with small fountains of lava jetting fifty to 100 feet. The sight was so mesmerizing that I stood there looking and saying "Pele, Pele, Pele," only later realizing that I had offended my own Christian god. When I awoke the next morning, the whole experience seemed like a dream. That was the only eruption of Kilauea to occur during our stay on the Big Island, but its memories are etched deep.

Not too long afterward, my father told us that he had been transferred to a base near Honolulu and that we would leave Hilo in about a week. Although Frank and I knew that we would have to move some time, the news was much harder to take than it was when we were younger. We had what every child dreams of—good friends, a wonderful environment, our own boat, a fishing paradise—and we were both very happy. A very special part of our life would soon be relegated to the past. Over the next week, Frank, my parents, and I said our good-byes to all of our local friends. In the early summer of 1935, the family and our dog, Kamehameha, boarded the inter-island passenger ship and the steamer made its way out into the harbor. With tears streaming from our eyes, my brother and I watched until the shoreline faded from view and then went to our staterooms wondering if we would ever see Hilo again.

The Return to the Navy Life

In Honolulu, we were picked up by a navy bus and taken to our new home. I had expected that we would live in or around the naval base at Pearl Harbor, but the bus headed south through town, past Waikiki and Diamond Head and along a large golf course. We finally turned off the main highway at a naval radio station, which was nestled in a small valley near a shallow coral lagoon that was protected by an extensive outer reef several hundred yards offshore. The base area was called Wailupe, and we were billeted in one side of a large duplex on a road that ran from the main highway toward the mountains. The curved road circled most of the government-owned property.

The base homes were built on the outer side of the road, and the inner area was filled with large trees, a nicely manicured lawn, a swimming pool, tennis court, a small movie house, and the single enlisted men's barracks. A small stream that originated in the mountains behind the base ran through the center of the property, neatly channeled through a cement rock culvert. The radio station was located across the main highway that ran north into Honolulu and south toward Koko Head Crater. The base was a small self-sustaining military community fenced off from the surrounding areas.

The day after our arrival, Frank and I scouted out our new habitat. There were about eight families located on the base, all of whom had children whose ages ranged from a few years to eleven or twelve (the same ages as Frank and me). We were surprised to find out that Jimmy Fraser, an old friend from our earlier years in San Diego, was living in the Wailupe enclave. Jimmy was the son of a chief petty officer, and our families had often gotten together for special holidays and other social events. We knew Jimmy's mother as "Aunt Jimmy." Jim's presence on the base made our transition from Hilo to Wailupe much easier. With Carl, the thirteen-year-old son of the base commander, we formed a band of four that played, explored, and frequently engaged in mischief together.

The base at Wailupe operated within an entirely different social and cultural framework than the one we knew on the Big Island. Status was based on the hierarchy of officers and enlisted men, and the class structure was passed down to the children. Thus, Carl was recognized as "king of the roost." This did not fetter Frank and me because he was a larger, more athletic lad, and a likable individual. The servicemen's children did not have friends among the locals who lived around the base. This was not entirely driven by class consciousness or racism, but to some extent the parents wanted their offspring to bond with the establishment.

The gloom of leaving Hilo was largely offset by finding Jimmy and because the base had a substantial number of perks. Frank and I had not seen a movie since the silent film at Tatoosh, and we were excited at the thought of seeing motion pictures with sound. The base was surrounded with interesting areas to investigate. To the south, there was a large valley occupied by hundreds of cows and breeding cattle owned by the Hinds Clark Dairy. At the lower end of the valley, the dairy operated a large milking facility that included bottling machinery, storage sheds for bottles and caps, and buildings for their many delivery trucks. Directly behind the station, the mountain rose steeply toward the center of the island, and seaward was a protected lagoon inhabited by a multitude of coral reef fish and invertebrate species.

Kamehameha arrived at the base in good shape and seemed to accept his new home and surroundings without evidence of stress. When we left Hilo, he was a little over three years old but still showed some puppy tendencies. At Wailupe, he seemed to decide to take his rightful place as a dog with a king's name. There were four other dogs at the base, and Kamehameha secured his position as leader of the pack by fighting and beating the two larger dogs and ignoring the smaller dogs, who barked a lot but didn't fight. One of the smaller dogs belonged to Carl, and Kamehameha seemed to know that he shouldn't harm or chase him, so he pretended that he didn't exist.

Shortly after our arrival, Frank entered Roosevelt Junior and Senior High School while I was enrolled in the sixth grade at Queen Lydia Lili'uokalani Elementary School. A small navy bus carried the children from the base to the various schools. My school was located just past Kimuki, a small shopping area just south of Honolulu. The students were a mixture of Hawaiian, Asians, and Caucasians, like at the school in Hilo. On my first day I was very upset, because my mother said that I should start wearing shoes. I was so upset, in fact, that I jumped off the bus and refused to go to school. My parents finally relented and let me go barefoot. When I got to school, I was quite pleased because almost all of the other students were without shoes. The older girls, however, tended to dress up more, and they wore shoes.

The Liliuokalani school had something I hadn't experienced before— an excellent cafeteria. For a dime, I could get a large scoop of sticky rice covered with chop suey, which I could eat endlessly. For an extra nickel,

I could top off my meal with an ice cream sandwich. Unfortunately, we didn't really have many nickels or dimes to spare, so any day that I went to the cafeteria was a real treat. Much to my delight, the school required that each student spend a day working in the lunchroom two or three times a semester. On those days the student was given a free lunch, which gave me the chance to feast on rice and chop suey. The ice cream sandwiches would occasionally have a little slip of paper under the chocolate cover that said, "free," which allowed the recipient to return for another ice cream. I was quite happy with the new school and joined in the recess football games. The one thing I hated was the noon nap period, a part of the school's agenda.

After school, our band of four would almost always head for the swimming pool, where we played tag and dove from the low and high boards. Carl, Jimmy, Frank, and I were all good swimmers and spent hours in the pool entertaining ourselves. When we got bored with the pool, we would organize expeditions into the mountains to explore the area above the base. The lower hills near the camp were covered by thorn trees and cacti, which made the going difficult. Several hundred feet up, the cacti thinned out and the trek became somewhat easier. On several occasions we camped overnight, perhaps a thousand feet up the mountain slopes above the base. It was difficult to maintain a campfire, however, because the trade winds blew endlessly and howled through the night. Thus, our treks into the higher mountains were infrequent.

One day, when we were exploring the foothills of the mountain behind our house, we discovered the entrance to a small cave hidden behind thick mesquite trees and heavy outcroppings of cacti. The cave mouth was about three feet high and perhaps five to seven feet across. We entered it and found that it expanded into a larger area, which averaged about eight to ten feet across and up to five feet high, and went back about twenty feet into the lava hillside. Although we were without lights, we could see some wooden objects along the side and bones strewn on the cave floor. We first thought that they were the skeletal remains of humans, but there was inadequate light to make be certain.

We returned to our homes to get candles and flashlights. We agreed not tell our parents about the cave because they likely would tell us to stay away, ending our exploration of what we believed was an ancient Hawaiian burial ground. Within half an hour, we obtained our supplies and returned to the cave. It quickly became apparent that the bones were much too big to be human remains and we decided that they were the remains of cows that had inhabited the adjacent valley. The real question was what they were doing there. We concluded that someone had stolen a cow and butchered it in the cave.

Continuing to explore, we examined some pieces of wood that were stored on one side of the cave—perhaps, we thought, the remains of an outrigger canoe. The search had been exciting, but also somewhat disappointing. Then when we were ready to leave, Frank yelled that he had found a small entrance to an inner cavern at the far end of the cave. The four of us converged on the small hole, which was just large enough to squeeze through and led to another open area. We discussed whether it was safe to proceed and concluded that it was, but Jimmy didn't wish to go any farther. The other three of us twisted and pulled our bodies into the inner cavern.

The second chamber was twice the size of the outer cave area and much higher—six to eight feet high. The sides of the cave had been formed from black lava; broken rock lay on the floor. It was pitch dark, and we had to use our flashlights to see. We moved slowly forward, farther and farther back into the cave. After about fifty feet, the cave suddenly took on an oval shape and its sides turned a deep red. In less than a hundred feet, we found ourselves in a perfectly shaped tube; we felt like we were inside a fire hose. Frank and I had visited lava tubes on the Big Island and recognized that we had discovered a major one just outside of Honolulu. We followed the tube back under the mountain for almost half an hour, but our nerve ran out long before we reached the end of the tube.

Worried about how long the flashlight batteries would last and afraid of getting lost without light, we started back. On the way, there was a split in the cave that we hadn't seen before. Unsure which way to go, we started to panic. Carl kept his cool, but I was about to cry, sure that we were lost. Fortunately, we chose the right path and soon found ourselves in the inner chamber. As we made our way through the small entrance, we noticed that the light from the mouth of the cave was much dimmer. Jimmy had decided to give us a scare by piling brush and ferns across the entrance. The bright tropical sun defeated his meager efforts, and we pushed our way out into the sunlight. Afraid of being punished, we never told our parents about our discovery, but several decades later I did report it to the Bishop Museum in Honolulu.

Life within the confines of a military base was a new experience for me, because I couldn't remember the time we had lived on a base in the Philippines. We had an obvious respect for rank, the flag-raising and lowering ceremony, the wearing of uniforms, and the tidiness of the grounds and buildings, which all projected a sense of order, belonging, and national pride. At times, however, the regulated ways seemed to be superficial nonsense, even to an eleven-year-old.

Movies were shown three times a week at a movie house housed in a building that was about twenty-five feet wide and forty-five feet long. Rows of foldout chairs ran, with an aisle down the middle, from the back of the room to within roughly fifteen feet of the screen. In the first row, there were three large wicker chairs with cushions—the seats for the commanding officer, his wife, and his son Carl.

The movie hall opened thirty minutes before show time, and servicemen and their families would get there early to talk and enjoy a treat from the snack bar in the back of the hall. Ten to fifteen minutes before the scheduled start, the attendees began to seat themselves, but the movie did not start until the commander and his wife and son had taken their seats. The commander never arrived on time so there was always a waiting period before the grand entry. He and his family would walk down the center aisle as if they were royalty. Still, Frank and I relished the films and the social opportunity, not to mention the treats from the snack bar.

My favorite treat was the Nestle's chocolate bar, which came with a one-and-a-half by two-inch color picture of various animals under the wrapper. The company gave out booklets for placing each picture; the goal was to complete a set. There was always a great surplus of some animals and several rare pictures for which everyone competed. Although we traded with friends to get what we needed, I was never able to complete my book. But I did eat a lot of candy bars and learn the names of some strange creatures.

Running down the south side of our property line was a fence that separated our home from the dairy next door. We had never ventured into the grazing area, because there were always a few large bulls roaming about, and they looked mean. One day, however, Carl and Jimmy decided that we should cross the pasture and visit the bottling area of the dairy. We armed ourselves with a few sticks and started to walk the several hundred yards that separated the base from the milk-processing plant. Two bulls were in the field, so we picked our way across, making sure to stay as far from them as possible. About two-thirds of the way, one of the bulls began to walk toward us slowly, with his head slightly lowered. As a result, the four of us covered the last fifty yards in record time. After we'd climbed the fences on the other side, we looked back; the bull was quietly observing our retreat. He had cleared the field of a herd of two-legged intruders with a slow walk and a nod of his head.

Having negotiated the pasture, we began to scout out the activities at the dairy. Before long, we found a large warehouse filled with boxes; closer inspection revealed that the boxes contained cardboard tubes that were about three feet long and filled with milk-bottle caps. When empty, the tubes made good and generally harmless clubs, or could be used to orchestrate musical tunes and odd sounds. We climbed up on the boxes, which were stacked fifteen to twenty feet high, to observe the goings-on of workers entering the building. After several minutes, a worker came over and told us to get our "asses down from the boxes and get the hell out of the building."

We did this on command, but when the worker left the warehouse we ran back into the building, climbed up on the boxes, and restacked them so that we couldn't be seen from the floor level. From our fortress, we were able to survey the warehouse for a couple of hours. Nothing very interesting happened, but knowing that we had regained our position after being kicked out was enough to satisfy our egos.

Later, we sneaked down and out of the building, each of us carrying a tube filled with bottle caps. We ran to the pasture, carefully crossed the bull-grazing area, and returned home. Our group was very honest and would have refused to steal a nickel, but absconding with four bottle cap tubes seemed the rightful bounty for pirates who had captured and occupied the Hinds Clark warehouse for more than two hours.

In Wailupe, a senior petty officer named Graves took a shine to Frank and me, and frequently took us on trips to various parts of the island. He was an avid surfer and one day suggested that we go to Waikiki, where he would teach us to surf. We enthusiastically grabbed our swimming suits and towels and waited for Graves to pick us up in his small Ford coupe.

This took twenty minutes, which included getting permission from our parents. During the mid-1930s, Waikiki was somewhat laid back and not as crowded as it is today. There were only three hotels on the beach, the Moana, Royal Hawaiian, and Halikalani, which were all very elegant. We were most impressed by the outrigger canoe and surfing club that, as I remember, was on the Diamond Head side of the Moana Hotel. There, surfers stored their boards upright in several long rows that stretched from near the sidewalk down to the beach. Ten to twelve feet in length and made from solid wood, the boards seemed monstrous to us.

Graves checked out his board and then one of us accompanied him on each surfing effort. We would kneel or lie on the board as he paddled it out past the inner *keiki* (baby) surf. I was surprised to see how shallow the water remained as we made our way over sand and broken coral. Several hundred yards from shore, we could still touch bottom. When Graves felt that the waves were large enough to get a good ride, we were probably a mile offshore. To catch a wave, we were instructed to lie slightly forward on the board and help paddle to increase our speed just before the wave reached us. As we gained momentum and moved down the face of the wave, Graves would maneuver into a standing position and then have us try to get up just in front of him. It worked once or twice, but most of the time I ended up in the water. If I didn't make Graves fall into the water as well, he would continue toward shore and I would swim around until he returned for another attempt.

Frank did somewhat better than I did, but neither of us ever became proficient surfers. We didn't own a board and seldom swam at Waikiki. If the family was in the mood for a beach outing, we usually would go to Hanauma Bay, one of the most enchanting swimming spots on the island. The bay had formed in the crater of an ancient volcano when its seaward side sank into the sea, leaving the shoreward side of the crater rim and a sandy beach, a protective outer reef, and several sandy areas inside the reef, including the popular "keyhole." Those who were timid could stay within the inner reef and enjoy fish watching. Those who were more at ease in the water could go beyond the reef and find quantities of large parrotfish, snappers, damsels, and butterfly fish. We didn't have facemasks and snorkels, but used goggles that were held on with a stretch band. When we wanted to body surf, we usually went to the north shore or one of the beaches on the windward side of the island.

It was at the Wailupe base that I first learned Morse code, which would enhance my career options when I joined the U.S. Navy some years later. Carl, the commander's son, and I told our fathers that we wanted to learn to send code. As a result, Commander Holden had a wire strung between our homes and a telegraph key located in our bedrooms. For several months, Carl and I would contact each other during the evenings and attempt to communicate some of the day's interesting happenings. We were never very proficient, but managed to get our primitive, sometimes garbled, messages across. It was a novel arrangement that was not available to many youngsters. My brother showed no interest in Morse code, finding it much easier to pick up the phone or walk down the road and talk to Carl.

During our years on the island, tension built between Japan and the United States, and Frank and I often heard our parents talking about aggressive moves of the Japanese military. We also were aware that there were increased war games and military maneuvers on the island. During one of the maneuvers, soldiers from the Schofield base built a machine-gun nest on the peak of the hills just above the Wailupe navy station. We frequently visited the location to chat with the soldiers. One day after we had returned home, three navy Grumman fighters swooped down from over the sea and made an imaginary attack on the machine-gun emplacements. The kids on the base were impressed and sure that our military was unbeatable.

In the spring of 1936, my father received orders to proceed to San Diego, where he was to be reassigned to a supply ship. Our honeymoon in Hawaii rapidly was coming to a close. We had left San Diego in the late 1920s, and Frank and I had attended five different grammar schools, from southern California to Hawaii. During our absence, I completed my primary schooling and was ready to begin junior high, while my brother soon would enter the eighth grade.

In the six years that we were on the road, we made many friends and interacted with a number of ethnic groups. Our hearts were still in Hilo, but we realized that our memories of the islands eventually would fade. In one sense, going home was exciting, but what were we going home to? Even though our mother told us that our friends Oscar and Gene still lived close to our old house, they had grown from children to young adults. We prepared for our trip by organizing our few belongings and buying shoes—a sure sign that we were mainland bound.

As we were moving from place to place, the Great Depression had spread across the United States and the rest of the world. Prohibition disappeared, and global air travel began with flights of Pan American Airway's China Clipper. Emperor Hirohito and the Japanese military flexed their muscles in Asia. A strange man named Hitler was making noise in Germany, to the discomfort of his northern European neighbors, and Mussolini, also known as Il Duce, was stirring up trouble in southern Europe and northern Africa. President Roosevelt was busy with a myriad of new social programs designed to improve the lot of the millions of Americans suffering the depravations of the Depression. Frank and I entered our teens, understanding little of the world.

In the early summer of 1936, we were driven, in the same navy bus that took us to Wailupe, down to the pier next to Aloha Tower in Honolulu. There, we boarded the Matson liner *Malolo* for the second time. We took Kamehameha to the top deck and placed him in a large individual kennel; then we returned to the passenger deck to watch as the ship departed. The band on deck played *Aloha Oe*, and we all had a good cry, but we were less emotional than we had been when we left Hilo. As a tug pulled the ship away from the pier, the crew threw streamers and passengers tossed coins to Hawaiian swimmers in the water. Two or three of them also were on the boat, collecting coins from the passengers as advance payment for making a somersault dive from the mid-deck of the steamer. After collecting all the money, one diver jumped into the water, creating a giant cannonball splash. The other two made the dives they'd promised.

After several loud blasts of the horn, the ship moved quietly out of Honolulu Bay, swung south around the tip of the island through the channel between Oahu and Molokai, and headed east toward San Pedro. As we passed through the channel, we could see into Hanauma Bay and, on the other side of the ship, the northern tip of Molokai. The only thing I remember about the voyage home, however, is going to the upper deck to visit our dog. The pomp and ceremony and elegant dining that were so impressive on the trip to the islands apparently had lost their charm. I was deep in thought about my new life in east San Diego, the place we called home.

Alverson home, Hilo, Hawaii; early 1930's

Breakwater Gang: Front Row: Lee Alverson, Buster's younger brother, Abraham (last name unknown). Second Row: Buddy Terrel, Buster Puehow. Frank Alverson. Third Row: DeDe Puehow

Lee and Frank Alverson with outrigger canoe

Chapter III
Stateside Again

The ship returned to the same wharf in San Pedro we had departed about five years earlier. Uncle Raymond and Aunt Kathryn and our two cousins, Kathryn Jo and Florence, stood in the viewing area on the dock just as they had when we left. My cousins had grown into young ladies, and it was obvious that five years had brought about significant changes in the family. We waved vigorously as the tugs pushed us into the berth and waited anxiously for the gangways to be put into position. We all, including the dog, joined our welcoming party on the dock. My aunt and uncle took us to their home in Maywood, California, a suburb of Los Angeles. They lived in a two-bedroom house on the outskirts of town, several blocks from the Los Angeles riverbed.

We stayed there for the night, and the adults spent hours talking about what was happening around the country and national politics. I remember that there was considerable discussion about the animosity between labor and management and the sometimes fatal clashes between union members and non-union workers. My uncle, a truck driver, was caught up in the conflict. Threats had been made on his life, and on occasion he had moved his children to temporary locations for safety.

The discussions did not hold our attention long, and Frank and I soon engaged in more mundane chitchat with our cousins. Our minds were on important things, such as when school would start, our other relatives, and our grandparents, who now lived in Idyllwild, a small mountain village in the San Jacinto Mountains southeast of Los Angeles. Florence and Kathryn Jo, called K Jo or Katie, for short, convinced us that Idyllwild was the place for fun. While our parents talked well into the night, my thoughts turned to San Diego and the friends who had more than doubled in age as we'd traveled to Tatoosh Island, Bremerton, Wilmington, Hilo, and Honolulu.

Settling In

The next morning, Uncle Raymond drove us to our house in east San Diego. The small, white, one-bedroom house was one empty lot from the corner of 35th Street and Orange Avenue. I don't remember where Frank and I had slept before, but it was apparent that we were short on living space for a family of four with two teenage boys and a dog. My father and mother soon made arrangements with a carpenter to extend the back porch and add a small room on the end of the house. In the interim, we slept in the front room and took turns using the sole small bathroom. When the 120-square-foot room was completed, my parents purchased bunk beds, and the place seemed luxurious to us. I was given the upper bunk, Frank the lower one, and the dog took up residence on the throw rug. In the midst of the Depression, we felt like kings.

After a few days, Frank and I began to reacquaint ourselves with our old friends. Oscar lived next door in a room his parents built above the garage. Oscar's add-on living space was in some ways like ours, but because it was both away from the house and rather large, it became a hideaway that was envied by all the neighborhood boys. Most of the houses in the area were built with one or two bedrooms and often were modified to accommodate growing families. Oscar was stocky and slightly shorter than Frank and me. He reintroduced us to Gene and Louis (Louie Black), who lived with his three sisters about a block up 35th Street. We slowly began to build a new network of friends.

Frank and I had never felt the impact of the Great Depression before, and it suddenly dawned on us that we no longer had the freedom and activities that had shaped our lives in Hawaii. We had lived close to the ocean and in Honolulu had access to on-base recreational facilities. We could scavenge tropical fruits and spent little on our clothing. In San Diego, however, families would show up on our doorstep from time to time asking for food, and Mom always packed them a dinner. Every now and then Frank and I wanted to see a movie or catch the streetcar to the beach, but money was not easy to come by. Our folks gave us twenty-five cents a week in allowance, but we quickly consumed it. We set out to earn a little money by doing yard work and selling magazines.

I soon got a job delivering and selling the *Saturday Evening Post* and *Colliers* magazine to save money for a second-hand bicycle. I received one-and-a-half cents for each magazine sold, so it took a lot of work to earn a decent weekly income. I built the route up to about twenty-five, sometimes as many as forty, magazines per week. The other sellers and I carried the magazines in cloth bags equipped with a sling that went over the shoulder, but when we carried more than twenty magazines, the weight began to take its toll. Yet, because selling magazines required stopping at every house, using a bike was not very convenient. Still, I found the job easy; it entailed giving a pitch about how you would win a prize if you sold several more magazines. My route ran about a square mile in all directions from our house and covered about five miles altogether. I soon knew most of the people within several blocks of our home and had a group of weekly customers, and could predict the likely response of most of the housewives. Nevertheless, squeezing a nickel or dime out of a potential client wasn't easy in the middle of the Depression.

Within a month, we were integrated into the group of young teenage boys who lived in our neighborhood. Four to six of the local boys would regularly join us in a game of sandlot football, Anti-Anti-Over, Capture the Flag, or riding bikes. Hide and Seek had given way to Kick the Can, and instead of Cowboys and Indians we played rubber-gun battles, antecedents of today's paint-ball wars. We made our guns out of wood, shaped like a pistol, but with a barrel that was up to eighteen inches long. We fit the handle with a clothespin, and left the barrel end flat. Then we placed a rubber band, cut from an inner tube, under the clothespin and stretched the other end over the front end of the barrel. We chose teams, two to three on each side, and each group attempted to eliminate the other by hitting an individual with a rubber band, fired by pressing down on the clothespin. We were supposed to aim below the neck, but occasionally there was a hit to the face. Games could last a few hours and range anywhere within our block. I don't know if this activity made us less sensitive to the rest of mankind, but no one was ever really hurt and no one ended up in the hoosegow. Our major debates were over whether or not an individual was hit, which led to endless squabbles.

Skateboards were not yet on the scene, but all of the neighborhood youngsters knew how to build skate scooters. Their construction involved cutting a two-by-four into two sections, one thirty inches long and the other about twenty-four inches. The wheels were harvested from skates and placed on the front and back of the longer piece of wood. To form the scooter's neck, the shorter two-by-four was nailed upright, at a right angle, to the forward part of the undercarriage. Then we inserted a short wooden brace between the two pieces, and formed the handle by placing a short section of a broomstick across the top of the brace. If we felt like adding carrying capacity, we nailed a wooden apple box to the lower two-by-four just behind the upright board. If this Cadillac style was used, then there was no need for the brace, and the handle was attached across the forward end of the box. We used the skate scooters to get to the movies, to race, and to perform tricks (curb jumping) up and down the street. Once in a while, we would venture several miles from home, taking along a lunch in the apple box. Each scooter had its own name and license number. Mine was "Jump-1."

The Mountain Hideaway

Shortly after we settled into our home, we planned a trip to visit our grandparents in Idyllwild. About three hours away by car and 120 miles northeast of San Diego, the town was nestled at about 5,000 feet in a valley below Tahquitz Peak. Frank and I were eager to go as our cousins had given Idyllwild such good marks. The road to Idyllwild ran northeast to Escondido, north to Fallbrook, east to Temecula, and then across the hot and dry semi-desert country to Hemet, which stood at the foot of the San Jacinto Mountains. By the time we arrived in Hemet, everyone was sweaty and ready for a cool drink, so we headed to the local ice cream fountain near the center of town—a location we remembered during every other trip to Idyllwild. A mile or two east of town, the road began to wind up the edge of a deep gorge cut by the various streams that drained the western watershed of the mountains. About a third of the way up the two-lane highway, the road crossed Strawberry Creek and made a sharp turn back across the face of the mountain.

The temperature on that summer day was well above 100 degrees in Hemet, and a number of cars had stopped by the side of the road because their radiators had overheated and boiled over. If they were lucky and made it up to Strawberry Creek, they could get water out of the creek bed. Our '36 Chevy didn't overheat, however, and we made it up to cooler country when we entered the pines at about 3,000 feet. At Mountain Center, the road split off from the main highway and continued east until it reached Palm Springs. The other road swung off to the left, cut across the face of the mountain, and rose up a steep winding grade to Idyllwild. Just before we arrived, the road made a sharp turn to the right and dipped into a somewhat level region, filled with large yellow sugar pine and a smattering of oak. In those days, Strawberry Creek, which drained the valley, had a good flow of water even in midsummer. It was a mecca, hidden away from the hot desert sands to the east and the arid Hemet area to the west.

Idyllwild was comprised of a large inn, several small stores, a swimming pool, and some cabins, most of which were on a circular road that followed the general direction of the stream. My grandparents ran a photo and souvenir shop in the front of a log cabin. From the back of the store, a trail wound down the hill to the meandering creek that was hidden by tall pines on both sides of its banks. The stream supported a population of brown trout and was well stocked each spring. It was our favorite place to fish and hang out with the other teenagers who were lucky enough to spend the summer in the mountains. There, Uncle Bob, my mother's brother, taught us how to catch trout with our hands. In the late evening or after dark, we used a light to spot trout lying motionless under the rocks. By moving hand and arm slowly under the rock and up toward the fish's gills, and then making a quick grab, we had ourselves a ten- to fourteen-inch trout. It was illegal, but there was a certain thrill to snatching a trout out of the stream with our bare hands. The hand-caught trout tasted better than pole-caught fish.

Idyllwild was a hideaway of which Frank and I never got tired. We could swim in the pool, hike about seven miles to the top of Tahquitz Peak, talk to customers in our grandparents' store, or ride down the hill to Hemet with our uncle to pick up the two Los Angeles papers, the *Times* and the *Examiner*, which were sold in the store. Once in a while we would go with our uncle to the surrounding campgrounds to sell the

Sunday papers. On a good day, we could sell more than two hundred copies as bacon cooked on campers' stoves or outdoor fire pits. Uncle Bob always made driving exciting. During one trip to Hemet, which was about nineteen miles down a winding road with a number of hairpin curves, we reached speeds of more than sixty miles an hour. We prayed all the way. We called Uncle Bob "Old Lead Foot." Fifty years later, although driving considerably slower, he hit a tree on the same route, so he stopped driving from Hemet to Idyllwild.

We took trips to Idyllwild once or twice a summer and occasionally during winter when there was snow in the mountains. Snow trips were special and were envied by all of our friends. Behind our grandparents' cabin, a toboggan trail wound down a steep hill and leveled off just before Strawberry Creek. Uncle Bob would take us down the hill, sometimes one at a time and sometimes together. Occasionally, we would shoot off the trail and crash into a snow bank, which only increased the thrill of the ride. The downhill run lasted only a minute and half, but the twists, turns, and speed made us feel that we would never reach the end.

Bob gave us a great deal of his time until he got married to our Aunt Jimmy. After that, they both spent hours working in the store, helping my grandfather. Crippled by rheumatoid arthritis, my grandmother was confined to a wheelchair from the time we returned from Hawaii until her death just before the end of World War II. She often told us about her younger years in Germany, her birthplace, and about living in California at the beginning of the 1900s. She said very little about Hitler and the German aggression, but I believe that she remained loyal to her homeland well into the war and until after the world press publicized the atrocities of the Holocaust.

Getting an Education of Sorts

In the fall of 1936, Frank and I entered Woodrow Wilson Junior High (now Middle) School. For the first time, I moved from room to room throughout the day to attend classes taught by different teachers. English, math, history, civics and woodshop classes were the cornerstones of the seventh-grade education. We also were introduced to something called "homework," which was a change from the laissez-faire approach in Hawaii and not a practice that I particularly embraced. Homework took up a lot of time that could have been used more productively to play sports. The school was about five blocks from our house, an easy walk. Behind the school, which faced El Cajon Boulevard, was a large playground where we fulfilled our physical education requirements and entertained ourselves during lunch and other breaks.

Marbles was the "in" game during recess. Most of the boys had marble bags with shooters, agates, laggers, and traders of various colors and sizes. Each player put an agreed number of marbles into the center of a two- to three-foot ring drawn in the dirt. The players then took the lagger and took turns tossing it to see who could come closest to a line scratched in the dirt. The winner shot first, and the order of the remaining players depended on how they ranked during the lagger toss. The object of the game was to take a shooter, usually a slightly larger marble, and knock as many marbles as possible out of the ring. As long as a marble was hit, the player could continue, but when he missed, another boy would take his turn. A player kept all the marbles that he had knocked out of the ring; the best ones built up a substantial collection of quality marbles. Some girls also were quite good at the game I was a rather mediocre player, however, and generally left the game with fewer marbles than when I started. A bag of good quality marbles cost between fifteen and twenty-five cents, which was a lot of money to invest.

After school, the boys in the area played touch football, baseball, Capture the Flag, or whatever the current sport was at school. On Saturday or Sunday, we often headed for the movies. In the late 1930s, western or cowboy movies were in high style. On Saturdays, we could see a double feature—two westerns, a newsreel, a cartoon, and a serial for a dime. For a nickel more, we could buy a Pepsi or Coke and then watch Roy Rogers and Gene Autry all afternoon. My favorite serial featured Buck Rogers and his battles with King Ming. The precursor of *Star Trek*, it had its own rocket ships, cities in the sky, alien beings, ray guns, and an evil empire probably made up of Klingons. We might see a particularly good movie twice and blow the entire afternoon and early evening.

At that time, toy guns were not considered politically incorrect or detrimental to the formative years of youth. Most of the neighborhood boys had a play six-shooter pistol and could buy a large supply of rolled caps for five cents. To play our version of Cowboys and Indians, we gathered after dark in one of the large vacant lots. By late spring, lots were overgrown with tall, wild grass that looked like fields of wheat.

We chose sides and began a shootout. The good guys and bad guys were not cowboys or Indians, just two armies shooting it out. We found our opponents by watching the flash of their guns when the caps fired. Sometimes the battle would rage on for more than an hour before we got tired or ran out of caps. No one ever won, but no one ever lost.

When these activities got boring, we took our armies of lead soldiers to Oscar's corner lot. We built the armies by collecting the lead foil that lined discarded cigarette packages and rolling it into balls. At times, we also purchased pieces of lead from a local hardware store. The toy stores sold molds of soldiers in different battle positions—standing, kneeling, and lying prone on the ground—as well as molds of machine guns and canons. We melted the balls of foil or lead pieces in a cast-iron pot, ladled it into the molds, and set it aside to cool. Later, we broke open the molds and retrieved the soldiers. Over time, every boy in the neighborhood had his own lead army.

At Oscar's lot, each player chose countries that he would defend by hiding his soldiers along the borders. The game was played out with dice, each army element having a different value. For example, a soldier with a rifle had less value than a machine gun and gunner. Each player rolled the dice and kept track of his score. If you reached the total value of your enemy's troops before your opponent reached your army's value, you won the war. If you hid your troops well, and your opponent miscalculated their value, you might win hands down. Sometimes the battles raged for hours with nary a soldier harmed.

At the end of the school year, my parents decided that it was time for another visit to our family in Michigan. As we were older than we had been during the first trip, Frank and I were allowed to discuss the route with our father and mother. We were eager to participate in the debate concerning our route of travel. We didn't know much about the highway system, but we could read a map and knew what we wanted to see. The plan was to go north to Los Angeles, then east to Arizona, then swing north through Utah, stopping at several national parks, then through parts of Wyoming into Yellowstone. From there we would head east through the mountains, across South Dakota, into Iowa and Indiana, and up into Michigan. We planned to camp when the weather and location permitted and stay in motels when it was more convenient to do so. Our year-old Chevrolet sedan was comfortable enough for a family of four and a dog, but it did not have air conditioning.

The trip started off well, although it was extremely hot crossing the desert. No one, not even Kamehameha, was comfortable. Mom had made some lemonade that we sipped throughout the day, which helped somewhat. I don't remember where we stayed the first night, but we covered a lot of ground and saw a lot of cacti, rabbits, and roadrunners along the highway. We arrived in Bryce National Park on the second night and stayed at the local campground. It was a beautiful area and the weather was great, so we slept out under the stars. The next day we stopped at Zion National Park, an experience that I have always remembered. The towering red rocks looked like a valley of statues and carvings made by the gods. We spent several hours looking at the canyon and then headed toward Salt Lake City.

As we moved north, we passed a number of freight trains and were surprised at the number of men riding the boxcars. Sometimes several hundred were hanging out the doors or sitting on top of the cars. They seemed to be moving in all directions, young, middle-aged, and old men searching for work. When we were in Hawaii, we had never perceived the depth of the country's economic situation. While several of our friends' parents were out of work, the effects of poverty were masked by warm and pleasant environment where wild tropical fruits and fish were abundant and families shared what they had. Seeing so many people out of work made us more appreciative. My parents had to scratch to make it, but I never felt impoverished.

In the rolling hills and pastures of Wyoming, we saw a good deal of wildlife, including deer, antelope, and several bears. We entered Yellowstone National Park from the south, circled Yellowstone Lake and stopped at the bridge where the river exited the lake, and visited Yellowstone Falls, Old Faithful, and the sulfur pools. Not too far from Old Faithful, people were feeding bears along the road. Kamehameha was asleep on the back seat next to Frank, but when we stopped and a bear approached the car, the hair on the dog's back grew stiff. He rose from the seat, showing his teeth, and let out a ferocious growl. At first he couldn't see the bear, but then he turned and lunged at the window. My father sensed the problem: neither the dog nor the bear were in a good mood. He quickly drove off, leaving the bear standing beside the road. It took twenty or more minutes to calm down the dog.

Once in South Dakota, we proceeded through the Badlands, and it soon became apparent how the name arose. The hills, buttes, and valleys seemed to be the result of organized chaos. We quickly lost interest and made our way toward the Black Hills and Deadwood—Indian and mining country. Then we headed south and east to Iowa, where some distant relatives lived on a small farm. East of the Missouri, the land flattened, and we could see the effects of the drought that had struck the Plains states in the early 1930s. The great dust bowl had consumed parts of Texas, Missouri, Arkansas, Kansas, Oklahoma, and other central states and ravished the once productive farm areas, leaving sand and wasteland. Even as far north as South Dakota, the land looked barren, dry and hot.

In Iowa, farms improved in appearance. When we arrived at our relatives' farm, we drove up to a large two-story home that sat in the middle of a green corn field. Much to Frank and my surprise, they had a young daughter who was about our age but whose name I can't remember. She entertained us by showing us around and allowing us to ride her ponies. In the evening, we played cards for some time before we were sent off to bed in a large screened-in porch on the second floor. It had been hot and humid during the day, and we welcomed the cool air of the evening. Soon after we went to bed, a large storm passed over, lighting up the home with continuous flashes accompanied by loud claps of thunder. I think Frank and I had a crush on the young lady because after we left the next day, we were both silent and lost in our dreams for most of the morning.

Burma Shave ads were once ubiquitous along roads all over the country, from coast to coast and from Mexico to Canada. Every motorist had seen them. Rather than billboards, they were small, orange signs with one or two words. The ads promoted Burma Shave in poems that appeared on a series of five to ten signs placed in the ground. One example: "The next time you shave, try the present rage, Burma Shave." I can't remember all the sayings, but they were fun to read. Frank and I always competed to see who could spot the next sign. We must have seen several hundred on our trip.

A day or so after leaving Iowa, we arrived at our grandparents' farm, and someone insisted on reminding us not to "bury the chickens." In spite of the ribbing, we had a good time with our uncles, who took us fishing on a local lake and let us ride on their wagon and do chores around the farm. The few days in Michigan passed quickly, and soon we were preparing for the return trip.

We had seen my grandparents and other relatives in Michigan only twice in our lives. They were my father's clan, but Frank and I didn't know that side of the family very well. We frequently visited, however, my mother's sisters and their families in the Los Angeles area, particularly our cousins Katie and Florence. Sometimes we would spend the night at their house after seeing a show together. We also drove to my Aunt Ernestine's to visit her children, Joan and Donald. During one of these visits, I joined Katie and some of her girlfriends at the local bowling alley. There I met a relatively tall, brown-eyed brunette named Ruby. We didn't talk to each other much, but she was very attractive and I fantasized that we might get married in the future.

In the eighth grade, Frank, Gene, Oscar, and I joined the local Boy Scout troop. We bought our Boy Scout handbooks and began to learn the oath and the other requirements for working our way up from tenderfoot to second- and first-class scouts. Scouting gave us the opportunity to camp, hike, and work together while gaining our merit badges. We located an excellent site for camping out near Padre Dam (also known as the Old Mission Dam) in the San Diego River Gorge. Mountains rose sharply on both sides of the river, which made for excellent climbing.

During spring break, there frequently was enough rain to create a substantial flow through the gorge, so it had the added allure of having swift rapids for swimming. It remained a favorite location well into high school. Like all good things, our visit there came to an end when a group of older boys began shooting their .22-caliber rifles over our heads, and we had to hide behind large boulders for protection. The shooting lasted almost an hour, and although the shots were not close to us, we could hear the slugs hitting on the bank a hundred or so feet above us. After things quieted down, we managed to get in touch with our folks, who came and picked us up. The identity of the shooting gang remained unknown, but there was a lot of talk about the incident at school and we all suspected a certain group.

By the time I reached the ninth grade, our neighborhood group—Gene Ryan, Oscar Olson, Louis Black, Eldon Crotts, Frank, and me—had become fairly close. We went camping together on Mission Beach, which was our favorite spot to body surf. We played countless games of touch football and baseball, or sat on apple boxes at the corner store, drinking Coke or Pepsi—discussing worldly and philosophical matters, naturally! When we were allowed to stay out later in the evening, we started to get into a little mischief. We built a hideaway on top of the Piggly Wiggly store (an early version of the supermarket) on El Cajon Boulevard. We climbed up a fire ladder behind the building and used a rope to haul up large cardboard boxes, which served as our camp. We supplied it with food, water, and playing cards. But things got boring, and we became bolder; we stupidly decided to drop water bags on passing walkers. We thought it was hilarious for about five minutes, and then someone called the police. They caught us coming down our "escape" rope. We all got in big trouble—some more than others.

Our friends seldom engaged in any sleazy activities. Once in awhile, however, we were overcome by the desire to demonstrate that we were macho. As the local farms were a considerable distance from our neighborhood, stealing melons or fruit from trees was not feasible. But one summer evening, we felt a touch of bravado and decided to harvest a few watermelons from the local supermarket, which was a block and a half from my house. The California Market displayed its fruit on one end of a row of stands, with canned and other dry goods at the other end. Our scheme was to send one person to the cash register located in the middle of the store to distract the worker while the rest of the group confiscated a few watermelons. Oscar went to the cash register and asked for something away from the fruit display. We picked up three watermelons and ran.

Victory was ours! About halfway down a dark alley, we stopped to eat our booty. There was far more than we could consume, so we ate only the sweet heart of the melons and discarded the rest. After bragging to each other about our caper, we left the area strewn with watermelon rinds. On reflection, we felt a little uneasy as the parents of a schoolmate owned the market, and it all seemed a little shabby. Besides, it was just too easy. We really didn't enjoy the payoff, and the consequences of being caught dampened our joy over the conquest.

Junior high school wasn't the highlight of my life partly because, for some strange reason, my body decided that it was not a good time to grow. Throughout grade school, both Frank and I had kept up with the pack and were of average height, but our growth hormones must have fizzled out in the summer of 1936. The discrepancy in height between us and others our age bothered us since we had never been the runts among our friends. Furthermore, it made life difficult. We were unable to participate effectively in many sports, and chasing girls was difficult and, at times, frustrating. There were no growth spurts later in junior high or high school to remedy the problem. Still, I managed to play on the ninth-grade championship softball team and was a member of a makeshift tumbling team. That was little solace, however, for a kid who wanted desperately to play on the school football team.

While school was uncomfortable, the summer months were filled with many great memories. Most of our friends, particularly Oscar and Gene, loved to go to the beach. Our hangout became Mission Beach, right in front of the roller coaster at the amusement park. We quickly mastered body surfing and would spend most of the day catching waves. Sometimes we brought bag lunches, but if we were flush, we squandered a dime to get a hamburger that would put a Big Mac to shame. For a nickel more, we would get a twelve-ounce bottle of Pepsi and eat and drink in style. On the way home in the late afternoon, tired and sunburned, we would brag to each other about the number of unbelievably large waves we had conquered. Several years later when we were in high school, we moved our beach site to Old Mission. The waves seemed bigger and easier to ride, and more of the Hoover High School crowd hung out there.

Getting to the beach could be difficult because our parents weren't always available to drive us the twelve to fourteen miles down Mission Valley Road and across Mission Bay. Nevertheless, we usually managed to persuade one of the mothers to drive us. Better yet, once or twice a summer, we were given enough money—fifty cents—to buy a weeklong streetcar or bus pass. This allowed us to go beach hopping to Ocean Beach, Coronado, Mission Beach, and La Jolla, which became our favorite spot for abalone diving. The weekly passes also gave us the opportunity to play bus and streetcar tag all over the city. The only unbreakable rule was that we had to stay on the streetcar or bus and go to wherever the pass would take us. If we were clever enough, sat in separate seats one behind the other, and moved the pass along quickly after showing it to the conductor, we could take along a friend who didn't have the money to buy a ticket. I often suspected that the conductor knew what we were doing, but took pity on us.

Our home in San Diego was situated on a plateau above Mission Valley, which was about a mile and a half to the east. To the west, a series of canyons scored the plateau running toward San Diego Bay. Both Mission Valley and the canyons were exciting areas to explore. Frequently, we would hike east to Mission Valley down Ward Street or just drop down the steep bluff into the valley. At the bottom of the hill, we passed a dairy, where we got our informal sex education by watching the bulls service the cows. Beyond the dairy was a small two-lane road that led to where the San Diego River ran into southern Mission Bay. The riverbed was usually dry, except after heavy rains, but even during the dry periods there were large ponds filled with sunfish, perch, and other fresh-water species that could be caught on a small hook and line. We used the valley ponds as local swimming holes and the dry areas as campsites for rabbit- and bird-hunting expeditions.

As we grew older, we were privileged to own .22-caliber rifles and .410 shotguns. We used them to hunt rabbit, quail, and doves. If we were lucky, we would bag enough for a bird feed and cook them next to one of the waterholes. Frank was very good with the shotgun and bagged most of the game that was shot. He always managed to get several quail, sometimes half a dozen or so, along with an occasional dove or rabbit. Our parents had given us strict gun rules, but we also had our own laws: no bullets in the gun chamber until we reached the hunting area, guns on safety until game was seen, and all hunters in sight when a gun was to be fired.

The canyon to the west ran from several blocks east of University Boulevard all the way to the bay just south of downtown San Diego. We explored the whole length of the canyon several times before graduating from junior high school. In a number of places the canyon had steep slopes, which made excellent sledding areas when they were covered with grass. We used large pieces of cardboard as sleds and raced down the slopes with a good deal of speed. After sledding for an hour or so, we would rest at the bottom of the hill among the sage and small trees.

Once, after we had gathered to rest in shade of the chaparral, I was suddenly hit in the back of the neck by a BB, probably shot from a Daisy Air Rifle. It hurt like hell. I turned around to locate the shooter but couldn't see anyone through the sagebrush. After a short time, we heard some boys laughing. I was mad, picked up a rock, and hurled it over the top of the sagebrush. Much to my surprise, a boy came running out, holding his head. Blood was seeping down his face, and he was looking for revenge. He was substantially larger than anyone in our group, as were his comrades. We scattered and headed in all directions. I ran as fast as I could; scared, I seemed to have more stamina than usual and stayed well ahead of my wounded pursuer.

I made it home, believing that I had escaped detection, but the next day the lad and his father showed up on our doorstep. The dad demanded that my parents pay the cost of stitching up his boy's wound. I had shown my folks the welt on the back of my neck and told them what had happened. My father showed the man my neck and argued that I had a right to protect myself. The boy admitted to his father that he had shot the BB gun but had not meant to hit me. His father didn't seem to believe him because he agreed to take care of his costs and acted very civil about the whole matter. I was told that the boy lost use of the BB gun for some time.

We also explored a third canyon, which ran east along Balboa Park and the zoo, from the navy hospital to University Avenue. It was relatively far from our house, requiring us to walk several miles northwest on University Avenue, then west for several blocks, and past the municipal swimming pool to the canyon. We would spend an hour or two checking out the canyon, where other kids usually were playing, and then head up the other side toward Balboa Park. On the way to the zoo, we walked under a large rail trestle that was used by streetcars moving to and from town. We would always cross the trestle to show that we had no fear (or brains) of being caught by the streetcar in the middle of the bridge. A merry-go-round at the zoo entrance offered a free ride to anyone who could grab the golden ring. We wasted more money reaching for that ring than we spent on lunch. The zoo, of course, was a great place to explore, check out the animals, and watch the Gibbon monkeys jump and swing through their cage.

After several years, Frank and I dropped out of the Scouts when our troop was disbanded. Later, I joined the Sea Scouts, which held their meetings on board the old *Star of India*, which was anchored just south of the San Diego–Coronado ferry dock on the east side of the bay. A decade or so later the ship was refurbished and moved north to the foot of Broadway, where today it serves as a marine museum.) The Sea Scouts attracted members from all over town. The troop owned an old navy landing craft with a hull like an oversized dory, which had been converted to a small two-mast schooner without a cabin or covered area. It was great for sailing in the San Diego Bay, and at times we would take it far offshore.

One summer, we sailed to Newport Harbor, about fifty-five miles north of San Diego, where a Sea Scout jamboree was taking place. The event ended with a dance. We dressed in our white uniforms and entered the pavilion, where a number of attractive young ladies were waiting to be asked to dance. Unfortunately, most of the girls were taller than I was, at least with their high heels on, and I felt uncomfortable. I stayed and watched for a time, felt lonely although I was in the middle of a large party, and left without asking anyone to dance. The fact that I was not much of a dancer probably also hastened my departure. While I was a Boy Scout, I had earned the rank of Star Scout, but I so enjoyed the sailing and camaraderie of the new group that I never attempted to acquire any more merit badges. I just hung out with the other Sea Scouts and had a good time.

My great triumph at this time was being selected to play a major role in the ninth-grade play, *Saturday Evening Ghost*. I played a twin; my twin sister was played by Gertrude Downs, a pretty blonde, who was perhaps an inch shorter than I was. . It wasn't a very serious endeavor, but it gave me a chance to interact with the school aristocracy. Gertrude treated me well enough, but it was obvious that she didn't want to be associated with a short kid. Nevertheless, the play was a success, I earned a little notoriety, and after the play, cast members were invited to a dinner dance. It was my first dance, and somehow I persuaded Betty Lee Anderson—later a cheerleader at Herbert Hoover High—to be my date. Unfortunately, I really couldn't dance, so she spent the evening giving me pointers.

When I entered the tenth grade, all sixty-one inches and 118 pounds, I felt like a dwarf among my contemporaries, and it became increasingly obvious that there would be no invitation to join the high-school football team. The coach probably thought I was too small to be the water boy. Nevertheless, I was determined to find a way to play football, even if it meant joining a sandlot team. With a little urging, I managed to talk Eldon, Gene, Oscar, Louie, and my brother to form a six-man team and challenge the other neighborhood groups to weekend contests. Our largest player was Eldon, who was just over 145 pounds, followed by Louie who was a few pounds lighter. Recognizing that we could hardly play power football, we decided to build our team strength on trickery. We practiced almost every day after school, perfecting a series of plays involving reverses, fake reverses, double reverses and fake hand-offs ending in passes.

I played the quarterback position, but seldom took the ball directly from the center. In most of the plays, the center threw the ball back so I could hand-off to either Gene, who was our passer, or to Oscar, who was usually our blocking back. After considerable practice, we became quite good at fooling our opponents. We bought sweatshirts and cleats so we'd look like a respectable team. We first challenged the kids in the next block but gradually expanded and began playing games on the high-school practice field almost every weekend during the fall. Despite our short stature, we won every game in September and October, and our reputation spread throughout the school and in adjacent neighborhoods. The competition became more formidable, however, as the size of our opponents increased.

The team remained together throughout high school, probably because none of us grew big enough to play on a varsity club. Our winning streak continued until our senior year. By that time, I had grown to a whopping five feet four and 132 pounds, while other members of the team weighed as much as 160 pounds. During the team's last year, some of our school companions decided to try to beat the 35th Street Gang. Their group included a few members of the high-school junior-varsity team, and its smallest player was larger than our biggest, Eldon. The game took place on a field next to the high school, and a number of our respective supporters, including a few JV coaches, gathered along the sideline to watch the show.

We had little difficulty scoring on the opposition, and at first their defense was utterly confused. Still, their size gave our defense a lot of trouble. For the little members of our team, including Frank, Oscar, and me, tackling their 170-pound back wasn't easy, and we were getting hammered. Our only chance was to throw a low block and knock the runner off his feet. It was a day for the offense. The score switched back and forth throughout the game. Near the end of the game, our opponents had a six-point lead. After they kicked off, we got a good run back, and on the next play we ran a fake reverse and then passed, scoring a touchdown and tying the game 34–34. If we made one more point, we would win the game. There were no goal posts on the field, so we knew we'd have to run or pass the ball.

By then our opponents had adjusted to our bag of tricks, so we decided to try something new. We put my brother in the quarterback position and Oscar on the line. Frank was our smallest player, and we were sure the other team would expect one of our trick plays. They probably would not expect Frank to run since he was so small. We told Frank to go up the middle. As soon as the ball was snapped, we all ran forward to block. They spread out their line to stop our frequent reverses and laterals, and their backfield sagged. They were caught off guard, and we easily crossed the goal for the win. That was the last game we played together and we went home bruised and tired, but proud. The next school day, the JV coach stopped me in the hall and told me that I had a lot of natural talent. He probably said it just to make me feel good, and it did, but I would have given up all these memories for one good day on the high-school varsity team.

High school, like junior high, was not easy for me. Because of my size—by graduation, I had reached the remarkable height of five feet six inches—I did not excel at most sports. The only sport I participated in was boxing, competing in the 132- to 135-pound class during my senior year. I got along well with the coach, who trained me well. I won three matches in a row and eventually the school finals. Girls were often out of reach, as well. I dated Eldon's sister once in awhile, but only had enough money to take her to the movies and have a coke. In addition, except for math, in which I got mostly A's and B's, I did not do well in school. I had an aversion to homework, was hyperactive, had a short concentration span, and was a daydreamer. Graduation exercises were held in Balboa

Park, just outside the zoo. Let it be known that I graduated without honors but happy to have completed twelve years of schooling. My grades were such that I had absolutely no chance of entering a college, or at least that's what I thought.

To earn spending money, Frank had a paper route with the *San Diego Union* and I worked for the *San Diego Sun*. The *Union* is still in business, but the *Sun* failed many years ago. On Sundays, we usually had a few extra papers, which we sold at the corner of 35th and El Cajon. On Sunday, December 7, 1941, we were selling papers on the corner when the news broke about the attack on Pearl Harbor. We sold our extras and returned home to be with our mother. Retired from the navy for several years, my father had already been called back into service and had left for his station at Point Loma. My mother, brother, and I sat glued to the radio, listening to every word, but I could not believe that the surprise attack had been so successful. We had such pride in our military service and the navy. It seemed impossible that the Japanese could have destroyed most of our battleships. I was overwhelmed and wondered whether we were receiving false reports. We did not know the true extent about how much damage the Japanese had done for several months.

Our family had been a part of the navy scene in Hawaii, and the attack on Pearl Harbor shredded my belief in the invincibility of our military. From December to the end of the school year in June, I was preoccupied with the news of the war, hoping each day to hear something encouraging about the battle at sea. After school, our group frequently gathered at the malt shop on the corner to discuss our potential involvement in the war. We knew we were bound to be part of the escalating conflict, either in Europe or the Pacific. Gene wanted to join the Air Force, Louie wanted to be a gunner on a B-17, Oscar was undecided, and both Frank and Eldon wanted to be paratroopers. I wanted to follow in my father's footsteps and join the navy. It wasn't long before we were all given a chance to volunteer or be drafted.

After graduation, my parents moved to Ocean Beach, and I began putting in a lawn and landscaping the yard around our new home. I went surfing at Old Mission several times, but it just wasn't the same. Many beach buffs had already entered one of the services, and I felt uneasy, knowing that the draft would soon call me to service. The chances of getting into the navy were less likely if I waited to be drafted. So in late

June 1942, with my parents' permission, I joined the navy reserves with the rank of seaman first class at the age of seventeen years and nine months. I received this rank, instead of ordinary seaman or second-class seaman, because of my service in the Sea Scouts, and perhaps because the recruiters knew my father.

Things Just Change

My description of life in the late 1920s, '30s, and early '40s may not seem that different from life in the 1990s or 2000s. To me, however, the difference is as obvious as the difference between day and night. When I was a very young lad in San Diego, the horse was rapidly fading out as a method of transport, but ice was still delivered along the street by horse and wagon. To ensure delivery, Mom would put a sign in the window that said "ICE." On hot days, we chased the wagon, hoping to scavenge a couple of extra pieces. Bread wagons delivered bakery goods, and a vegetable truck came by about twice a week. We had no concept of supermarkets. We listened to the radio for news and, occasionally, to some comics who performed on Sunday evenings, such as Jack Benny and Fibber Magee and Molly. A few of our friends had radios in their bedrooms, where they heard the adventure serials, such as the Lone Ranger; Zorro; Jack Armstrong the All American Boy, and the Shadow. We also listened to the music of Glen Miller, Benny Goodman, Arty Shaw, Tommy Dorsey, Les Brown, and other great bands and the singing of Bing Crosby, Frank Sinatra, Patty Page and the Andrews Sisters.

For the most part, however, electronics did not play a major role in our lives; there were no televisions, computers, or Nintendo or other video games. There were also no McDonald's, Burger King, Kentucky Fried Chicken, or Taco Bell, although drive-in restaurants did open in the late 1930s. With the exception of baseball, professional sports had not entered the scene in a big way, and the PAC 10 had not yet signed an agreement with the Big Ten for the Rose Bowl. Air transportation was for the rich; in any case, most of us felt it was not very reliable. Most of the population was poor according to modern living standards, but we kids were not aware of any material shortcomings.

For all practical purposes, drugs were unheard of in school. A few kids smoked Joe Camels or Lucky Strikes behind the gym or other less obvious areas. Most of the students knew about the few girls who would "make out," but not about those who were more discrete in their sexual affairs. I suspect that they were better known to the "in group," which included the athletes. Pregnant classmates were a rarity, but we all knew one or two girls who had to drop out of school to marry and have a baby.

On average, the students were not deeply into the sex scene, partly because we had so little money—in those days, the boy paid for the dates. No more than two dozen students owned cars, and it was difficult to take a girl on a date if she had to ride on your bike's paper rack or handle bars. Finally, most of us were not very sophisticated, perhaps because there was no T.V. to "educate" us and promote premarital sex. Teenagers were more naïve then; there were no free condoms, no pharmaceutical drugs to cure sexually transmitted diseases, and no sex-education classes (beyond a film about venereal disease that was shown to seniors). Sex, however, was not completely removed from our thoughts, and a few of the more adventurous students traveled to Tijuana, Mexico, to buy their own sex education.

This was a period of rapid change in global politics. The Depression, which had held this country in a tight grip during these years, had also had an impact most of the other technically advanced nations. Germany had become increasingly powerful, and by the time I graduated, Hitler had invaded Austria, Poland, and the Rhineland. War broke out throughout much of Europe. The great German air raids on England (especially London) and the submarine attacks on shipping in the North Atlantic were daily news. The war seemed to go poorly for the British after the Allied Army made its great escape from Dunkirk. Mussolini had fought his war with Ethiopia and was flexing his muscles in Greece and Albania.

U.S. relations with Japan had slowly eroded during the late 1930s following the latter's invasion of Manchuria and later China. Little trust remained between our two nations. I listened to the accounts of these developments over the radio with interest, but they all seemed remote from our lives, that is, until the surprise attack on Pearl Harbor and our entrance into the war. Most high-school seniors recognized that they were next in line for the draft and that there was little purpose in planning a life after graduation. Thus, I never gave any thought to getting a job or other post-graduation plans. The men in the senior class just knew that they would end up in the military service, unless they were 4-F (physically unfit).

Before December 7, 1941, the United States had managed to stay out of the growing global conflict, although President Roosevelt, with the help of Congress, was busy shipping arms, petroleum products, surplus navy ships, and other war materials to England. After war was declared, gasoline, select foods, and other scarce materials were rationed. From time to time, San Diego had blackouts because of Japanese raids on coastal facilities. A Japanese submarine shelled Santa Barbara, and balloons carrying bombs landed in Northern California and Oregon. Unfortunately and unnecessarily, some of the public feared Japanese Americans living along the coast, and they were sent to special camps. History would later stamp its disapproval on this decision, but at the time it was, for the most part, a popular move. The military presence in San Diego grew rapidly as the number of recruits arriving at the navy and marine boot camps expanded and the armed forces on Point Loma were reinforced. By the time I entered the service, the streets were crowded with men and women from all branches of the military.

CHAPTER IV
THE MAKING OF A SAILOR

My entrance into the military came as no surprise to the family. In fact, they expected it. It was just a matter of time before all the men in our family would be in the nation's armed forces. My father had been recalled to duty right after the attack on Pearl Harbor. During my last semester in high school, my attention was centered on the struggle in the Pacific. There was not a lot of good news in the early months of the war. In the aircraft carrier battle of the Coral Sea (the first major sea battle following Pearl Harbor), we seemed to come out reasonably well considering the circumstances (each side lost one major ship). But we had lost the battle for the Philippines. General MacArthur said his farewell to his troops on Corregidor, and Japan took Hong Kong and Singapore from the British. During this period, the United States organized the military branches, developing its industrial arm to build warships, transports, air craft, tanks, small arms, and the other equipment needed to wage a war.

Upon entering the navy, I was sent to the 11th Naval District Headquarters in San Diego for a physical, given the normal round of shots, and then sent off to boot camp. Later, I was assigned to a radio school, which was located at the edge of Fort Rosecrans on Point Loma. There I joined about twenty other students in a small makeshift classroom, where we learned first how to use a typewriter and then Morse code, by listening to tapes that fed into earphones. The code was played slowly at first and then at greater speeds as time passed by. The goal was to train the students to read about twenty words of code a minute, which meant that our typing and reading abilities had to increase simultaneously. School lasted from about 8:00 a.m. until 2:30 p.m., after which we trained in fire fighting, use of weapons, marching, and military conduct. On Saturday evenings, if our conduct had been satisfactory during the week, we were allowed to go to the movies or the canteen, which was three blocks or so from the base.

My salary was about $56 a month. Since the navy provided room, board, and clothing, that amount seemed like a fortune. There wasn't much opportunity to spend it, however, because we were not allowed off base during the training period, except to go to the canteen. Within about three weeks, I had increased my code-reading speed to over fifteen words per minute and hoped to reach twenty minutes within the next two weeks.

Most of the men in the class were recent high-school graduates, although none were from Hoover High. A few were more than twenty years old, but most were still in their teens. I made friends with a young man named Chris Vesper who had come down from Whittier, part of greater Los Angeles. Chris was a stocky young man in his early twenties; he had played high-school football and was engaged to a girl named Cameron, who also lived in Whittier. We went to the movies together and played football on the marching field in the late afternoon. Chris was also progressing well at the navy school, and we thought that we would probably get shipped out around the same time.

By early September, Chris and I were reading more than twenty code words a minute, and we soon expected to be moved to a duty station at sea or perhaps to a land-based combination facility. Thus, we weren't surprised when our training petty officer called us up to his office and told us that we were being considered for a special assignment that would require extensive background security checks. After the meeting, Chris and I laughed and wondered if they would contact our schoolteachers. We soon learned that the check was a little more formal and comprehensive than that. Our parents told us that the security people had questioned almost all our friends and their parents, teachers, classmates, and probably the local police. We wondered how much dirt they had collected on each of us. I was even concerned that the Piggly

Wiggly affair, which I assumed had been documented in a police report, might jeopardize my new assignment.

Imperial Beach

A week later, we found out that we had been cleared temporarily and would soon move to a new duty station. Several days later Chris and I, along with our sea bags, boarded a truck and departed Point Loma with no information on the character of our new assignment or where, when, and how we would reach it. We expected the truck to take us to the waterfront, where we would transfer to a ship, or perhaps to the airport or train station for transportation to some other part of the country. We were surprised when the truck turned south, passed through San Diego and South Bay, and continued towards the U.S./Mexico border. We began to speculate and joked that we were going to a secret location in Baja California. The truck turned toward Imperial Beach, which was located at the southern end of a long sand spit that sheltered the outer side of San Diego Bay and then ran north to what we called Coronado Island (actually the broad end of a peninsula).

The town at Imperial Beach consisted of a gas station, several bars, and a dance hall/bar located next to an old fishing pier. We turned up a small road, which followed the beach to the north, and stopped at a fenced compound. A Marine seated inside a small guardhouse asked us for our papers, examined them, and then let us through. I saw a large building that housed the men on the base and, just north of that, two small buildings, one at the end of a boardwalk and another at the end of a walking bridge that crossed a large fresh-water pond next to the sand spit. Several larger buildings were under construction east of the pond. Our sleeping quarters comprised a room about twenty-five by forty feet attached to the main building; we were shown our bunks and lockers and told to report to the main office in the morning. The living quarters were heavily curtained so that the men could sleep during daylight hours, which suggested operations continued around the clock.

The next morning Chris and I reported to a small office in the same building where we'd slept. The office, enlisted men's sleeping area, galley, and eating facilities were all under one roof. The lieutenant in charge told us that we were to be assigned to the 500 kHz emergency frequency until our full clearance had been approved. This meant standing watch on the frequency set aside for distress signals from military or civilian ships. At that point in time, the emergency frequency was used to send out two basic Morse code signals—SOS: three dots, three dashes, and three dots, or SSS: three dots repeated three times. SOS was the intentional distress signal for ships in trouble due to submarine attacks, bad weather, or any other problem that impeded vessel safety. SSS was used to denote the sighting of a foreign submarine. The assignment did not require a great deal of experience, as the signals were generally transmitted slowly and carefully, and the information was repeated. On one hand, it seemed like an important task. On the other hand, it was suitable to our skills.

My first shift began that day at 4:00 p.m. and ended at midnight. Chris was my replacement and would work from midnight until 8:00 a.m. The watches were scheduled to be eight hours on and eight hours off over a forty-eight-hour period, with a twenty-four-hour break before the series was repeated. Thus, each of us was assigned one evening, one day, and one midnight watch with sleep intervals in-between before we had a day off. This meant that the only time Chris and I would see each other was when we replaced each other on duty and for part of our twenty-four-hour leave. It also meant that we were on duty for about sixty hours a week. The extended work hours were required because there was a shortage of operators but many communication demands. It was easy duty compared to combat.

That afternoon, I walked to my station wondering why my security clearance was taking so long and what was so secret about the work at an isolated base on the beach. A number of other men worked in the building on the other side of the pond, and we were told that the base staff would increase significantly after the new buildings were completed; several new radio towers also were being erected around the compound. At the radio shack, I was shown to a seat, given a set of earphones, and told the job requirements. First, it was essential to copy the data from the vessel in distress, regarding its identification, position, and other relevant information. Second, we had to take a direction-finder reading on the signal to help identify the position of the ship in difficulty and/or the enemy ship. When this was accomplished, the copied message was relayed to a central location where the information was evaluated and action taken.

Taking a reading involved turning a loop on top of the radio direction finder (RDF), exposed on the top of the building, until you reached a null (i.e., where the signal disappeared or was very low) and exercising a procedure to check the signal's direction. After about an hour of instructions, which included directions on what forms needed to be filled out (this was the military, after all!), I was left to fend for myself. I was a little nervous and sat listening carefully for any distress signals, but the radio was silent for the first several hours. Off and on, I changed the volume on the radio, thinking that it was not turned up high enough. Then around 8:30 p.m. an SOS signal burst into my earphone. I carefully copied the vessel's call letters and position without any difficulty, checked it against the ship's repeat message, and managed to get a good bearing of the RDF. Without hesitation, I radioed the message to headquarters using the transmitter key. After receiving an "R," designating "message received," I heard other RDF stations sending in similar messages and bearings on the ship. Of course, we had no idea of what was done to assist the vessel beyond that point. Later that evening, I copied two SSS messages, and all went well.

I felt quite good about the first night on the job and told Chris about my experiences when he relieved me at midnight. Then I went to the galley for a cup of coffee, where about five other radiomen, all petty officers, were making small talk, so I joined them. They were friendly, but it was obvious that they changed the topic when I sat down. They asked me a few questions about my first night's duties, and as my work was not classified, I told them what had happened. The conversation shifted to social affairs, comments on the war news, and the expected staff increase over the next several months. Before long it was 1:00 a.m. and time for me to hit the sack, if I was to be alert on the morning watch that was less than seven hours away.

Morning came fast, and by 6:30 a.m. I was preparing for breakfast. At 8:00 a.m. I relieved Chris at the RDF shack. He had a rather quiet night, just a few SSS reports, and suggested that some reading material could help pass the long intervals between distress and SSS messages. I'd had a similar thought and brought my radioman third class handbook. Both Chris and I would be eligible for promotion early in 1943, so we began our studies immediately.

We cycled through five or six watches before going on leave. We spent the first few twenty-four hour breaks on the beach, and although it was almost October, the weather was warm and body surfing was good. Few girls, however, went to Imperial Beach, and those who did most likely had been told by their parents to stay clear of sailors! Thus, there was not much chance of developing a relationship with the opposite sex. Over the summer and early fall, I had added three inches to my height and was slightly over five feet nine, and swimming and working out on the beach had not hurt my physical appearance. As San Diego was not the greatest town for finding a date, I got in touch with my cousins in Los Angeles to see if they could help. Katie told me that she had a number of friends who would like to go out with me, including some whom I had met in the past, so we mulled over a number of names. I decided to check out Ruby Lane, an attractive girl that I'd met at an L.A. bowling alley when I was about fourteen.

It was early October, and I had planned to take the 10:00 a.m. train at the San Diego depot, after completing my night shift. The navy bus took me into Coronado, then I jumped onto a streetcar in front of the Del Coronado Hotel, took the ferry across San Diego Bay, and arrived at the station around 9:40 a.m. The ticket to Los Angeles cost $2. A large crowd wanted to get on the train, but military personnel were given priority, and I entered a coach with about ten minutes to spare. The weather had cooled down somewhat; while it was still in the seventies during the day, the temperature dropped into the high fifties at night. I had my navy pea coat and toiletries in a small carry-on bag. As we traveled between San Diego and San Clemente, the summer crowds gave way to long empty beaches with occasional walkers. In my hurry to catch the train, I had not eaten breakfast, so I walked to the dining car and enjoyed ham, eggs, and coffee. I looked out the train window, trying to reconstruct Ruby in my mind, but my vision was of a fourteen-year old girl and she was now eighteen. Back in my seat, I dozed off, and suddenly the conductor was calling the stop for Whittier. I got off and hitched a ride to Maywood, which was about fifteen miles away.

I arrived at my Aunt Katherine's home around 2:30 in the afternoon. Katie, who was in junior college, and Florence, a high-school senior, were expected home soon. While we waited for them, my aunt asked me where Ruby and I were going that evening. The initial plan was to go to a skating rink on Atlantic Boulevard. Around 4:00 p.m., Katie arrived; she had plans to go dancing that evening with a young sailor named Bill Ming and wanted to know if we could join them. It sounded great, but since my dancing abilities were next to zero, I declined the offer. After dinner, Katie told me how to get to Ruby's house, which was about half a mile down Atlantic and one block south. I walked down Atlantic, still wondering what my date looked like. Katie had told me that she was very good-looking, but then she was Ruby's friend and had a different perspective on looks and personalities. The house was two-bedroom stucco on a corner lot. I knocked on the door, and Ruby let me in. I was pleasantly surprised. She was about 5 feet six inches tall with relatively long brown hair, brown eyes, and a nice figure.

While Ruby completed some last minute beauty preparations, her father and mother gave me the once over. What they saw was a young seaman sailor without rank or transportation who was nervous and anxious to get out of the house. I doubt I was what they had in mind for their beautiful daughter. Nevertheless, within a few minutes Ruby and I set off for the skating rink, which was about two miles away. As we walked, she asked me about my role in the navy and other generalities about my life. I did likewise and found out that she worked as secretary at one of the Sears stores. In a short time, we reached the rink, rented our skates, and began skating to the music. Ruby held my hand while we skated, and it felt warm and comfortable, but she was quick to let go when we stopped skating. Regardless, the evening went well. She was friendly and had a great sense of humor, but was also very proper—too damn proper! Back at the door of her house, I tried to kiss her on the cheek but she moved away, said, "Goodnight, call me again sometime," and entered her home. That was the end of the date.

I walked back to my cousins' house trying to evaluate the evening, but there wasn't much to consider. We'd had a good time, said a few departing niceties, and it was over. No kisses, hand-holding, or even sweet talk. A total bomb by today's dating standards. Even though I was attracted to Ruby, I didn't have the slightest idea how she felt about me. I got back to my aunt's home around midnight and went straight to bed since I had to return to base the next morning for an evening watch at 4:00 p.m. I left for San Diego around 8:00 a.m., hitching a ride to the Whittier train station. In late 1942, there were hordes of military personnel hitching rides all over the country, and if you were in uniform, you usually were picked up immediately. In fact, I probably could have thumbed my way to the base without any trouble, but I boarded the train and arrived at the San Diego depot just after noon. From there, it was a short walk to the ferry, a fifteen-minute ride across the bay, and another ten-minute ride from the ferry terminal to the Del Coronado Hotel, where I could catch a navy bus back to the base.

We were still working the RDF station in mid-November and had heard nothing about our secret mission. Both Chris and I continued to study for our radioman third class rating and hoped to take the test early the next year. Several other radio operators joined the base later that month, and we were put on a forty-eight-hour-on, forty-eight-hour-off schedule. This gave us considerably more time off and that I could go to Los Angeles on a more regular basis. Katie continued to serve as my dating service, introducing me to a number of her friends. They were good-looking women and I liked them all, but for some reason I kept going back to Ruby. She was more fun and better looking than the others. She also had a sense of grace and was easy to talk with but after four dates, we were just beginning to hold hands and we hadn't even kissed—this was unheard of! By Christmas, it was obvious that I wasn't making any headway with her even though she sent me an unexpected present (a box of candy). I abandoned the chase for Ruby and began to look for other female companions.

In early 1943, Chris and I were told that our full clearance had come through and that we would be reassigned to a new task the next day. When I asked why the clearance had taken so long, the officer shrugged his shoulders and said, "It always takes longer than you think." We were to be part of a top-secret group that was intercepting Japanese radio messages and attempting to break their code. The base also was part of an RDF network stretching across the Pacific that monitored transmissions from Japanese submarines. The radio intercept station plotted Japanese submarine positions to aid antisubmarine warfare efforts. A large group of radiomen from Bainbridge Island, Washington, trained in copying Japanese code, would arrive at the base soon to begin an expanded radio-intelligence effort.

In the interim, we were to train ourselves to read the Japanese code and be ready to man the RDF intercept facility in the next several weeks.

The work sounded exciting; Chris and I, as well as several newcomers to the station, were soon hard at work learning the code, which had almost twice the number of characters as our alphabet. At the beginning of the self-training, we were given a list of Japanese characters and the number of dots and dashes associated with each one. Instead of A, B and C, the Japanese characters were represented by two-letter sounds such as ho, he, ha, and hi. To fit all of the Japanese characters on a typewriter keyboard, it was necessary to assign two sounds or characters to each key. When we heard one of the two characters contained on a particular key, we either hit the shift for he or ha, which were written on the keys of the typewriter, and copied the upper designated letters (e.g., ho), or left it down and typed the lower character (e.g., ha). It took a good deal of skill to copy the code rapidly because there were so many characters and we had to keep shifting and lower the typewriter keys.

Within several weeks, our watch officer felt that Chris and I were ready for our assignment to the RDF intercept station, which was located just east of the large pond near our living quarters. This two-story building was larger than the RDF shack on the beach and guarded by an armed marine. We were shown an operational table with a telephone link system that would allow us to contact a network of stations from Imperial Beach to Bainbridge, Washington State. Japanese radio signals were broadcast via a speaker system simultaneously through the RDF network, along with information on the frequency of the incoming signal. When an incoming signal was on the speaker, it was our job to run upstairs to the large RDF unit, tune to the given frequency on the radio, match the incoming code signal with the one coming over the phone link, take a bearing and differentiate it from its reciprocal, meaning that the signal could be coming from either of two directions, 180 degrees out of phase. We then pinpointed its origin by plotting the bearing we received from the network. Our major task was to listen for the signal NA-RA-E that denoted a Japanese submarine and take bearings on its radio transmissions.

Chris and I soon learned that our promotions to radioman third class had been approved. We were now part of an elite radio intelligence team that had been organized before the war and had decoded at least parts of Japanese radio messages relating to the attack on Pearl Harbor. Since that time, radio cryptology (code breaking) had made rapid advances.

The effort was assisted by an extensive network of intercept stations and a new computer facility in Washington, D.C. The radio intelligence work included identification of incoming signal sources, radio fingerprinting (identifying a Japanese operator by the characteristics of his key signals), and code busting.

The first great success of the radio intelligence group was in May 1942, when it decoded messages about a planned attack on Midway Island, which allowed the United States to surprise the Japanese flotilla. The information, along with a good deal of luck and brave pilots, resulted in the sinking of four Japanese aircraft carriers. The battle was considered a turning point in the war in the Pacific. Radio intelligence would play a major role in other naval and air battles in the area, including the defeat of the Japanese submarine menace and the shooting down of Admiral Yamamoto, the man who led the attack on Pearl Harbor. Everyone associated with the effort felt a sense of pride and accomplishment, even though we could not discuss our role with our friends and families.

The RDF station usually was manned by a single individual, but the coverage was around the clock. This required four operators: Chris, two newcomers, Jim and Bill, and myself. Jim had come down from Bainbridge and was well trained in copying Japanese code. Bill came from Point Loma and, like Chris and me, had attended the local radio school and trained himself to read and copy the code. The work was far more interesting, exciting, and demanding than watching over the emergency frequency. Numerous NA-RA-E signals came over the telephone link. I always tried mentally to position the submarine, based on the bearings I took and those transmitted across the network. The strength of the signal gave me a sense of it was close or far, but even that was a pure guess. To pinpoint an actual location, we needed to know the bearings taken from the other stations on the net, but that wasn't our job. By the spring of 1943, my code speed had improved, and I had no difficulty copying the routine Japanese messages that were transmitted on a number of frequencies.

The base personnel increased rapidly in number, and by late spring most of the new facilities had been completed and routine activities were operating smoothly. The RDF crews moved to a new building with technology (automatic direction finders or ADF) that allowed us to take almost instant bearings on incoming signals without turning the large-winged RDF. The incoming signals were intercepted from a series of antennas forming a circle, and I believe the bearings were calculated from the differential strength of those signals. The job became much easier. The new ADF activity increased the number of operators on each watch to about ten. This allowed us to take bearings on intercepted submarine transmissions and at the same time search for other broadcasts from submarines or ships on the known range of frequencies used by the Japanese. During this period, Jim and I became good friends and at times were assigned to the same watch.

During the first nine months of service, I had been very thrifty with my income and put most of my checks in the bank. Since the logistics to and from Imperial Beach were time consuming and difficult, I decided to buy a car. Having personal transportation would make it easier to drive around the south bay and into San Diego. After visiting a number of used-car lots, I settled on a 1936 Ford Coupe. It had a reworked engine, a new head and retreaded tires, and the interior looked good. I paid about $450, which was a lot of money. I drove the car back to the base, reaching about sixty miles per hour on the open highway. It had a slight shimmy, which I suspected was a result of a much-needed wheel alignment. I thought the car would give me added attraction as a date. Many of the girls I had gone out with weren't thrilled about having to walk or take a streetcar or bus.

After getting a booklet of gasoline ration stamps, I called a girl I had known in high school and asked her out on a date. The first thing she asked me was whether I had grown since graduation and how tall I was. Fortunately, I had had grown to a little over five feet ten since my enlistment in the navy. My height passed muster, and we arranged to meet several days later. Betty lived about two blocks from our house in east San Diego. She had always been very shy in high school; thus, I was surprised to find that she had blossomed into an and outgoing attractive girl. She immediately took my hand and let me know that she loved dancing and just happened to know where a band was playing in town. I wasn't very enthusiastic because my dancing skills, although improved thanks to Ruby's efforts, were still primitive. With a few squeezes of my hand, however she convinced me to take her to a dance in Chula Vista. After spending the first hour attempting to teach me how to fast dance, she resigned herself to my incompetence. Later, she asked me if she could dance with some of the other men. When I said yes, she excused herself and began to do the Lindy and other fast dances with several other sailors and marines, all of whom were eager to dance with a good-looking blonde. Betty returned throughout the evening to dance the slow tunes with me and to assure me that I was her date for the evening.

Following the dance, we went to a drive-in and had a coke and hamburger. Betty was friendly, sat close, and squeezed my hand as we talked. She even suggested a quiet place at the end of a dead-end street where we could have a good view of the city. I'm not sure what the view was, but I know that she knew a lot more about making out than I did! When we returned to her house, she invited me inside to meet her folks. They asked me a lot of questions about my work at the station and whether I expected to be shipped out in the near future. Such question-and-answer sessions, which were not infrequent, made me uneasy; they forced me to build a reasonable story and keep it consistent over time. Betty seemed anxious for another date, but I wasn't keen on the idea of going back to another dance. I suggested that we go to the Mission Beach Fun Zone, but she was in love with dancing and didn't like the alternatives. So ended the Betty affair.

By early 1943, my immediate family was totally immersed in the war. My father had been transferred to the cruiser *Honolulu*, which was engaged in activities in the western Pacific; Frank had joined the army and was training to become a paratrooper in the 11th Airborne. Mom was working for Consolidated Aircraft. Like most homes with sons or fathers who had gone off to war, ours had a banner with three stars in the window. Every member of our sandlot football team also was part of the war effort. Eldon, like Frank, had joined the paratroopers and was serving in the 82nd Airborne; Oscar had joined the army infantry and was training for the invasion of Europe; and Louie was in the air force, a tail gunner on a B-17 making flights over France and Germany. Gene had opted for the U.S. Army Air Forces and was somewhere training to become a pilot. Our group was spread across all the major services except the marines. I suspect that each of us wondered if we would return unscathed or if the war would take its toll from our ranks. Until mid-1943, my part had been easy, and I remained out of harm's way. My promotion to radioman second class came through as soon as I had nine months' experience as third class. The work was demanding, but passing the war at Imperial Beach was a gravy train compared with the challenges that would confront of us in the future.

I had not seen much of my brother or our friends since I joined the service. Most of the others completed their basic training away from San Diego, but when they returned, we met to exchange stories about our experiences. Each would stay a week or two at most before he shipped out again. Frank came home for a few days and then returned to continue training as a paratrooper. The war in Europe had heated up, the Allies had invaded North Africa, where General Patton and his tank corps engaged the famous German warrior Rommel, the "Desert Fox." The Allies had taken Sicily. By mid 1943, U.S. and British planes were carrying the war deep into German-occupied territory in Europe. Louis was smack in the middle of the B-17 raids over western Europe.

In the Pacific, the U.S. position had improved steadily since the Battle of Midway, and the navy was engaging Japanese fleet in the western Pacific. The battle for the Japanese-occupied Guadalcanal had begun in early August 1942, and fearless fighting continued into 1943; the islands were not completely secured until February. A series of raging sea battles had taken place around islands that were unknown to most of us. During one of the engagements, my father's ship, the *Honolulu*, was hit by a Japanese torpedo and lost a hundred feet of its bow. The ship's watertight bulkheads held and the vessel was towed back to Pearl Harbor for repairs. Dad had been on the bridge when the torpedo hit; he felt the sickening heave of the ship and assumed the vessel would be lost. The bow sagged, in front of a major bulkhead, ripped clear of the rest of the ship, and sank into the ocean's depths.

Dad's quarters were in the bow section, so he lost all of his personal belongings. He came home for a short leave while the ship was under repair and gave me a cursory update on what he knew about the battles in the Pacific theater. We were making advances and generally winning the night engagements, but were still losing a lot of ships, men, and aircraft. Dad soon returned to his ship, which was involved in a number of battles as U.S. forces made their way across the Pacific and prepared to retake the Philippine Islands.

I had owned my car for about four months, using it to visit my mother and date several local girls. One evening while on watch, a mate asked if he could run into Imperial Beach to get hamburgers for the watch members. After checking his license, I gave him permission to make the short run, expecting him back in fifteen to twenty minutes. After about an hour, it was obvious that he was not going to return. When I ended my watch at midnight he still was missing from the base, and the next morning the base commander sent for me. I was stunned to learn that my friend took the car to San Diego, picked up a couple of girls, got drunk, drove down the wrong side of a freeway, and ran head-on into another car. The two girls in my car had been killed. Two of the three people in the other car died several days after the accident.

The police came to question me about the loan of the car, but since the driver was properly licensed and had my permission to take the car, he was held legally responsible for the deaths. The insurance companies, however, tried to hold my parents responsible since my friend was broke. I suspect the fact that our whole family was participating in the war and that I was over eighteen years of age had some bearing on the court's ruling, which limited the damages to the maximum extent of my insurance. My car was now a burnt hunk of metal. Since I had no collision insurance, I had lost the full $450 I'd paid for the car. My ex-mate was in the brig. The deaths bothered me for a number of years, and I made no attempt to replace the car.

Chris had been going out with Cameron for some time, and he married her in the spring of 1943. They found a small apartment about three miles from Imperial Beach and invited me over for dinner shortly after they moved. Cameron invited her sister Betty, who lived in Los Angeles, to join us. The matchmaking was okay with me because I wasn't dating anyone at the time. The apartment was small, but Cameron made it attractive and homey. Her sister was an attractive blonde with deep-blue eyes. She looked quite young but was seventeen, a senior in high school, and we got along well. The evening was full of laughter, and we planned to get back together in the near future. But that didn't happen. The next day I stopped by the army exchange at Point Loma, purchased a silver cross and chain, mailed it to my cousin Katie, and asked her to give it to Ruby—with no card or note. I don't know what she thought when it arrived, and I didn't hear anything from her.

Throughout the early months of 1943, the Imperial Beach base had been a beehive of activity. Our watch schedule remained heavy, requiring twenty-four hours on over a forty-hour period and then twenty-four hours off, so there was little time to visit Los Angeles. As a result, I sort of lost touch with my cousin's girlfriends, although I received letters from several of her companions —but not from Ruby. During my second summer at the base, I spent a good deal of time on the beach and helped out as a lifeguard. All of my prewar friends were either overseas or training for duty elsewhere in the country. San Diego had become a center for military activity. The navy and marine boot camps were turning out classes as fast as they could. North Island, adjacent to Coronado, was busy refitting carriers and training new pilots, while ship repair and construction was going on around the clock in the south bay area. The downtown was filled with swarms of military personnel, and the local bars and the seamy side of life flourished. With the mix of services and the number of men eagerly looking for entertainment and companionship, it was often difficult to get through town without someone challenging you to a fight or trying to sell you flesh.

As fall approached, the United States and its allies continued to move westward in the Pacific, as battles raged in the Solomon Islands. The United States lost two carriers, but its naval strength in the Pacific grew with the addition of pocket carriers, destroyers, revamped cruisers, better planes, and submarines, as well as an expanding capacity to supply its forces in the Atlantic and Pacific. News was censored, and the full story of U.S. losses seldom was told at the time, but reality would leak out several months after events occurred. Given my West Coast background and my father's participation in the naval engagements in the region, my attention tended to focus on the Pacific theater. Everyone read the newspapers and listened to the radio to keep abreast of the changing tide of the war. By late 1943, it was obvious that we were beginning to gain the upper hand in naval engagements and were rapidly gaining air superiority over the Japanese. I wondered about the role radio intelligence would play in the final outcome.

Sometime in the early winter, the watch schedule switched back to forty hours on and forty-eight hours off, giving us more time for social engagements. A couple of weeks before Christmas, I made a trip to Los Angeles to visit Uncle Bob's wife Jimmy. We met in Pasadena and went to a local theater. Surprisingly, the theater manager recognized me; he had been the manager of the North Park Theater in San Diego. We chatted for a few minutes and then he seated us in the loge, a perk resulting from our earlier encounters and my service in the navy. Jimmy asked me whether I was still dating Ruby, and I said, "No, but it would be nice to date her again." It had been almost a year since I had seen her, and I had no idea what she was doing or whether she was dating someone else on a steady basis. Jimmy convinced me to give her a call and see if she was interested in renewing our acquaintance. When I called, she seemed friendly, and we agreed that during my next leave we would go dancing at the Hollywood Palladium, where Harry James and Betty Grable were performing. She loved to dance so it seemed like the thing to do. Besides, my dancing skills were improving.

The following week my watch ended on a Friday morning, after a midnight to 8:00 a.m. stint. I showered, put on my dress blues, and took the train to Los Angeles, where I planned to see Florence and Katie. Florence had graduated from high school and was dating a sailor. Since my last visit, Katie had become engaged to Zollie, also a sailor. They were in style! I arrived at Aunt Katherine's just after noon, and since I had been up all night, I took a short nap. By the time I got up, my cousins had returned home from work. Their boyfriends had also come for supper. During the meal, I quizzed Katie, but she hadn't seen much of Ruby; junior college and her young sailor friend kept her busy. She told me that I would have to wait until our date the next night to get caught up on Ruby's social life. That Saturday felt like the longest one on record. Ruby and I had agreed to meet about six in the evening, get a bite to eat, and then go to the Palladium. Around 5:30 p.m. she called to say that her father had agreed to let me drive his car, which was great because it's difficult to be romantic on a bus or streetcar!

As I walked down Atlantic Boulevard, I reminisced and fantasized about Ruby—what she looked like and how the date would work out. I knocked on the door of her house, and Ruby invited me inside. She had on a new dress that had a V-neck, a tight-fitting top, and a skirt that blossomed out to just below her knees. Her long, brown hair fell down over her shoulders. We talked to her folks for a few minutes before her father gave me the keys to his car. As we left the house, Ruby placed her hand in mine and gave it a gentle squeeze—things were off to a good start. We stopped at a drive-in and then headed to the Hollywood Palladium, arriving around 8:30 p.m. As we expected, there was a mob waiting to get inside. The dance floor was crowded, but there was still room to move around. The Palladium always mesmerized me due to the quality of the great bands playing in the spectacular dance hall, its elegance, and its size. I believe that it covered an entire city block. We danced for about an hour, holding each other close during the slow dances. During the faster tunes, Ruby tried to teach me to jitterbug and the other "in" dance steps. When Betty Grable came out to sing, we crowded up against the stage along with everybody else and just swayed to the music.

We had a wonderful time and the evening passed quickly. As we drove back, Ruby moved over to my side of the car, and somewhere between the Palladium and home, I ask her if she would wait for me until the war was over, and she said "Yes, yes!" Before we arrived at her house, I had kissed Ruby's ruby lips not once, not twice, but to tell all wouldn't be nice. We became engaged that night; all I needed to do was find a ring to confirm that commitment.

Over the next several weeks I dated Ruby every time I had leave, and once or twice she came down to San Diego for the weekend and stayed at my mother's house. Her folks were not overjoyed that their daughter was engaged to a sailor, but they also didn't make any demeaning comments to me, which wasn't the case with Ruby's Baptist minister. About a month after we got engaged, Ruby asked me to go to her church, where she taught Sunday school. I suspect Ruby wanted to introduce me to several of her girlfriends, who also attended the church, and parade me around a little. During the sermon, however, the minister lashed out at the morals of servicemen and warned the girls in the audience that they shouldn't trust sailors. He reminded them about the saying that sailors had a girl in every port. I wanted to slip down into the pew and disappear, but I sat quietly until the service was over, then quickly retreated. Ruby introduced me to several friends whom I had not met earlier, but I could see that she was very uncomfortable.

By the end of 1943, the United States had made a major move in the western Pacific by landing the marines on Tarawa. The island would be used as a base for an intensive bombing campaign against the Japanese navy bases in the Marshall Islands. The long-term U.S. strategy was to form a ring of airfields around the heavily fortified Japanese navy bases on Rabaul. The radio intelligence group was very familiar with the Japanese bases in the western Pacific bases because we had copied radio traffic from all of their strongholds. We felt that we were part of each landing that moved the Allies steadily westward. By February 1944, U.S. troops had landed on Kwajalein and Eniwetok Islands. Meanwhile, in the European theater, there was considerable talk about the extensive buildup of arms and troops in England and the potential of an Allied invasion of France and Germany. The war seemed to be progressing on a more positive note, and the horrible losses at Pearl Harbor were being offset by the country's growing military might.

I became eligible for promotion to radioman first class in March 1944. I spent a great deal of time studying for the exam, which was a prerequisite for promotion. The course work for the test was more difficult than what I had previously faced and demanded an extensive understanding of simple physics, radio theory, and radio maintenance. I spent hours pouring over the manuals and began to think that I would never pass the test. Still, when the next opportunity came, I applied and took the exam for RM1C and chief in early April. The test took the better part of three hours to complete, and I had missed a number of questions. Nevertheless, two weeks later, I learned that my score was adequate for promotion. During the following duty period, my watch officer recommended me for first class. I hoped that my upgrade would be approved within a month.

Assignment Unknown

Late in April, Jim Murphy and I were summoned to the commander's office. Neither of us knew what to expect, though we remembered that the last time the entire base staff had been called to attention, they had requested volunteers to pick strawberries. There was an apparent shortage of farm workers east of San Diego, and local farmers had asked for help from the local servicemen. This time, Jim and I attempted to volunteer each other by pushing the other forward. The commander told us that since we had been on base longer than any other petty officers, he wondered whether we would be interested in volunteering for an important and perhaps dangerous mission. He wouldn't, or was unable to, give us any other details. At first we thought that he was kidding and was talking about the strawberry detail, but it soon became obvious that he was dead serious. We both said yes, although we knew only that we would be shipped out in the next ten days. We began to ponder where we would end up. There had been some speculation that a group would be transferred to Australia, but we were not sure what could be potentially dangerous about an assignment "down under."

During the next few days, I packed my gear in my sea bag so that I could leave at a moment's notice. I also checked with my mother to see if she knew anything about my brother and father and our neighborhood friends. Frank had completed training for the paratroopers; Eldon and Oscar were in England awaiting the invasion of Europe; Louis was flying bombing missions over Germany; and my dad had been involved with several naval battles in the western Pacific. I had heard nothing more about my new assignment, but expected to leave sometime in early May. Ruby came down to San Diego for the weekend, and I told her that I was expecting to ship out within the next week. Two days later, Jim and my orders came through. We were to leave the next day by train for Washington, D.C., for special training, and then we would be transferred to our new duty station.

Early the next morning, Jim and I were driven to the San Diego train depot and to catch a train for Los Angeles. My mother came down to say goodbye, and I could see tears in her eyes. Little wonder: my father was smack in the middle of the Pacific naval conflict, my brother was en route to the front lines in New Guinea, and I was being shipped to an unknown destination. We climbed on the train with our sea bags, waved goodbye through the window, and in a few moments we were headed north. I had served less than two years with the radio intelligence group, received my upgrade to first-class petty officer, and was still only nineteen years old. The time at Imperial Beach had gone by fast and Jim, Chris, and I had become the old timers. Ruby knew we were coming through Los Angeles, and she was there to meet the train. There were only had a few minutes before we had to board the *Santa Fe Chief*, bound for Chicago. Ruby and I kissed hello, I gave her an engagement ring, then we kissed goodbye and promised to write. With a long whistle, the train pulled out. As we moved out across the desert, my thoughts drifted back through my service career and my courtship of Ruby, and I wondered what the future held, with so many of my high school friends engulfed in the war.

Jim and I were given a sleeping room that was quite comfortable. The train carried several men and women from all branches of the service. Some were heading home on leave, others were going overseas, and others had new stateside assignments. The train took a route that was similar to our last family trip to visit our grandparents in Michigan. We headed northeast to Salt Lake City and into Wyoming, and then east across the Plains states toward Chicago. It was my first overnight train ride, and both Jim and I were excited. The diner was a pleasant change from navy food, even though the meals at Imperial Beach had always been quite good. We were treated very well by the civilian passengers, as were all service personnel. They questioned us about our role in the war and where we were going, and some asked us to be their guests in the formal dining car. Within a day we knew most of the people on the train and were playing cards with our military comrades.

The train arrived in Chicago about two days later, and we had to take a taxi to another train station, although it was probably quite close, and we could have walked if we had known the way. After a few hours in the Windy City, we headed east for Washington, D.C. As I recall, we arrived there early the next morning. Since the Department of the Navy didn't open until 8:00 a.m., we found a nearby restaurant and enjoyed a good breakfast. At that time the department was just in front of the Reflecting Pool that runs from below the Washington Monument toward the Lincoln Monument and the Potomac River. It was a warm spring day, and we were impressed by the city's beauty. When we arrived at the department, we were given our schedule. First, over the next several days we would receive a number of shots, including vaccinations against cholera, yellow fever, and plague. Somehow these didn't sound like shots needed for Australia, but then I knew that the northern part of that continent encompassed parts of the tropics. Jim and I speculated that our potential assignment could be anywhere from the South Pacific to India.

That afternoon we were sent to a building on the outskirts of town, I believe somewhere on Wisconsin Avenue, to visit the navy's top-secret radio intelligence facility. A lieutenant commander welcomed us, took us into his office, and told us that we were being sent to join a special navy group in China. Our assignment and the character of our training in radio intelligence was not to be discussed with anyone, including our parents, relatives, or close friends. When we received our order to ship out, all of our clothes and belongings, except for one set of working jeans, would be sent home, and we would receive new clothes before our departure. During our short stay in Washington, we would receive our shots and special information and training to help us adapt to our new environment, including a few key expressions and words in Chinese, lessons in using chopsticks, and tips on avoiding potentially lethal diseases.

After this briefing, we were shown the new navy computer facility. It housed an early computer, about thirty feet in length, that was made up of a mass of radio tubes and circuits. It looked impressive, but it didn't have as much memory as a modern desktop computer. At the time, however, I was amazed at its size and its ability to break enemy codes.

Navy housing facilities in Washington were bursting at the seams, and we had to find accommodations in the private sector. We began shopping for temporary housing that afternoon and found a furnished basement apartment at 19th and G Street. It was ideally located for our needs and was next door to a United Service Organization (USO) facility that provided free breakfast and other amenities for servicemen. It was perfect for sightseeing, a short walk from the Reflecting Pool, the Washington Monument, and Pennsylvania Avenue and the White House. And it was just around the corner from a YMCA and one block from a White Spot restaurant, where we could get three hamburgers for a dime.

The following morning we reported for our shots and then went to a briefing on how to survive in China and the nature of the navy's intelligence work. Almost nothing was said, however, about our specific duties or eventual destination in China. Later I found out that the navy group in China had been arranged by General Chiang Kai Shek and the secretary of the navy, Frank Knox. For more than a week, we followed the relaxed routine, took our slew of shots, and in the afternoons toured our nation's capital. Once our shots were completed, we were issued passports with visas for entrance into China. This seemed an odd requirement during war time, but we were told that passports were necessary because of the special entry arrangements between the United States and China. We were given special training for about three weeks and then, to our surprise, told that we could go on a six-day pass. Upon our return, we would receive further orders regarding transportation to China.

Jim and I examined our options. We didn't have the money to fly to California, and the train would take about three days each way. It was obvious that a visit home was out of the question. But Jim had relatives in Omaha and the Alverson clan lived in Michigan. We made arrangements for train tickets to Cleveland and planned to hitchhike to Detroit and on up to Rochester, Michigan. We departed the morning after we received our leave papers. Riding the train was a treat for young servicemen because people treated us like kings, and we almost always were invited to meals in the dining car. We were in luck on the trip to Cleveland; a middle-aged couple sitting across from us quickly engaged us in conversation. The husband was the vice president of a company that manufactured explosives for the war effort. They were returning home after contract negotiations in Washington and offered to buy our lunches. We agreed, knowing that it would cut down on our trip expenses. We arrived in Cleveland sometime in the afternoon and got to Detroit before evening.

We had planned to proceed to my grandparents and other relatives in Rochester, but it was getting late and we decided to spend the night in Detroit. We tried several hotels, but they were booked. One of the desk clerks, however, told us that the Salvation Army had a shelter just a couple of blocks away. When we arrived, a young woman greeted us, asked our names, and told us that she could put us up for the evening free of charge. We were given a couple of cots and small lockers in a large room, which must have accommodated close to a hundred people. It was filled with servicemen; most were from the army, but there were perhaps a dozen sailors in the room. Later, we were given a free meal, and then we went to a show at a local USO. The show featured comedians, dancers, and singers and included several entertainers from Hollywood. By our standards, it was first rate. Except for the train ticket, we hadn't parted with a dime during the whole day. It was a great time to be a man in uniform serving the United States.

Early the next morning, we began our trip to Rochester. We caught a ride with the second car that we thumbed. The driver was a farmer, probably in his late forties, with a son in the air force, a mechanic based in England. He was probably serving in the 8th Air Force in a reasonably safe situation, but he was also in an active combat zone and there had been broad-scale Luftwaffe attacks on British airfields. At any rate, the farmer was pleased to give us a ride and, although it was a little out of his way, he offered to deliver us to my uncle's door in Rochester.

I wasn't quite sure of family protocol, but I went to my father's brother's home because Uncle Omer had a large family and I suspected that he could accommodate us. Uncle Omer and Aunt Katherine had six children—Charles, David, Phyllis, Gerald, Melvin, Shirley, and Barbara. Shirley, the oldest, was about two years younger than I was. Before leaving D.C., I had called the various members of the Alverson clan to let them know that we were coming. Still, I had some apprehension about the character and size of the welcome mat that we would be shown. My concerns melted away shortly after we arrived, as the whole family treated us like royalty. Shirley and the older boys were bursting with questions about the nature of our jobs and our destination, none of which we could answer truthfully. Our answers probably sounded a little reserved, and it must have been obvious that we were avoiding some of the questions. There was, however, still plenty of room for conversation about family, my engagement to Ruby, the war, and my uncle's work in an arms plant.

After lunch, my uncle took us for a ride around Rochester and out to see the old farm that my grandparents had owned. They were now living in a small house in Romeo, Michigan, where we planned to visit next. That evening, my aunt and uncle put on a great feed that included fried chicken, mashed potatoes, and all the trimmings. Jim and I spent the rest of the evening bonding with my cousins, particularly Shirley, who was a junior in high school and closest to our age. The next morning, Uncle Omer took us down to the plant where he was foreman and showed us the production lines. We were especially interested in a new grenade-launcher ram device that could be loaded into a .30-06 rifle. The ram went down the rifle barrel, almost to the rifle chamber. Uncle Omer allowed us to test the device, which threw the grenade several hundred feet. We both greatly enjoyed our visit. Sadly, several months later, my uncle was seriously wounded when an explosion ripped through the plant. He remained in the hospital for almost two months and never returned to that job.

The following morning we spent some time with my grandparents in Romeo, and then visited my Uncle Russell, who had no children and lived in Romeo, not far from the old farm. He loaned us his car and suggested that we check out the amusement park in Flint. I remember riding everything that went fast, twisted, whirled, and made riders scream.

We even tried our luck at a dart game and won a well-endowed clay doll. We were going to give it away to some young lady when the evening was over, but instead decided to take it on our travels and scratch the name of each major city that we passed through into its clay flesh. We had no idea what path she would travel or the number of cities that would ultimately adorn her body. The next day, we hitched a ride into Detroit and grabbed a train to Omaha, where we spent one night with Jim's relatives and took in a stage show. The next morning, we headed back to the capital a little worn out. But we had had a good time, and the trip had given me the opportunity to learn more about my father's family.

China Bound

Sometime in late June, we were issued a set of green khaki work clothes, several pairs of dress khaki pants and shirts, army shoes and socks, and a standard GI knife. We were instructed to keep our dungarees. We sent home our navy blues and whites, pea coat, black socks, and dress shoes. During the next few days, we finished getting our shots and were trained in the handling of the knife, hand-to-hand combat, the use of small arms, and key Chinese phrases and words. Jim and I were puzzled. The training wasn't exactly what we expected for two members of a highly secret radio-intelligence team. We were told to say absolutely nothing regarding our training, but at that point it didn't take an intelligence wizard to know that we were being trained for a range of activities other than radio intercept. The next morning, we were told to tell our parents not to expect communications for some time. Each of us was issued a passport, given $200 in cash, and told that we would need the money en route to our destination. Later that afternoon, Jim and I received orders to proceed to Newport News, Virginia.

That evening, I called Ruby and my mother to let them know that we would be leaving the country in the near future, but couldn't reveal the location of our assignment. Mom told me Eldon had been killed shortly after D-Day when he had parachuted behind the lines at the start of the invasion of France. It was quite a shock, as he was the first of our sandlot football team and school friends to die in the war. I wondered about Louis Black who was making B-17 flights over Germany. There was no news of Dad or any of my other friends. After telephoning, I went to look for a small gift to send Ruby as a going-away token. She had a silver charm bracelet that she cherished, so I checked out several jewelry stores and found a clever silver Egyptian mummy charm with a secret lock. It had no significance; I chose it because it was attractive. I wrapped it in tissue and sent it along with a letter telling of my departure. But when Ruby received it, she decided that I was giving her a clue as to my overseas destination. She discussed it with my mother and both came to the conclusion that Jim and I were headed for Egypt. I never knew that they had cleverly determined my future placement until after the war.

The next morning, shortly after daybreak, we were on a train headed south for Charleston, Virginia, along with three other radio-intelligence petty officers, Benedict, Stringfellow, and Estes. We disembarked from the train and were trucked to a ferryboat, which hauled us across the bay to the navy receiving station in Newport News. There, we were housed in a corrugated metal Quonset hut with about twenty other sailors, who presumably were headed to the same destination. There was considerable discussion regarding our past experiences, but almost nothing about our current assignments because we had all been told to keep our mouths shut. We stayed in Newport News for two days before we received orders to board a navy transport docked in the harbor.

Into the Atlantic

During the first week of July, we packed our sea bags, climbed into a six-by-six truck and headed off for the ship. When we arrived at the dock, there were several thousand troops from all branches of the service waiting to board the navy transport ship, the *General Anderson*. We were told that the large ship could carry five thousand passengers. It was painted Navy gray with camouflage-black, zigzag stripes down its sides. A band on the dock played all the service songs, including *Anchors Away*, *From the Halls of Montezuma*, *Off We Go Into the Wild Blue Yonder*, and *As the Caissons Go Rolling Along*, etc. As the song of each branch was played, the servicemen being honored let out a loud cheer. It was a time when we felt proud to be a part of the war and stimulated by the ranks of other servicemen surrounding us.

I don't know if it was done on purpose, but our small navy band was loaded first. Jim noted, "The navy takes care of its own." We climbed up a long gangway and were led through a hatch down into one of the ship's holds. The compartment, which was about ten feet below sea level, held about two hundred to three hundred men and was equipped with bunks that were four levels high, showers, and heads. Being so far below the sea surface was a hideous location to be in if the ship was torpedoed. We threw our sea bags on two adjacent bunks and returned topside to watch the departure. The waiting servicemen were embarking on two gangways, and the deck was rapidly filling with troops waiting for the ship to cast off. When the vessel was finally loaded, the horn let out a loud blast and several tugs began to pull us out of the berth. The servicemen and a few navy and army nurses let out a cheer. Most of us stayed topside as we cleared the harbor and headed out to sea. Those on deck became increasingly less boisterous, and we could sense a change in the mood as the passengers began to realize that they were unlikely to see their families and friends for a long time.

Jim and I stayed on the deck for a while before we returned to our hold. As we came down the stairwell, we noticed that some of the men had begun a poker game and were gambling away the money they had been given to support themselves during the trip. When we first arrived, the pots amounted to $20 or $30 but gradually increased to several hundred. Asked if we wanted in, we declined. Someone had loaned me a Zane Gray cowboy western—*The Riders of the Purple Sage*—which gave me an excuse to return to my bunk. About three minutes later a fight broke out between some of the poker players; Jim and I and several others tried to break up the confrontation. One of the fighters had a bloody nose and the other a cut lip, so it ended in a draw. We later found out that one of the players had lost most of his $200, causing him to blow his cool.

Chow time started in the late afternoon. Each compartment had a scheduled time to go to the mess hall. Given the large number of men, feeding the troops was a continuous process. Men were cycled in and out for several hours before dinner was completed. The eating area had a number of parallel tables and fixed stools. The metal tables were just wide enough, perhaps two and a half feet, to accommodate two individuals across from each other. The line to the mess hall seemed eternal. You waited your turn, grabbed a tray, took what they served, and found an empty spot at the tables. There was time to eat, get a second cup of coffee, return your tray to the collection area, and then make room for those waiting. The food didn't compare to the grub at Imperial Beach, but under the circumstances it was more than adequate.

After breakfast the next morning, I was anxious to go topside and get some feeling for the route the ship was taking. Besides, I was a little claustrophobic. When Jim and I arrived on deck, we walked forward toward the bow, where there was ample open deck space. The weather was clear and a comfortable breeze blew across the ship. From watching the bow waves, it was apparent that the ship was making a good speed, probably more than twenty knots. We had lost the destroyer escort that had accompanied us during the first day out, but I suspected that we still had air cover. By checking the time and the position of the sun, I came to the conclusion that our course was roughly south and perhaps a little east. I told Jim that it appeared that we were headed for the Panama Canal, and it was unlikely that we would cross the Atlantic. I felt somewhat relieved. I was of the impression that there was less danger of a submarine attack in the Pacific than in the Atlantic or Mediterranean or in rounding the Cape of Good Hope into the Indian Ocean.

I didn't want to tell Jim or the other members of the radio-intelligence group that I was nervous because our compartment was below the waterline, so I decided to spend as much time as I could up on deck. Sitting and watching the ocean wasn't always stimulating. A few albatross followed in our wake; from time to time, a school of dolphin would come alongside and play in the bow waves; and we saw an occasional flying fish. After a couple of hours, Jim suggested that we get out of the sun and check out the radio room. As we were wearing our navy dungarees, we were given a fairly free run of the ship. We asked one of the ship's crew where the radio shack was located and how to get there. We made our way up several ladders and entered the communication center. There I offered to stand watch, copy the incoming news, or do other chores as needed. I was given the job of summarizing the news coming via radio, copying it, and preparing a newsletter for the passengers. For me it was a godsend in that it allowed me to stay out of the hold for the greater part of the day.

Later that afternoon, someone announced over the loudspeaker that we would arrive at the Panama Canal the next morning. Each section was to limit its sightseeing to one side of the ship, apparently to ensure that the ship didn't list and endanger safe passage through the canal. The temperature increased as we moved south, and while it was hot on deck, it was even hotter in the hold. The smell of human sweat below decks became increasingly obvious, regardless of how many salt-water showers a day we took. About a day out of Colon, Panama, we were introduced to salt pills that we were required to take for a considerable portion of the trip and warned to drink lots of fluids to prevent dehydration.

Across the Isthmus

The next morning, the *General Anderson* approached Limon Bay, located adjacent to Panama and the locks. The crew reminded us again of the need to prevent the ship from taking a sharp list. Each section of the hold was instructed to view the transit across the canal from designated port or starboard sides of the ship. The open deck area was filled with the men and women. However, we were caught off guard when the ship slowed and made its way to a pier in Colon Harbor, Panama. There were either too many ships waiting to make the crossing or the ship's captain had decided to take on water, fresh vegetables, and fruit. Whatever the reason, we stayed there overnight, and the crew and our navy group were given shore leave. Jim and I, along with Stringfellow, Benedict, and Estes, went ashore and found a lively restaurant with a loud band and thinly clad waitresses. Each of us ordered a large T-bone steak, a baked potato, and a sliced tomato salad. The meal cost us forty cents apiece.

Later, we looked over the town and returned to the ship around 10:00 p.m. When we got to our bunks, Jim got out the clay doll we had bought in Flint and scratched Colon on her body. As the ship got underway the next morning, we heard that some of the crew and several members of our intelligence group had gone to a bar to see a rather crude girlie show, got in a fight with some local marines, took their share of hits, and were on report. We never did hear the outcome of the hearing, but assumed that the incident was dismissed as an energy burst due to cabin fever. We were sort of envious of the perpetrators.

The ship began its transit through the canal in the early morning with passengers crowded on every spot of open deck—forward, aft, along the sides, and on top of the ship's house. All morning long we moved, step by step, from one lock to another. Although it was hot and muggy, no one left his or her viewing station unless it was for a pit stop. We gradually made our way through Gatun Locks and into Gatun Lake. Along the lake, we could see almost nothing but the thick jungle that crowded the shoreline. I used my Sea Scout training to check the ship's course, and it seemed like we were sailing to the east, which didn't make sense to me. Later, I checked a map in the radio shack and saw that the transit of the canal from the Atlantic to the Pacific ran southeast to Panama Bay. Thus, when we arrived in the Pacific Ocean, we were east of where we started. As we departed the Canal Zone, we again were escorted by two destroyers, which I suspected would accompany us across the Pacific. I thought that we might make our way around the south of Australia, but since the Allies had been advancing west across the Pacific, perhaps we would be routed north of Australia through the Strait of Malacca. Later, I realized that that part of the world was still under Japanese control.

The Blue Pacific

We saw quite a bit of sea life as we moved out of Panama Bay into the Pacific, including numerous birds diving for bait and what appeared to be a large school of tuna or mackerel. We also were tailed by several albatross, which glided back and forth across our stern, apparently looking for leftovers. To avoid leaving a trail, however, the ship discarded garbage and trash in weighted bags and only at night, so there were few pickings for the albatross. Signs of the productive coastal sea diminished as we moved away from land, and by late afternoon we were crossing the deep blue ocean water, seeing little more than the occasional flying fish. Sometime during the first night, our destroyer escort left us. I had hoped that it would stay with us for the remainder of the trip but concluded that the ship's speed and course wouldn't place it in great jeopardy. As the *General Anderson* moved into the vastness of the Pacific, Jim and I spent more and more time in the radio room, copying news or playing cribbage. We bet the awesome sum of two cents per point, losing or winning a dollar or two a day.

We started preparing to celebrate the crossing of the equator soon after we left Panama. Navy tradition called for all those who had not previously crossed into the Southern Hemisphere to submit to a series of taunts and hazing before they could become selected members of the Society (or Kingdom) of Neptune Rex. The experience was confined to navy personnel, including members of the *General Anderson's* crew and navy passengers in transit, such as Jim and I. On the day the ship crossed the equator, there was great excitement as the crew decorated the ship's stern, or "fan-tail" area, which was decked out with flags and balloons along the hazing course. The non-navy passengers were invited to watch the shenanigans from the decks above.

Each applicant to the Kingdom of Neptune Rex started with a "wack'um" course. That is, we walked between lines of men, who used wooden paddles to wallop our backsides. Then we had to eat some greasy, foul-smelling food, which made us want to vomit. I managed get through everything until I was put into a dunking pool—a shallow pool on deck that had been constructed using wooden frames and canvas. The hazers would dunk us under the water for a short time before allowing us to come up for air. Growing up on the beaches of San Diego, I became a good swimmer who could hold my breath for a considerable time. I decided to have a little fun. As they turned me loose so I could surface, I grabbed the legs of one of the men doing the dunking and tossed him out of the pool—a bad decision. They evened the score by holding me down a little longer for three successive dunks, but each time they let me surface, I tried to toss another mate out of the pool. After the third time, everyone watching gave me a standing cheer. Several days later, each participant was issued a beautiful Neptune Rex certificate showing the mythical sea king rising out of the ocean.

After about a week in the Pacific, all hands were told to keep the deck chatter down, as there was a very sick GI in the ship's hospital. The word quickly spread that he was suffering from an attack of spinal meningitis, creating an uneasy feeling among the crew and passengers that it could be very contagious. Jim and I didn't know if it were true, but we were all caught up in the same rumor. The man died two days later, and there was a burial at sea. Solemn sailors fired several gun salutes; the chaplain spoke a few words; the ship's bugler played a soul-wrenching taps; and then crew members slid the corpse, wrapped in canvas, into the sea. With the deck filled with military personnel standing at attention, it was a sobering moment.

After the equator celebration and the burial at sea, the days began to run endlessly together. Going to chow was the only excitement to break the daily routine. Jim and I gathered with the other radio intelligence members in the radio room to listen to or copy the news, play cards, and drink coffee. Still, we had it better than most of the GIs, who had to stay on deck or down in the hold. The officers bunked in cabins in the ship's main house and had their own mess. We heard over the radio that

American marines had landed in Saipan, which would be used after its capture as a major base for the attack on Japan. The war in Europe had been bitterly fought since the landings in France on D-Day. The Germans had mounted a number of counterattacks, but General Patton swept down through France to help free the 101st Airborne trapped in Bastone. Out in the South Pacific, the *General Anderson* plodded toward New Zealand and Australia.

One morning, we spotted a large island off the starboard bow. It was the Northern Island of New Zealand. As we moved closer over the next two hours, we could pick out the Southern Island. The weather had turned cool. It was midwinter in the southern hemisphere, and we could see a number of snow-capped mountains to the south. Some of us believed that the ship would stop at Auckland or Wellington for supplies, but we moved into Cook Strait without slowing down. The ship transited the strait and headed out into the Tasman Sea. Australia lay several hundred miles to the west, so there was still a chance we could put into a port on its southern coast. All of us were eager to get off the ship and walk on terra firma. Jim and I stayed on deck to watch New Zealand fade into the ship's wake and then headed to the radio shack. The wind off the bow had a chill, and we were getting cold.

A day or so later, the *General Anderson* headed into a large dock in Melbourne Harbor. Jim and I thought we would be given liberty but that was reserved for the ship's crew. Our hearts sank when we realized that we wouldn't get on shore until we reached Bombay, which we now knew was our final destination. We watched as the ship was secured to the dock, supplies were taken on, and trash was unloaded. Some of the dockworkers began trading local coins, magazines, etc., for American cigarettes, and there was considerable chatter between the men on the dock and the GIs watching over the side. Jim and I volunteered to join the work crew that was unloading garbage so that we could set our feet on the land down under. For about an hour, we carried garbage cans down to waiting trucks, dumped them, and returned to repeat the operation. After we loaded the last can on the truck, we walked up the dock to check out the waterfront, staying perhaps half an hour before returning to the ship. Those in charge, however, had already noted our absence, but only "ripped" us with some foul language and sent us to our quarters. From our perspective, we had visited Australia, so Jim etched Melbourne into the Flint doll.

The Indian Ocean

Before the sun rose the next morning, I felt the ship's engines come to life and knew that we were pulling out. We headed west along the south coast of Australia and later north up through the Indian Ocean, and two destroyers (from Perth, I think) joined us and remained with us until we reached Bombay. About a day after the ship entered the Indian Ocean, I began to feel pain in my lower stomach, which lasted overnight and into the next morning. With some urging from my mates, I checked in at sickbay. There, a young doctor diagnosed appendicitis. I argued that my pain was more likely due to constipation, but the doctor had spoken. (I've never believed there was anything wrong with my appendix but thought that the doctor was looking for some at sea experience.)

The doctors scheduled my appendectomy for the next morning. As I walked up to the ship's hospital I could see large swells, about twenty to twenty-five feet high, running from the stern. They were hitting the ship on the starboard side, causing it to roll heavily. The doctors were undaunted, however, and proceeded with the task at hand—to give me local anesthesia via a spinal injection. I was concerned that the ship's rolling would lead to injury of my spinal cord, but the doctors were up to the challenge, and the local worked well. I was conscious and had feeling above my chest. There were several large lights above me, some on and some off, and I could see what was happening in the light's mirror reflector. The doctors cut into my abdominal cavity and moved some of my gut on to gauzes on my stomach, and then I vomited. The operation did not take very long, and it seemed like only minutes before I was in a hospital bed. A nurse showed me the appendix, and it looked quite normal to me, not that I knew what one looked like. I still think that the operation was a training experiment and hope that it helped build the doctors' confidence.

I slept for several hours before waking sometime late in the afternoon. The local had worn off, and I checked to ensure that there was feeling in all of my appendages. The ship was still rolling heavily, which made me feel a little uncomfortable. I asked to see my appendix again, but it had already been disposed of in the ocean—my first contribution to the field of marine biology. That evening, Jim and several other members of the radio intelligence group came to the sickbay for a visit. We chatted a while, and then Jim asked me if I wanted a Coke. Although the doctors told me not to eat or drink after the operation, my will was weak and I eagerly drank it all. That night, I was in a lot of pain, but who was to blame? Jim, maybe, or the idiot who drank the Coke!

We arrived in Bombay two days later. Everyone in sickbay was taken off the ship in stretchers, which was embarrassing for me because I could walk quite easily. I was put into an ambulance, taken to a British hospital on top of a hill somewhere in the outskirts of Bombay, and examined by a doctor who prescribed a few more days of rest. Jim and the other members of our group, en route for China, would leave by train for Calcutta the next morning; my transportation would be arranged after my release. The hospital was filled with "limeys," most of whom had been wounded in the jungle battles underway in Burma. They seemed genuinely interested in talking to a "Yank" and thought it peculiar that a navy lad was headed for China, but then Jim and I hadn't unraveled that mystery either.

Getting to Calcutta

The realization that I would have to cross India without my companions made me uneasy, but the solo journey also would give me a chance to see Bombay and learn a little bit about the country. After three days or so, I was feeling good and could walk without pain. The doctor said that I would be released within a day or two and could take liberty that evening. I asked a young English corporal, Thomas, who was recovering from a bullet wound in his shoulder, to join me in a tour of the city. We decided to see a soccer game in the afternoon and work out the rest of the schedule later. We left the hospital with passes that allowed us to be absent until 10:00 p.m.

I had never seen a soccer match before, but Tom was a former soccer player and totally consumed by the game. The match was between a group of New Zealanders and a local Indian team, and I don't remember who won, but it was interesting and fun to watch. After the match, we went to a restaurant recommended by one of the nurses. We still had plenty of time, so I suggested that we go take a look at Bombay's notorious "cage girls," whom I had read about in *Time* magazine. Known as the "two ana [four cents] girls," they had been sold into prostitution at a very young age, and lived, and took care of their business, in cages in a rather scruffy area of town. Tom and I talked it over for a few minutes, jumped into a taxi and headed for the city's "educational center." When we arrived, the taxi driver told us not to spend too long in the area, but there were a number of other servicemen on the street, and it looked safe enough to us.

The cages, which ran for five to six blocks mostly down one side of the street, were rectangular cubicles, about ten to twelve feet deep and eight feet across, with bars along the front section and one girl to a cage. Drapes were pulled when the girls were occupied. They appeared to be mostly Indian, with a sprinkling of other nationalities, and for four cents were said to provide sexual pleasure to the local population. Almost all the girls had sores on their arms or legs. We walked the length of the cage section several times and then went to a bar, where we each had several beers (no one asked for identification) and listened to a local girl singing popular American songs—she was good. By the time we left the bar, it was nearly midnight; we were already two hours over our leave time. Most of the other servicemen had already left the area, and Tom and I grew a little nervous.

We caught a cab several blocks away, but we couldn't communicate effectively with the driver. He didn't seem to know where we wanted to go. perhaps we pronounced the hospital's name (which I can no longer remember) incorrectly. We decided to catch another cab, but luck was not running with us. It was another thirty minutes before we found a taxi driver who knew where we wanted to go. It was then well after midnight, and we were becoming increasingly panicked about getting into big trouble. The taxi started off for the hospital, but it was obvious that the driver was in no hurry. We tried to encourage a faster pace with no success. Neither of us knew the way back, and we had no idea where the driver was taking us. We decided that when we came to a well-lit area we would jump out and get another taxi. Fortunately, we didn't have to put the plan into effect because we soon saw the hospital up on top of a hill and sighed with relief.

The next morning, a British major called me into his office and observed that if I could stay out until 1:30 in the morning, I probably didn't need further medical attention. I was released from the hospital at noon and a U.S. Navy bus came to pick me up. I was driven to an apartment house downtown that housed half a dozen navy petty officers. I was introduced to the group and told that I would spend two days there before taking a train to Calcutta. The next morning, one of the men offered to show me around town. He owned a Chevy sedan and had a driver at his disposal. We cruised along the waterfront, which looked quite clean. Small fishing boats were anchored offshore, including a variety of Arab dhows. Away from the waterfront, the city was less attractive. A large number of beggars were working the streets. Some well-versed children yelled, "Boxies Sahib, no Mama, no Papa, no brothers, no sisters and no per diem." All in all, my visit to Bombay went better than I had anticipated, but I saw only a thin slice of Indian society.

Just after 7:00 a.m. the following day, I boarded the train for Calcutta. There were about twenty cars joined to a steam engine, which appeared to be occupied primarily by black troops from West Africa. My sleeping area was in a small room with four bunks. The three other men in the compartment included two white and one black British officer. The black officer and one of the white officers spoke English while the other officer spoke French, as did the remainder of the troops. I threw my sea bag on my bunk, introduced myself to my English-speaking roommates and acknowledged the third via translation. As I remember, one was John, one was Charles, and the French-speaking officer was Gene. All were dressed in Army fatigues, armed with rifles, and had English-style helmets at their sides. I was dressed in brown khaki shirt and pants, army boots, an Eisenhower hat decorated with a bronze USN pin, and armed with only a knife. No one commented on or questioned the United States Navy's presence in India. They either didn't know the significance of the pin or didn't really care.

The train engine let out a shrill whistle, and we started moving out of the station. The petty officer who had brought me to the train had departed already, and all I could see was a mob of locals peddling fruit, sandwiches, and other merchandise. Other black soldiers were boarding a train on the adjacent track. I had my orders stuffed in my shirt pocket. If everything went as planned, I would arrive in Calcutta in about three days. The petty officer told me not to eat any raw fruit or drink the local water and to stay away from fresh salads. He also had said that my best bet was to eat in first-class restaurants and stick to soups, tea, and C Rations. I found myself in the company of the British Army, headed for a destination a few thousand miles to the east. In the midst of many, I suddenly felt very alone.

After chatting for a short time with my English-speaking companions, I offered them most of a carton of Camel cigarettes, which seemed to install me in the clan as a member of the group. An hour or so later, I went looking for the toilet. It was a slit in the floor, with no running water, no shower, and no paper. I also discovered that there was no dining car on the train. At this point, I decided not to ask questions or complain and just do the best with what was available. The train made a number of short stops at stations, where there were stalls selling fruit, candy, and various local foods. Some stations also had first-class restaurants, but I limited my purchases to tea and boiled water. I was nervous about leaving the train, as I seldom understood what was being said about the layover time.

In addition, I found it difficult to deal with the hundreds of beggars whose bodies were deformed in every imaginable way. They included children whose arms, legs, and faces had been purposefully deformed to aid their begging. Just before sundown, we entered a rather large town. As we slowed and came to a full stop, beggars swarmed along the sides of the cars. For several hundred feet, we could see the type of pain and suffering that Jesus must have seen in Galilee—there were blind, crippled, and tortured bodies and lepers of all ages. The sight cut deep into my conscience, and I thought a long time about the tremendous advantages of being born American.

The first evening, I found out a little about my traveling companions. The two English-speaking officers had grown up in Kenya. The white officer's father owned a small plantation where they grew pineapple and avocados; the black officer, had been educated at a mission school in Kenya and an English university. Both had joined the army in the early 1940s, some time after Hitler's mission had become apparent. The third officer was of French descent and had grown up in Morocco. All three had fought in North Africa and were being reassigned to the India/Burma region. They all looked like they had seen their share of war and were older and generally quiet. While they knew that I wasn't a combat veteran, they treated me with respect. I spent most of my time listening to background chatter and train noises and watching India slide by outside the window. After dark, we played poker for cigarettes for a few hours before retreating to our bunks. It was not a great night for sleeping as the train continued to change speeds, lurch, and stop at various stations, and the track seemed rather bumpy.

I woke up feeling somewhat sore, but realizing that I was much better off than the hordes of people along the train track who were struggling to make a living. Around 8:00 a.m., the train pulled into a small town, where I purchased some hot tea. The countryside had changed overnight; here, there was more vegetation and the air seemed more humid. The first-class restaurant in this was clean and equipped with a bathroom, which I used. I bought a couple of chocolate bars from a stand and climbed back on the train, pleased that in a day and a night I would be in Calcutta. The second day seemed endless, as we went from one town to the next occasionally stopping because there were cattle on the tracks or to make rest and food stops. I shared the candy bars with the three British soldiers, and they offered me some bottled water. Unused to the hot, humid weather, I sweated profusely, developed a heat rash around my waist, and was generally uncomfortable, but I did not complain. By day's end, I was glad to see the sun fall below the horizon and looked forward to the morning reunion with my navy buddies.

Over the Himalayas

After a second night of broken sleep, we arrived in Calcutta. I said goodbye to John, Charles and Gene, who had been good travel mates. We shook hands firmly, knowing that we were unlikely to ever see each other again. Our lives had touched for a few days, and I had learned a lot from my comrades who had been involved in the war for more than four years. I didn't have a lot to offer in terms of experience, but at least they enjoyed my contribution of cigarettes and chocolate bars. Much to my relief, Jim Murphy was waiting for me at the station. He and a local driver helped carry my belongings to a jeep, and we drove through the streets of the sprawling city to a rather nice two-story building that housed the radio-intelligence unit on the outskirts of town. I was tired but relieved to rejoin my mates. After a short conversation about my experiences in Bombay, what my friends had encountered, and what we had to look forward to, we went to lunch—the first decent meal I had since leaving Bombay.

The next morning a lieutenant laid out the schedule for the five of us who were bound for military intelligence work in China. We were to spend several days in town training on a field direction finder and would be issued arms and new clothing. We started off the morning by walking about a mile to the field RDF. It was about 9:00 a.m., but it was already hot and humid. Within an hour, a heat rash had broken out on my stomach. The four of us each took turns practicing on the RDF. It was somewhat similar to the one that we had used at IB, but it had a better signal-resolution device that allowed bearings to be more rapidly resolved. Several hours later we returned to the headquarters building and were each issued a .45-caliber pistol, .30-30 carbine with a belt to hold its ammunition, Thompson submachine gun, and a new jacket with Chinese letters and a flag on the back. The jacket would supposedly put us in better standing if we were shot down or landed in an area under control of Chinese nationals.

The sudden endowment of weapons placed our upcoming mission in a somewhat different light. We were given a short lesson on disassembling and reassembling our arms and instructions on their maintenance. Then we were told that we could take leave that night, but must be ready to ship out to a transfer station the following morning. Jim and I decided to have dinner at the famous Hotel Green and check out the city. The doll we had brought from Flint, Michigan, now displayed the names of Newport News, Virginia; Colon, Panama; Melbourne, Australia; and Bombay, India. That afternoon we carved Calcutta, India, along one of her legs.

That evening, Jim and I and three other friends made our pilgrimage from the outskirts into the famous city located on the upper Bay of Bengal just south of the Ganges River. It was teeming with people, and the streets were crowded with cars, trucks, buses, and carts, all trying to weave their way through the heavy traffic. Young children raced along side, successfully persuading us to part with a few anas. The Hotel Green felt like a veritable oasis, and its furnishings and elegance reminded me of the Del Coronado in San Diego. At the restaurant, we selected several dishes of curry and were well satisfied with the food and service. After dinner, we did a little more sightseeing and had our pictures taken in a photo shop. We noticed a great number of individuals sleeping on the outcroppings of buildings or on the sidewalks. Poverty was everywhere, and we grew increasingly depressed. We just couldn't believe that so many people had to live with so little. Our education had only begun.

At 7:30 the next morning, we stacked our gear in the back of a six-by-six truck, said goodbye to a few of the men who were permanently stationed in Calcutta, and headed out of town to the transfer station. We were taken to Kanchrapara, which was out in the country, about forty miles from Calcutta. The camp comprised half a dozen or more permanent buildings scattered among grassy fields and occasional clusters of trees. The main facilities for transit personnel included hundreds of semi-permanent tents on wooden platforms a foot or two off the ground. Jim and I were dropped off at one of the tents, which would be our home until we got orders to proceed over the Himalayas into China. It wasn't exactly high living, but it was more than adequate for a brief stay. We choose our cots, stored our arms, and waited for further instructions.

There wasn't a great deal to do at the camp, so most of the men laid on their cots, read, and waited for chow call. Jim and I took a walk around the camp, which was surrounded with a barbed-wire fence. Outside the compound, we could see a few cattle in the fields and several small children walking a large water buffalo across a meadow. Food for the camp residents was prepared in a large wooden building about two hundred yards from our tent. At lunch time we lined up, waited our turn, entered the mess hall, picked up a tray, took what we wanted of the available chow, and returned to our tents to eat. Since the weather was extremely hot and humid, it was better to eat in the tent than in the mess hall, which housed the stoves on which the food was prepared.

Shortly after lunch, Jim and I noticed smoke coming out of the mess hall and ran to see if there was a problem. All of the cooks and servers had abandoned the building, and a small fire was burning on one side of the large cook stoves. We ran in and found a bucket of fuel oil sitting close to the fire and an opened fifty-gallon drum of stove fuel about fifteen feet away. We put the lid back on the drum and wheeled it out the nearest door. A fellow ran in to assist us, but did not see the small bucket of fuel oil on the floor. He tripped over the can and spilled it across the wooden floor, retreating out the door just before the fuel burst into flames.

The fire quickly expanded, and the flames reached the grass ceiling of the mess hall. Suddenly Jim and I found ourselves trapped between the fire and the doors. In an instant, we realized that we had to run through the flames or let the building burn down around us. In a split second, we covered our faces with our hands and dashed through about six feet of flames. We ran through the door and rolled on the ground to snuff out any fire. Luck was with us and, although the hair had been singed off of our arms and our eyebrows were burned, we were basically unharmed. The fire that had flashed up so quickly was soon put out.

Both Jim and I were annoyed at the would-be helper who had spilled fuel on the floor. As we approached him, he made a smart remark, saying that if we had just stayed put, we wouldn't have had any problem. He did not see that we had been trapped behind the fire. This turned out to be true, but we couldn't have predicted it at the time. At any rate, my adrenaline was running high and I wasn't in the mood for a debate. I threw a quick right and hit him on the chin. He went down on his back with a thud. Jim told him to keep his butt on the ground, but others quickly arrived and ended the scuffle. But there was a problem: the downed man was an ensign. Jim and I were in big trouble. We were required to stand mast, i.e., go before an officers hearing. There was no discussion, just the reading of the findings. I hit an officer, and Jim swore at him. Luckily for us, he was not wearing his officer's bars on his hat, which reduced our punishment to five days in the brig on bread and water. Furthermore, if we stayed out of trouble for the next six months, the incident would be removed from our records and we would still be eligible for our good conduct medals.

We were placed in the brig, but the guards all knew what had happened and believed that we were the heroes, and the officer the dunce. As a result, they passed us sandwiches and candy through the bars. Our stay in the brig lasted only two days because our orders came through to move on to China. We were released and told to secure our belongings and be ready to ship out the following morning. Later, we found out that the officer who caused the explosive fire was given a medal for his heroic efforts to save government property. I don't know if this is true, but anything is possible in time of war.

Before sunrise, we were taken to Dum Dum, the airfield outside Calcutta, where navy personnel separated and assigned us to different aircraft. Jim and I were on different planes. A lieutenant called my name and briefed me on the flight. He told me that I was the senior ranked non-commissioned officer (non-com) and that I would be in charge of getting approximately twenty-two men to Kunming. We were to fly first to a field at the foot of the Himalayas, spend the night at the base camp, and then make the "jump over the Hump," as it was called during the war, into China. All of the petty officers were dressed in khakis and armed with carbines, Thompson submachine guns, and .45-caliber pistols. Somehow, they just didn't look like the U.S. Navy. We were given the order to board a two-engine C46 Combat Cargo Command plane. I called roll, we boarded, secured our parachutes, sat in the bucket seats that lined the sides of the fuselage, and snapped on our seat belts. The plane taxied to the end of the runway, turned, revved up the engines and, in less than a minute, took off. It was the flight of our lives, and we were on the edge of our seats. After twenty to thirty minutes, the roads and buildings began to disappear, and all I could see was a matted green jungle.

The first hop of our flight took us into the northeast corner of India. After about three hours, we landed in an airfield that I suspect was not too far from the border of Nepal. The runway was a long paved stretch carved out of the tropical foliage, and it seemed immense to me. I could see several cargo planes nestled along the sides of the runway, while others taxied in or out on the runway. Our plane came to a stop, and we disembarked, loaded our gear into a six-by-six truck, and drove to a tent city near the field. It was early afternoon, and the temperature and the humidity were trying to outdo each other. Within minutes, I began to sweat profusely and in less than half an hour I developed a bad heat rash around my waist. An army lieutenant met us in front of one of the more permanent structures at the transit base camp. He handed me a sheet of instructions, which included tent assignments, feeding arrangements, and orders to be ready for pick-up and departure at 5:00 the next morning. I read the pertinent information to our group, dismissed them, and we all headed to our respective tents.

After settling in, I went to the first-aid tent to get something for my heat rash. A nurse took a quick glance, gave me a small bottle of some kind of white lotion, and told me to shower frequently and apply the lotion after drying. I showered three times that afternoon, and each time felt sticky and sweaty minutes later. I applied the medicine as instructed but got almost no relief. I began to worry that Kunming would have a similar climate and that I would spend my overseas duty scratching my waist and putting on lotion.

Around 5:30 p.m., I went for a walk around the camp, which was called Chabwa. About a quarter of a mile from my tent, I located the post exchange, the watering hole at the base camp, and was delighted to find that several of my shipmates had already arrived. I joined them for a cool beer—again, no one bothered to ask my age. Our group was the center of attention. The quantity of weapons alone was enough to stimulate comments (we were told to keep them with us at all times), but the gold USN pins on the collars of our shirts fired up a heated discussion between the local army personnel and our transit navy team, who looked like army raiders. After we heard a few smart remarks like, "Are you swabs lost?" and "What the hell did you Sea Scouts do to deserve being assigned to this part of the world?" we quietly talked over the situation. If we got in a fight, we weren't likely to win, and if we went before an army court martial, we would probably end up in the cooler. I didn't want another black mark on my record. We decided to finish our beers and retreat.

The soldiers' desire for response from us must have been overwhelming because within a few minutes a sergeant, who must have been six feet two and two hundred and twenty pounds, walked across the room, stood next to me, and asked, "Come on, guys, what the hell is your story?" Everyone looked at me so I knew that I had the floor. If I stood up, he might think I was challenging his masculinity, so I invited him to sit down and join us for a few minutes. He did, we put a beer in front of him, and I gathered my thoughts. "To be honest," I told him, "we don't know where the hell we are going in the middle of Asia. But I think there must be more navy recruits than ships and suspect that we are going to be assigned to Chinese sampans." He laughed and said, "Yes, I know you're on some sort of secret mission, everyone in this theater is on a secret mission." It might have been a lousy story, but it broke the ice and we parted friends. Interestingly, several decades later a friend of mine, Clayton Mishler, wrote a book about our war effort titled *Sampan Sailor*.

As I left the post exchange, a woman in army fatigues stopped me and introduced herself as a *Stars and Stripes* reporter. She had seen several of our heavily armed comrades in the camp and asked if I could give her a story about our destination in China and our mission. The camp was a transit stop for people going into China, so she knew our general destination. "Yeah," I said, "a few navy men are on their way to

China and if you want details, you will have to ask the navy brass." She asked if I would join her for a cup of coffee, but I knew that she wanted to get something more for her story. I excused myself and returned to my tent lost in other thoughts. About a month later, Ruby sent me a strip from the comic *Terry and the Pirates* showing a group of well-armed navy men boarding a flight headed over the Hump into China. I always wondered if the reporter had leaked the information to the comic strip author, but it could have happened a thousand other ways.

That evening, I sat on my bunk and thought a lot about the morning flight, which would be over Japanese-held territory without fighter escort. In 1943, the Allies had lost about five percent of their planes going over the Hump, but there were fewer losses in 1944. But planes were still being shot down, and others crashed due to mechanical problems. We would be flying very high, and the passengers would have to use oxygen for a good part of the trip. There was some comfort in the fact that we would have lots of company. Part of the survival strategy was the sheer number of flights launched each day, and the varied courses tended to ensure that a good percentage would reach their destination.

At 4:30 a.m., someone came into the tent and gave me a wake-up call. I got up, put on my heat-rash lotion, dressed, and put on my recently issued winter jacket with the Chinese flag and Chinese writing on the back. The temperature was in the high seventies and would get even hotter, but we were told to dress for a cold flight. I jammed a navy watch hat in my pocket, put on my army helmet, grabbed my sea bag, and headed for the pick-up location. Most of our group had gathered already, and at 4:50 a.m. I called roll and we were ready for departure. The truck showed up at 5:00 a.m. sharp, and within a few minutes we were dropped off next to a C-46. Those planes always looked like pregnant guppies to me, but along with the C-47s (DC-3s), they were the workhorses of the Hump flights into China. I suspect that a hundred flights or more a day carried men, arms, and supplies into China. Since the Japanese controlled all the coastal cities and had choked off access to the country via the Burma Road, the Allies reached China primarily by air transport over the Himalayas.

One of the pilots asked us to form a semicircle on the tarmac, while he briefed us on the flight and gave us survival instructions in the event the plane was shot down or had to be abandoned during the journey.

Before he began his spiel, a crewmember gave each of us a survival kit, which we were told to open. It contained a compass, small knife, map of the region we would cross, pills to purify water, a small amount of food, first-aid paraphernalia, sulfa powder (an antibiotic used in case of an injury), and a note requesting assistance in local languages, which was to be given to the inhabitants if we made it to a friendly village. The pilot then pointed out that part of the flight would be over enemy territory and explained in detail what steps to take if we had to ditch the plane. These steps included proper use of the parachutes; routes to follow, which depended on the drop area; and where and how to seek help along the escape routes marked on the maps. It was about that time that I decided war was a very serious and risky matter; all my mates were also quiet and attentive.

The pilot concluded with, "Are there any questions?" It was quiet for a few seconds and then, after scaling the potential distance to safety on the map, I asked, "How long does it take for those who have to bail out en route to walk back to friendly forces?" He answered, "From several days to months, if you're lucky," and then gave the order to board the plane. We mounted the ladder, carrying all of our weapons and personal belongings, stored our gear in the center of the cargo area, put on our chutes, sat in the bucket seats, and placed our rifles and submachine guns between our legs. As we waited for take-off, I could see a steady line of planes taxiing toward the end of the runway, and cargo planes lifting off every few minutes. The captain came on the intercom and told us that the flight would be close to five hours and that we would need to use oxygen for more than half the flight. We would fly at about 20,000 feet when the plane crossed over the Himalayas. "It will be cold in the cargo area, and the skin of the plane will be freezing," the captain said. He told us to stay seated and buckled up, and that he would alert us if there was trouble.

The pilot revved up each engine, gave the engines full throttle, and then we were on our way to challenge the Hump. The runway faded, and soon we were over a sea of undisturbed green jungle. It all looked the same, and the ground was hidden by the jungle canopy. The plane continued climbing for the better part of an hour before I made out the outlines of mountains, with broad valleys draining the surrounding highlands. A crewmember came back from the cockpit area and told us to hook up our oxygen masks and keep them on until we were told that it was safe to remove them. He remained long enough to ensure that we had all followed orders and then disappeared into the forward cabin. As I looked through the small ports, it became obvious that the valleys were narrowing and the peaks were getting higher, until the latter finally rose above the plane. For almost two hours, our entire group strained to watch the passing scenery and to look for signs of enemy planes. Thank god, they never made an appearance.

About halfway through the flight, my apprehension about having to use the survival kit waned, and I fell asleep listening to the drone of the engines. When I awoke more than an hour later, the pilot announced over the intercom that we had cleared the dangerous part of the trip and would land in Kunming in about forty minutes. During my sleep, my hand had fallen against the skin of the plane and when I sat up, I pulled a thin layer of skin off the back of my hand. It was bruised and oozing blood, so I took a bandage packet out of my survival kit, taped it to my wrist, and chalked it up to experience. We landed in Kunming around noon without incident, veterans of the flight over the Hump. Several officers were in the passenger unloading area to receive us. One asked for the passenger manifest, and I gave it to him. We were immediately broken up into several groups, and I was whisked away, loaded into a jeep, and driven about ten miles to a building at the foot of the mountains that overlooked a broad valley covered from side to side in rice paddies.

It was late August, and I had finally arrived at our "unknown" destination. Jim had already landed and was waiting for me at our new base. We had been on the road, discounting the training period in Washington, for more than fifty days. We sat down in the galley, poured a cup of coffee, and recounted our journey. We had traveled across the country by train to Washington, D.C., taken a short train ride to Charleston, Virginia, and a ferry to Newport News. From there we sailed across the Caribbean to Panama, crossed the breadth of the Pacific, turned north in the Indian Ocean to Bombay, and crossed the heart of India to Calcutta by train. We had been trucked to Kanchrapara and flown to northeast India and then over the Hump into China. All in all, it was a trip of about 18,000 miles. Our Kewpie doll, with a few new names scratched on her body, had arrived with us unscathed.

Lee Alverson and Jim Murphy (SACO members) with unknown WAVE. Washington, D. C. USO. May 1944.

CHAPTER V
DRAGONS, WARLORDS, PEASANTS, & SACO TIGERS

After a good evening meal, Jim and I retreated to our new quarters and chatted with acquaintances. That night, for the first time since leaving the ship in Bombay, I slept very well. In the morning, Commander Joyce, the CO, called all the newly arrived radio-intelligence petty officers to his office, and we finally were given a detailed explanation of the navy's mission in China. I have since read a number of accounts about the origins of the Rice Paddy Navy, and they all seem to support the story given by Commander Joyce. Apparently General Chiang Kai Shek had no great love for General Stilwell, the U.S. Supreme Commander in the China, Burma, and India theaters, and was reluctant to have U.S. Army ground forces enter and engage the Japanese forces in China. But he did get along with General Chennault; hence, the 14th Air Force, whose members included the soldiers of fortune who were the original Flying Tigers, was actively involved in China.

Chiang, nevertheless, looked for a way to strengthen his forces fighting the Japanese and the nationalist position in China; perhaps he also foresaw the post-war conflict with the Communist elements in the country. In a strange quirk of fate, he had become close friends with a young American admiral, Milton Miles. Working with the famous General Tia Li, Chiang's head of intelligence, the men formulated a scheme: U.S. Navy personnel would enter China to organize an intelligence network, train Chinese guerrilla fighters, aid the escape of allied pilots and crews who had been shot down by the Japanese, and generally obstruct Japan's war effort in China. I will not comment on the wisdom of selecting the navy for this job, which seemed to be a convoluted way of circumventing personality conflicts. I suspect that President Roosevelt faced strong resistance from several branches of the armed forces before he signed the order creating the Sino-American Cooperative Organization (SACO) in 1943. The group later was nicknamed the SACO Tigers. Its pennant had three question marks, three explanation marks, and three stars, which roughly translated into "What the hell is this?"—a fitting commentary.

The agreement gave Admiral Miles and the U.S. Navy the authority to establish a network of intelligence and training bases for an expanded Chinese effort, working largely behind the Japanese lines. By the time Jim and I arrived in Kunming, Miles and his Chinese friends had already established a number of camps across China, from the Gobi Desert south to Hong Kong and from the interior to the near coast facing the Pacific. The war in China, unlike most of the other theaters, had not gone well for the Allies. Slowly but surely, the Japanese had been advancing, closing down vital 14th Air Force bases and capturing strategic cities. The week we arrived, several radio-intelligence men returned to Kunming after being evacuated from Kwelin. Captured by the Japanese, Kwelin had been a long-time operational base for the Flying Tigers even before the United States entered World War II. Earlier, Nanking had been lost. For all practical purposes, the Japanese had backed Chiang and his allies against the Himalayas and forced them to walk into western China. The SACO mission was an attempt to encourage greater internal resistance in areas of the country lost to the Japanese. It also would give the United States much needed meteorological and intelligence information on Japanese movements in China and along its coasts.

Until 1944, the focus of the radio-intelligence activity was in Kunming and Chunking. The Kunming facility was located about twelve miles out of town, just east of Lake Dianchi (Dian). It was situated at the base of the steep rise of the eastern Himalayas and consisted of a large old building with thick earthen walls. In typical Chinese style, the tiled roof tilted up sharply at each corner end. Inside, the interior had been modified to house a galley, a large radio-intercept room, showers, and quarters for the officers and the other men. Work on the radio towers and antennas had not yet been completed; a new building with expanded quarters and cooking and eating areas was also under construction. The main building faced out on a very large fish pond that had been used for carp culture and retained a residual population of koi (large colorful goldfish, a species of carp). The compound was enclosed in back by the steep rise of the mountains and a wire fence. A large stream gushed from an artesian well at the foot of the mountain and ran down one side of the grounds to the road. Rows of trees and barbed-wire fence closed off the other side. A guard post sat just inside the road that led to the base.

After being briefed on the navy's mission in China, we were put to work finishing the construction of the radio towers and antenna, standing watches to intercept Japanese radio messages, and do whatever was necessary to get the base fully operational. We were all kept busy taking turns on various tasks. I was assigned to decode the messages coming into the camp and code the messages originating from the camp and then transmit those messages to Chunking, Bombay, or the United States. Jim and I also were asked to operate an RDF that had been installed right in the middle of the rice fields, about a mile from camp. From the station, we could see miles of rice paddies filling the valley floor. No wonder we were tagged the Rice Paddy Navy. To approach the RDF site from our base, we had to walk down a rice-paddy levee; if it was at night with no moon, the walk was in pitch blackness and difficult to do even with a small flashlight. The equipment was exactly the same as what we had trained on in Calcutta, so there were no real operational learning requirements other than getting familiar with the local facility.

About two weeks into September, Jim and I decided to take liberty in Kunming to check out the sites and a fancy Chinese restaurant. We were given a jeep ride into town and dropped off at one of the city's large gates, which looked like an elevated temple; it had a large access area cut through the long sides of a rectangular-shaped base and a lofty tile roof that arched up from the middle of the gate to peaks at each corner. Unlike many other cities in China, the streets in Kunming were relatively wide, perhaps because the city came into prominence during the war as the terminal of the Burma Road. Hundreds of shops lined the streets, selling everything from Western cigarettes to hand-carved teak chests. Dance halls, theaters, and restaurants were even decorated with a few neon lights. The streets were crowded with local and U.S. military buses, trucks, and jeeps as well as carts and street vendors and flesh peddlers.

Using a local map of the city, we headed toward the Li Li (Come) Restaurant, known for the quality of its food. As we turned a corner, we ran into a parade of dragons, with colorfully dressed men and women beating drums and cymbals and playing various local instruments. In the middle of the procession was a highly decorated wooden carriage carried by twelve men. We could not see who was inside because white silk curtains were drawn across the windows. Several hundred people followed the carriage, singing, waving, and apparently having a good time. We were told that it was a wedding celebration but never saw the bride. A few blocks later, we found the Li Li Restaurant. It was several stories high, and we could see diners through large windows on each floor.

It was obvious that the restaurant catered to foreigners and was a favorite of the American forces in Kunming. Almost all the waiters spoke English, and the menus were printed in both Chinese and English. Jim and I were a little disappointed because we were looking for something more traditional. It was, however, probably a good starting point. We could try what little Chinese we knew, and the waiters could correct our efforts and add a few words to our vocabulary. We soon learned that very few of the dishes offered resembled the Chinese food we had consumed stateside. With a little help from the waiter and nearby patrons, however, we managed to order a variety of dishes, which were not served together, but brought to the table in sequence. To demonstrate our maturity, we started off with mulberry wine, a reddish, sweet wine that satisfied our palates. We drank several glasses before our first course was served. The restaurant had a small three-person orchestra that played a mixture of Chinese and American tunes and a local woman singer who did her best to imitate Betty Grable, and, accompanied by other band members, the Andrew Sisters.

City gate, Kunming, China

We ate a casual dinner with a number of courses, some of which we passed over quickly and others that we found delicious. We found the entertainment good and the American servicemen in our vicinity friendly. Our impressions and the camaraderie of our fellow GIs likely were enhanced by our consumption of several more glasses of mulberry wine, which we used to toast the adjacent tables and partake in a Gom Bay (bottoms-up) exercise. By dinner's end, my brain finally kicked in and told me that I had had too much to drink, but it was too late. My stomach was protesting, I was getting nauseated, and my surroundings seemed less than steady. Jim also had consumed more than his share, but showed no signs of becoming sick. He just got happier and louder, suddenly becoming quick-witted and a stand-up comedian. After he had entertained the crowd for half an hour, I suggested that we hit the road and check out more of the city. We left after toasting all those around us, the waiters and waitresses, President Roosevelt, Chiang Kai Shek, and anyone else we wanted to toast. We were well on the way to being smashed, and the party was rapidly ending for me. I was sick and needed air and a place to upchuck.

Before leaving the restaurant, Jim purchased another bottle of mulberry wine. We caroused around town for about an hour. As Jim consumed the last bottle of wine, I expelled the wine, dinner, and whatever else was in my stomach. I was miserable, but Jim just kept getting louder and falling "in love" with everyone he saw. He talked to anyone who understood a little English, including the Chinese military, fellow Americans, and girls on the street. I wasn't surprised when a couple of MPs confronted us, asked for our identification and passes, and then told us that it was time to retire. I think they said, "Get your asses back to your base and don't give anyone a bad time." As a matter of fact, we had been very gregarious with all hands.

We managed to get a ride to the main road that led to our compound and then had to fend for ourselves. We tried to hitch another ride, but there wasn't much traffic. Jim's mood began to change, and then he became sick. He vomited so much on the side of the road, he began to have the dry heaves. I had never seen anyone in that condition, so I didn't have a clue what to do. We just sat there hoping that a ride would come along and that Jim would recover. He had the dry heaves for the better part of an hour and was so weak he could hardly get up. We lucked out, though, as an army six-by-six pulled up. I jumped in the back and, with the help of the corporal who was driving, got Jim onto a seat under the truck canopy. We had been baptized in China!

It was difficult for me to make it through the morning watch, but somehow I managed. Jim also struggled, hating every minute of it. When the day was over, we went to our room and discussed the "mulberry wine night." Jim remembered everything until after dinner, and then it all became dim. We jointly decided that we had to find some other beverage, something that would stay down. Those who had been around Kunming for a while just laughed, and I took it to mean that they had already put a similar experience behind them.

The next time that Jim and I had leave, we decided to do something more educational. About a mile from the radio station, there was a broad trail leading up into the mountains that edged the valley. We had no idea where it went, but we noticed that coolies—a term used at the time by Chinese and U.S. forces to describe workers who undertook tedious and underpaid work—came down the trail laden with various vegetables, fruits and paper goods. We packed a lunch and set off to explore. After crossing a few rice paddies, the trail climbed steeply for a mile or so and then opened into a small valley. We passed several groups of walkers moving down the trail and a few coolies carrying goods swinging from the ends of their yo-yo sticks. The trail soon climbed again, following a small stream that wound through the surrounding hills. We continued to climb for over two hours and began to see larger and more frequent trees and denser foliage along the trail.

We were perhaps five or so miles from base camp when we began wondering whether we would encounter a village or rice paddies in the uplands. We stopped and tried to ask one of the coolies how far it was to a settlement, but I suspect that our Chinese was too flawed. He smiled and headed on down the trail. We still had no idea where we were headed. After about three hours, just after noon, we decided to find a good spot for lunch. But before we found a suitable location, we saw a village on the banks of a large pond crowded with white ducks and a few geese. Various fruits, eggs, and rice cakes of some sort were for sale in a small earthen building with an open front. Two old men sat inside, having tea and cakes. We found a stone bench, sat down, and ordered tea. We were served a kettle of steaming tea, which was poured into two tea bowls, and we unwrapped our sandwiches and began to enjoy our meal.

Our arrival generated a good deal of interest among the villagers, particularly the children. Within minutes, twenty to thirty children, ranging from three to fifteen years of age, had formed a semicircle around our bench. The youngsters giggled and chatted among themselves, while the older children just smiled and watched. We suspected that the villagers hadn't seen many, if any, Caucasians in their community. The shopkeeper and a woman, probably his wife, tried to sell us some fruit, eggs, and rice cakes, but Jim and I were concerned about sanitation and turned them down. Later, they tried to give us the food as gifts, but we refused, knowing they had little to give. We did accept a bowl of rice, which we ate with chopsticks to the great delight of the surrounding children.

After lunch, we paid the shopkeeper about a hundred CNC, or fifty cents, and he gave us about forty CNC back. When we left, the shop owners and a number of the adults waved at us and some of the children followed us for half a mile or more. We had consumed no mulberry wine, although the storeowner offered us something called "chiew," a white rice wine that we later learned had the kick of a mule. We left untainted and returned down the trail that had been worn into the mountainside by thousands of peasants and other people moving products to and from Kunming for almost five centuries.

Old man drinking tea, small mountain village, 1944

On the Move Again

Jim and I settled in and became comfortable with life at the base. Early October brought a twinge of cool fall air. One day, I had just finished breakfast when someone told me that the skipper wanted to see me in his office. Commander Joyce asked me sit down and then said, "You have a lot of experience with the RDF." Then he explained that the radio-intelligence group wanted to establish an RDF unit in Japan's backyard. He pointed to a spot on the map that appeared to be right on the coast of China. "We have just opened up Camp Six here," he said. "It is located in the mountains behind the enemy lines about forty-five miles from the coast. We plan to install an RDF and weather station at the camp. You have the experience, would you like to go?" I don't think my brain had kicked into gear, because I didn't ask any questions, and just said yes. I assumed that Jim and others would accompany me. Then Joyce then explained that I would go alone and that sometime in the future they hoped to add enough men to carry out a twenty-four-hour watch. I was still eager to go, but inside my brain I kept hearing the words "alone, alone, alone."

That afternoon I sat down and talked over the new assignment with Jim and Verne Benedict. I was not happy about leaving Jim and the others who had been with me on the *General Anderson* and on into China. Still, there was something challenging and exciting about the adventure into the unknown. I put my gear in order and collected a few things that might come in handy in an isolated location, including a sewing kit, aspirin, and cigarettes. I didn't smoke, but cigarettes were always good trading stock. Jim and I decided that the Flint doll should come with me, so I wrapped it in a T-shirt and stowed it in my bag. After I'd packed, I wrote long letters to Ruby and my mother, advising them not to expect mail for long periods. That evening I was given my travel orders; transportation had been arranged and it would take a week or two to reach my destination.

The sun was just creeping into the sky the next morning when I threw my gear into a jeep and was driven to the airfield. In a Quonset hut that served as an office, I met Parsons and May, two second-class petty officers from the general communications group who were also en route to Camp Six. In a few minutes, the pilot, who looked to be in his early twenties, showed up with the crew and told us to bring our gear to the plane. It was another C-46. This time, the pilot did not offer any general instructions on survival. He took me aside, however, and said, "You know we will be flying about 800 miles over enemy territory with no fighter escort." He then looked at the Thompson submachine gun that was slung over my shoulder and said, "Do you guys know how to use those things? We might need some protection." I smiled and replied, "We know how to shoot the guns, but we would be lucky as hell to hit anything moving in the air." (I thought we would be more likely to shoot ourselves down than a Japanese Zero.)

The pilot seemed a little uneasy about the flight, but then I'm sure he had lots of company. We tossed our gear into the plane's loading bay along with some mail and other freight, which was secured with cargo nets. Before takeoff, one of the crew came back and introduced himself. He told us that the trip would take about six hours and that we would be flying at 10,000 to 12,000 feet for most of the flight. "Oxygen," he said, "probably will not be needed and if it looks like we're going to have company (enemy planes), we will let you know." He concluded by saying, "Stay buckled up and don't do anything unless the pilot gives an order."

As we gained altitude, we could see the great high valley that surrounded Kunming and, to the east, more mountains. South China was made up of range after range of mountains interspersed with valleys that were, for the most part, planted in rice. The plane droned on hour after hour, and there was always another range of mountains ahead. About five hours into the trip, the plane began to descend as we approached our destination, an airfield in Fujian Province. Changting was about a hundred miles east of the China coast and the only 14th Air Force field in east China that had not been closed by advancing Japanese forces.

The plane began circling at around 1,000 feet. I could see rice fields, but not the landing strip. Then, without warning, the pilot suddenly gave full throttle to the engines, turned sharply to the left, and began to gain altitude. It was obvious that the plane was struggling to climb into the sky. We were flying up the wall of a valley, and the mountain crests were above us on both sides. I could see the valley floor narrowing and rising up toward the plane. The pilot maintained full throttle and attempted to climb over a crest at the head of the valley. We had no idea why he had aborted the landing and why we now were flying within several hundred feet of the ground.

Within about ten minutes, we approached the pass at the upper end of the valley. The mountain cliffs on both sides had closed in and were as near as several hundred feet on either side. The valley ended at a ridge several hundred yards ahead. The plane seemed to be flying at the same height as the crest. Seconds later we swooped across the top of the ridge with perhaps forty feet to spare and I could see the grass on the hill flatten in our wake. Looking around the plane, I realized that the three of us all had been straining to get the plane up and over the crest. We soon put the ground well below us, and the mountain fell steeply away into another valley.

The plane soon descended again, and a crewmember came back and told us that we were about to land at Changting. He then shook his head and said, "You will be briefed when we're on the ground." As the plane circled the field, we could see the runway and supporting buildings. After landing, we taxied to a Quonset hut at one end of the field, where there was a hurried discussion between the pilot and the officers who met the plane. After a few minutes, the pilot came over to me and said, "We almost made a horrible mistake. We were all about ten minutes from being shot down or Japanese prisoners. The field we were about to land on was a Japanese airfield about sixty miles from here. I was attempting to make ground contact when I spotted two Japanese Zeros landing. They either didn't see us or were out of fuel. We got out of there as fast as we could and stayed low so it would be difficult to spot us. I guess we didn't have much time to spare when we crossed the ridge." I suspect that the crew was more nervous than the passengers because they knew what was going on—we didn't.

The pilot told us to unload our gear as fast as possible because the Japanese probably had seen us and would raid the field. We got our belongings, mail, and the equipment that was bound for Camp Six and were off within minutes. The pilot then said, "Goodbye and good luck. I'm going to get debriefed, have the plane fueled, and get out of here." We carried our gear into our assigned quarters, and the three of us returned to the mess hall for some coffee. We hadn't been there five minutes when the air-raid siren went off. We ran for the slip trenches where we remained for fifteen minutes or so, while a Japanese recognizance plane circled high over the field. After it departed, the C-46 took off and headed back for Kunming. We assumed it made it.

May, Parsons, and I discussed our future travel plans over dinner, but it was the blind leading the blind. The only thing we knew was that a bus would take us from Changting to our next destination and that we would leave either the next day or the day after. The navy apparently felt that was all we needed to know. How far and how long we would travel was considered unnecessary information. After a good night's sleep, we were told by a Chinese military officer that we would not leave until the following morning and that, if we wished, we could visit the city. The three of us jumped in the back of a six-by-six truck that was going into town, about five miles from the air base. The corporal driving the truck told us that he would pick us up after lunch.

Changting was somewhat larger than I'd anticipated. The buildings were mostly two stories, neatly decorated, but nothing like Kunming. There was no visible foreign element. The local 14th Air Force crew was small, and its members were not very obvious in town. We browsed through a few shops, bought nothing, and then stopped to purchase some tea. We were pleased that our Chinese was adequate for the small purchase. The local tea was good and, as usual, a crowd of children surrounded the large, white foreigners. There wasn't a great deal to see in town, but after meandering around for an hour or so we stopped in a shop for lunch. Ordering lunch was somewhat more difficult than ordering tea, and we were not making any progress. We looked around, saw some of the locals eating noodles, and pointed. The waiter smiled and trotted off to the kitchen. We received three large bowls of noodles with some shrimp on top, chopsticks, and a big grin from the waiter. We knew that the other customers were watching and were probably amused by our attempts to consume our meal with the wooden sticks. The waiter provided soup spoons, but we stuck to our chopsticks. We were a little sloppy, but we managed to get a fair share of the noodles in our mouths. After that day, chopsticks would be the only eating utensils available to us.

The next morning, we were up and dressed by 5:00 a.m, had a quick breakfast, and hauled our gear to a waiting bus. It looked something like an old school bus and was powered by charcoal, which generated some sort of gas that fueled the engine. Such engines were common in China in areas where gasoline was difficult to find. The three of us boarded the bus along with our gear, three bags of mail that were bound for our base, and an assortment of boxes, crates, and miscellaneous cargo. An army sergeant told us, "You will be on the road for a couple of days. Keep your guns close at hand. Sometimes bandits ambush the buses. They won't give you a bad time if they see you are well armed. Don't eat anything that is not well cooked and drink only boiled tea. Good luck."

With those parting words, the bus headed out of the airfield and slowly gained speed. We headed east across rice-paddy fields and then up and over the mountains. The bus struggled along at about twenty-five to thirty-five miles an hour depending on the road conditions and street traffic in villages. As we passed through the rice fields, we could see farm workers bending down or squatting in the paddies and, from time to time, a worker plowing a field with a water buffalo. There was absolutely no powered farm equipment to be seen. The paddies were irrigated from adjacent rivers or streams. At times, we could see water wheels operated by the leg power of the workers. The wheels scooped water from the river and dumped it into a slough that ran into the paddies.

There wasn't a lot to differentiate one valley from the next, but on occasion we would enter a valley where the workers had been organized. In one valley, all of the workers wore light blue peasant hats straw hats, blue blouses or shirts, and blue skirts or pants. The workers looked clean and neat. We suspected that the landowners or war lords that ruled the region mandated the dress. Although Chiang had nationalized many regions of China, some areas still retained largely autonomous control and maintained their own armies. From time to time, we would pass through villages where the bus had to slow to a crawl. Sometime after noon we stopped at a roadside restaurant where we enjoyed tea, sticky rice, and a rice cake. We stretched our legs and made a pit stop, although there were no bathrooms, just old fashion outhouses, without paper. The toilets were designed to collect excretions in large buckets (referred to as "honey buckets"), which then were used to fertilize the fields.

Everywhere we stopped, children would gather around to check us out and, if we decided to take a short walk, a number of them would follow us down the street. The villages generally involved a series of earthen huts two stories high that housed a few homes, shops, and a wealth of children. The town and village names on our rubber maps were all in Chinese script so it was impossible to know where we were. Animal husbandry, as far as we could make out, was confined to raising ducks and geese and the occasional water buffalo, which was often led by a child. The trip was difficult, and each time we had to climb a steep grade, the bus would slow down so much, we would have to walk for several miles as the driver slowly made his way to the top. On the steeper mountain roads there were gangs of workers who would push or occasionally pull the bus up the incline using a hefty rope. The driver would pay their boss some money for the help.

Around sundown we pulled into a relatively large town whose name escaped me, but modern maps tell me it was Hsinchuan. The town's buildings were on both sides of the road, which wound down the edge of a hill. We were put up in a hotel with surprisingly clean rooms, but rather spartan furnishings. One of the owners spoke a little English and told us that there was an American mission a block down the road. After dinner, we walked down to the mission—which may have been Methodist, but I'm not sure— and chatted with one of the priests and his helpers. I had thought that the Christian missions had been abandoned after the Japanese took control of China, but learned that a fair number were still operational in areas not occupied by the Japanese. We had a good talk, and they seemed genuinely interested in what we knew was going on in the outside world. They had radios but seldom got to talk to anyone from the United States.

By sunup, we were back on the road, going over one range of mountains and down into a valley. We had been traveling for a couple of hours when we began climbing up into another mountain range, but this time we just kept going up. Frequently, the bus had to back up and maneuver carefully around the bends. At such times, the driver asked us to leave the bus and help guide him around a sharp curve where the outer side of the road fell off steeply into canyons. Throughout the morning we got off and on the bus as workers helped to push or pull us up steep grades. The old charcoal burner just didn't have a lot of steam. After two more hours, we entered a stand of conifers that appeared to be some kind of pine or fir. The first ones we saw weren't very tall, but as we twisted up the mountain, they got larger and larger. They were bigger than any of the yellow pines or other evergreens that covered Idyllwild valley. I was amazed, as I had always been told that the major forests of China had long since been cut down by loggers. As I think back, the trees had the look of redwoods, but to this day I don't know what they were.

We reached the mountain crest and then spent much of the afternoon slowly winding down the other side. We finally broke out of the mountains into stretches of rice paddies and small farmhouses. Late that evening we pulled into Lungyen, a large city made up of buildings several stories high. There, we were met by several Chinese military officers who were trained SACO guerrillas. One was an interpreter who told us that Lungyen was about forty miles inland from Amoy, a major Japanese base at the mouth of the River of Nine Dragons. Before the war there was a relatively good road to the coastal area, but all the bridges had been destroyed to slow the Japanese advance. We would spend the night in small bungalows on the side of a hill. The rooms were clean and had indoor plumbing—what a treat! We would spend the next day in town and leave for Camp Six early the following morning.

Herding Ducks- "Chinese animal husbandry"

The officers and the interpreter picked us up at about eight that same evening and took us to dinner. The food consisted of fish, duck, rice, soup, and several drinks of chiew. It was, of course, essential that we toast President Roosevelt, Chiang Kai Shek, Tai Li, and other notables. After dinner, we made our way back to the bungalows. As I climbed the stairway, I felt something drop onto the side of my face. As I reached to brush it off, I saw that it was a large rainbow-colored spider. I knocked it to the ground, but not before it bit me just above my eye. Within an hour my left eye had swollen shut and I was feeling a little woozy, but it could have been the rice wine. I showed the swollen eye to Parsons, who said, "I don't know what the hell we can do for it but I have a half bottle of chiew left. It might help to kill the pain." I took a couple of swigs of the rice wine and retired to my room. When I awoke the next morning, the swelling had gone down and I felt great. There is a remedy for spider bites.

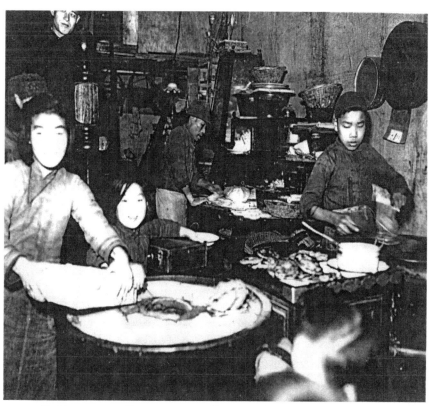

Cooking area in small restaurant on road between Changping and Linguine, 1944

We spent the next day exploring Lungyen with a group of children who took it on themselves to be our guides. We entered a few shops and looked at carved teak chests, jade, silver bracelets and beautifully embroidered silk gowns. However, we all agreed that it wasn't a time to buy. According to our Chinese advisors, we still had three days on the road, one by truck and foot and two by sampan. We would buy artifacts before we left China. That evening we had a simple dinner of chopped chicken, rice cooked with greens, and tea. On returning to the bungalow, I pulled out my rubberized map of Fujian Province and attempted to evaluate our progress since we'd left Changting. After two days on the bus, driving about twelve hours a day, we had only managed to travel about eighty-five miles as the crow flies. While the road mileage was probably more than double that, eighty-five miles in two days was hardly walking speed. The map indicated that the origin of the river we would float on for several days was about twenty miles from Lungyen. We were to do a good share of boating and walking the next day, so I turned in early, free of spider bites.

Early the next morning, the truck took us as far as about an hour out of Lungyen and dropped us off on a trail leading to the upper point on the river where sampan operations were possible. We hiked most of the day, with the Chinese workers carrying the mailbags and some other freight for Camp Six. When we finally made it to the river's loading area, it was about an hour before sundown and we had been on the trail for about seven hours. The village had a dozen or so earthen buildings. It was dirty and rundown, with no electric lights, hotels, or restaurants. We were taken to an abandoned building where we were to spend the night. Since it would turn pitch dark an hour after our arrival, the three of us placed blankets on the floor, made a makeshift arrangement to hold up our mosquito tents, and broke open C rations for dinner. When the sun set, we had little to do other than turn in.

I quickly fell asleep, but suddenly woke up frightened by something. My heartbeat was racing and I realized that a small animal was under the mosquito webbing with me. I grabbed the flashlight, jumped up, and turned on the flashlight in one quick movement. As I swung the light around the room, a half-dozen large rats scampered for cover as one raced away from my bed. How it managed to get under the mosquito webbing, I don't know; perhaps I had not tucked the net under my blanket. The incident so shook Parsons, May, and me that we abandoned the building and sat by the river for several hours. However, we soon got so tired that we returned and slept through the night—rats or no rats.

Just after sunrise the next morning, we loaded our gear onto a sampan that was about twenty-two feet in length with about six feet of beam. It had a small rounded roof made of straw matting over the stern and was manned by a father and his son, who was about fourteen. The father used a large stern oar to steer the boat, while the son operated a smaller one. Before we departed, our Chinese SACO officers told us to keep our eyes peeled for possible bandits and to swim for shore if we lost the boat in the rapids. The boat operator would then guide us to the next village downstream.

We started down the river around 6:30 a.m. The first half of the day we crossed several steep rapids and watched as the boat operator skillfully worked his way around the large boulders that broke through crashing white water. As the day passed, a number of streams and rivers added volume to the river. Just after noon, the boatman brought each of us a bowl of rice and several bananas. We were getting used to chopsticks, so we managed to eat our rice as we moved down a deep and slow section of the river. The riverbanks seemed to get steeper and higher as we moved through the mountains. We seldom saw any signs of life, though from time to time, there were sampans nestled along the riverbanks or boat "pullers" helping other sampans over the rapids. The sides of the mountains were covered with small trees, including some evergreens. We saw no wild life but were told that occasionally a tiger was seen along the riverbank. The boat operator worked his way downriver for most of the day, taking great pains to stay clear of the rougher white water that at times stretched across most of the river.

After almost eleven hours on the water, we pulled into a small protected dock area where about twenty boats were tied. We had arrived at Changping (not Changting), which was located on the river. The town was clean and seemed prosperous in comparison to the village we had left. We were met by several SACO-trained guerillas who spoke no English. They took us to a small hotel that was a lot better than the previous night's accommodations. We joined the Chinese soldiers for dinner, which included sweet-and-sour pork, rice, chicken soup, and sweet fresh watermelon. Despite their poor English and our feeble Chinese, we made it through the meal, hoisted several toasts with chiew, and turned in for the night. We expected to arrive at Camp Six the next day. We were a long way from Kunming and in the real heart of rural China. The thought that we soon would join our navy comrades helped stave off the loneliness.

We were off again just after sunrise the next morning. The river had become broad and deep, running between mountains that sloped up from each bank. Appearing to rise 1,000 to 2,000 feet above the riverbed, the mountains were covered with dense foliage, which included small and moderate sized trees, grass, and vines. The rapids became less frequent and, for long sections, we drifted slowly seaward as father and son positioned the vessel well away from the riverbanks. For great stretches, we saw no signs of life, not even small villages, just an occasional sampan making its way upriver. Sometime after noon we approached a major tributary that joined the River of Nine Dragons. The branch entered from the north side, and on the lower side there was a bluff that rose sharply from the point where the two rivers met. Several hundred feet up the mountain sat a beautiful temple. There were no obvious roads or paths to the building, or people near the temple and, unfortunately, we couldn't ask our boat operator any questions about it.

Cape Six, after construction. August, 1944.

Camp Six

Late that afternoon we pulled into the small village of Huaan, where we were met by several SACO soldiers and an interpreter. Camp Six was just a few miles downstream, and we would arrive in about an hour. At Huaan, we changed to a smaller open sampan and said our farewell to the boatmen who had brought us safely from Lungyen. We then began our drift down to the camp. The pace of the river had quickened, and we crossed several rapids. After moving through the third set, the boatman turned the boat sharply into the left bank of the river toward a small trail that climbed steeply for about twenty feet. Five or six Khaki-clad Americans came down to the boat and helped us offload the sampan. Of course, they asked if we had the mail. Commanding Officer Birthright briefed Parsons, May, and me on camp activities and our assignments. May was assigned to the communication center, I'd been assigned to radio intelligence in Kunming, and in a day or two Parsons would join a coast-watch group stationed in Changchow, about twenty miles downriver and next door to the Japanese base on Amoy.

At that time Camp Six was comprised of twenty people. Eighteen were navy and two were marines, and they were volunteers from all over the United States. The camp was situated next to the river, with several rice paddies in-between and large hills surrounding the back of the compound. At Huaan, the mountains opened up into a valley several miles long, which also was crowded with rice paddies. An eight-foot bamboo fence, with each stake sharpened at its upper end, circled the camp. The enclosed area was about a hundred yards long and sixty yards wide. Two entry gates facing the river were guarded by Chinese SACO trained soldiers. Inside the compound, a large earthen building—an ancient temple—housed the galley, mess hall, radio room, and sick bay. Against the mountains sat an L-shaped building that housed the enlisted personnel. Near the river, a smaller rectangular-shaped building housed the camp commander, the doctor, one ensign, and a lieutenant. The toilet facilities consisted of a four "holer." Water was brought to the camp via a bamboo piping system that tapped into a spring high up in the hills behind the camp. The makeshift shower did not have warm water; it was okay during the warmer months, but was unused in the winter.

Downriver from the camp was a small village occupied by Chinese soldiers who were being trained by the American camp personnel. Across the river and downstream, there was a clumping of farmhouses amid rice fields. The site had been chosen as a training ground because the river was impassable below Huaan. From there, it dropped sharply though a coastal range of mountains and a series of steep white-water rapids and small waterfalls prohibited boat movement—and Japanese attacks—via the river. Furthermore, the trails that came from the coastal areas dropped suddenly down the sides of the mountain and were easily defensible, even with small arms. Everything that moved up the trails from the coast had to be transported by water or carried on foot or horseback. It was next to impossible to move heavy weapons inland.

Camp Six was one of a series of navy facilities built in a no man's land between the Japanese coastal forces and those that occupied much of central China. In cooperation with Chiang's forces, the U.S. Navy had managed to establish guerilla units from north of Shanghai to south of Hong Kong. In west China, the navy worked with Chiang's forces in Kunming and Chunking and, to the north, in the Gobi desert. The number of Americans in each camp was small. In total, there were perhaps six hundred to eight hundred SACO American navy men scattered behind the Japanese lines, training Chinese guerilla forces, monitoring the weather, collecting intelligence data, rescuing downed pilots, and harassing Japanese forces wherever possible. In addition, the SACO units deployed coast watchers that spied on Japanese coastal garrisons and reported on troop movements and shipping activities in the major Chinese harbors. Parsons had been deployed to one of these units.

Although there weren't many people at Camp Six, it was great to be back in the company of fellow Americans. It had taken six days to get from Changting to a mountain base near Huaan. I was initially quartered in a large room with five other men, including the camp cook and Rainnie, a marine who, along with Captain Daine, was training a group of about one hundred Chinese commandos to attack bridges, communication centers, and Japanese coastal installations. Our quarters were somewhat crowded, but more space would become available as soon as the new addition at the end of the compound and the officers' quarters were finished. Each room was serviced by a Chinese houseboy, who cleaned, brought us warm water, and kept a small charcoal stove burning during the winter months.

My first assignment was to put in a power supply to the RDF unit located about seven hundred and fifty feet up a nearby foothill. The building, which sheltered the radio equipment and the rotating RDF, had been constructed by the Chinese, but it lacked power or communications with the camp. It was surrounded by a circular bamboo picket fence and a guard was posted at the single gate. Aided by several Chinese workers, Chief Newell, a radio-material specialist, and I attached a lead wire from the small gasoline generator that provided power to the main building and radio shack and led it to a twelve-foot pole inside the camp and then to one we set outside the compound. From there, the power line would run up the hill attached to posts positioned between the camp and the RDF station.

For the next three days, with the help of the Chinese, we dug with hand tools and erected a series of poles along the trail up to the RDF building. Then we ran the power and telephone lines up the hill to the RDF shack. At the top of the hill, we dug an approach trench across the flat area and brought the power in underground. On the fourth day, we should have been operational, and Chief Newell hooked up the RDF and checked it out. Everything seemed to be working, except that the set had trouble differentiating the direction of the incoming signal from its reciprocal (180 degrees out of phase). Newell and I discussed the problem, but he was very sick with malaria and decided to look at the matter when he was feeling better. Unfortunately, he became so rundown and weak that he was taken back to the States a few months later before he could complete his work on the RDF.

Nevertheless, the RDF was operational. I had been given instructions about which frequency to communicate the bearings I took back to Chunking, but no instructions on what frequencies to monitor or hours to guard, which seemed strange. I decided to initiate evening watches from 4:00 p.m. to midnight, taking my carbine and .45 pistol, record pads, and chronometer to the RDF shack with me. I had no difficulty locating and taking bearings of a number of Japanese stations, but I was never quite sure whether the RDF was properly sorting out the incoming signal from its reciprocal. Each evening I would take forty to fifty-five bearings, but had no idea if they involved frequencies that needed guarding. In desperation, I sent a long communication to Chungking headquarters requesting information about which frequencies to cover and whether or not the bearings I'd taken were true or reciprocals or both. While I waited for a response, I continued to maintain the evening watches.

In the mornings, I helped with the training activities on the rifle and pistol courses. After a week or so, I was an expert on both courses and had learned the mechanics of breaking down the weapons for repair and cleaning. Approximately twenty Chinese a day trained on the pistol range, and several hundred were on the rifle range. Others were trained in the use of explosives, Composition C (a plastic explosive), and dynamite. The Chinese seemed to pick up things quickly, but we had difficulty with our interpreters because the trainees came from different regions and the various dialects caused confusion.

Shortly after my arrival at the camp, rumors spread that a Japanese force was going to come up the river, take over Changchow, and move on to Camp Six. We were all alerted to the possibility of a rapid evacuation. Birthright didn't think that the Japanese would come up the river to Huaan and the camp because it would be too costly for them. To ward off a surprise attack, a large contingent of Chinese SACO soldiers monitored the trail coming up from the river. Additional guards also were posted around the camp. Birthright was apparently correct, because no Japanese raid on Changchow or Camp Six ever materialized. The camp stayed on edge for more than a month, however, and everyone slept with their guns loaded and ready for quick retrieval.

By early November, it had turned cool, and the houseboys lit the charcoal stoves each evening to keep the rooms warm. I had been in camp a little over a month, and the lack of bathing facilities and warm water showers and the repetitive food menu started to wear on me, as they did on everyone else. The camp was so remote and isolated from logistic support that we had received no mail since my arrival, food supplies were provided by the local Chinese, and medical supplies consisted of whatever Doc Coleman had brought with him. The last included sulfa drugs; cholera, yellow fever, and plague shots; atibran (which was used to ward off malaria), and some form of aspirin. At that time Fujian Province was considered one of the most disease-ridden areas of the world, with an abundance of malaria, plague, dengue fever, cholera, leprosy, scabies, open ulcers, stomach flukes, etc. I had already had a bout with dysentery and scabies, and several members of our group had come down with malaria and leg ulcers. I would entertain all of these before left China.

Our daily food intake consisted of duck eggs, fried rice, boiled rice, rice with pork, rice with greens, and rice with fish. Rice was occasionally replaced with an alternate carbohydrate—sweet potatoes. There was always adequate and nutritious food, but the menu and preparation might be best described as boring. Still, the cook baked pretty good rice bread and, although there was no butter, it went well with coffee and green tea. The houseboys brought washbowls of warm water each morning and night, which allowed us to wash our faces and hands and lightly scrub down the rest of the body —the traditional "spit bath."

Four weeks after I transmitted the message about the potential problem with the RDF, I still had heard nothing from Chunking or Kunming. There was no word either about the questions I had posed or the request for instructions regarding frequencies that should be covered. My messages with accompanying bearings on enemy transmissions were acknowledged each day, but otherwise all I heard from Chunking was silence. Thinking the original message had somehow been filed in the circular basket, I repeated the message, adding that I needed guidance and help if the Camp Six RDF was to contribute to the war effort.

Sometime in early December, the camp woke up to the news that our commanding officer had disappeared. There were a number of rumors, but the most common was that he had become mentally disturbed or "had gone berserk"; China had the highest number of Section Eight cases during the war. Somewhere in the navy records, the story probably is more complete but, whatever the circumstances, we were left under the command of Lt. Lowell until a new commanding officer, Helprien, arrived on the scene later in the month. Helprien, who was an ex-pro football player, never said a word about our departed CO. He just took charge and sped up the training activities. He seemed to have an "in" with the supply officers back in Chunking. More and more guns arrived at camp, including small rocket launchers and .50-caliber machine guns, augmenting our few World War I Lewis .30-caliber machine guns.

Just before Christmas, we received word that Parsons had been captured during a coast-watch mission along with Captain Lin, a Chinese SACO member. It had been just over two months since we had come downriver together, and everyone was in shock. It might just as well have been me. He had been sent to a small island (Whale) in the estuary of the river to observe Japanese shipping activities in Amoy harbor. The site had been used previously since it afforded an excellent view of the whole estuary. The Japanese had somehow learned about the spying activity and were waiting for Parsons and his Chinese companion. As soon as the two men reached the observation point, Japanese patrol vessels armed with a couple of dozen men landed at several points on the island. Parsons and Lin were surrounded, captured, and taken to Amoy for interrogation.

At the time we did not know the fate of either man, but long after the war Parsons wrote a letter to the U.S.-based SACO veterans group and told his story. It was even worse than we had imagined. In Amoy, he was told that he would be treated as a spy and shot. After being beaten and tortured for some time, however, he was transferred to a prison camp on Taiwan, which was held by the Japanese. There he was again beaten, tortured, and held in confinement. Later in the war he was transferred to another prison camp, this one in Japan, where he received similar treatment. Parsons' faith and internal strength held on, and he eventually was released after the Japanese surrender in 1945. The Chinese SACO agent was released from Amoy, so both would survive their grueling internments.

After Parsons' capture, there was a sense of uneasiness among the men at camp. Word had spread, and was apparently true, that the Japanese had placed a bounty of $100,000 a head on the American forces operating with the Chinese. None of us were quite sure of the loyalty of the Chinese and the senior petty officers had a discussion about our dependence on our fellow Chinese SACO personnel. In the course of the debate, it soon became obvious that we had little choice. We lived among, were supplied by, and depended on our hosts for intelligence information. If they were not loyal to us, we were dead ducks. There was no evidence that the Parsons event had anything to do with the Chinese forces, and one of their members also had been taken captive. We decided to encourage our comrades to recognize the special ties we had both with the Chinese SACO soldiers and also the community of farmers and villagers who lived around us in Huaan.

From then on, the camp residents seemed to go out of their way to be respectful and friendly toward the Chinese people who lived in and around the camp. As we prepared for Christmas, we learned that there was a small Christian church in Huaan. The Sunday before Christmas, about ten of us decided to walk to the village and attend the services, even if they were all in Chinese. We were armed with our traditional weapons, side arms, and carbines. When we arrived at the small church, we were suddenly caught in a dilemma. We didn't want to take the guns into the church, but we never left them far from our sides. We were standing outside, talking about leaving our weapons outside with one of the group, when one of the Chinese interpreters came up to us and said, "Take them inside, the preacher knows your concerns and doesn't feel it matters."

We slowly moved into the church and sat on one of the benches about two-thirds of the way back from the pulpit, placing our guns on the floor at our feet. The church was far from elegant; in fact, it could be described as rundown and somewhat shabby. Regardless, the parishioners walked in talking and smiling and filled all of the available seats. It didn't take long for them to spot the group of American servicemen in their midst. When the preacher came to the pulpit, he gazed across his flock, looked at the foreign participants, smiled and continued to scan the congregation. He then motioned for the congregation to rise, and the faithful began singing "Onward Christian Soldiers" and then "Silent Night," all in Chinese. Our group joined in, singing the English versions, and it all seemed to blend together. Whether or not the song selections were for our benefit, we made it through the rest of the service very much at ease. Afterwards, some of the congregation came to greet us and, as usual, the town's children were watching as we departed the village.

Christmas Eve was clear and, in the dark of the mountains, a myriad of bright stars stretched like a canopy over Camp Six. We gathered early for a sermon from Father Shannon, who had recently joined our small family, and then we went to an early dinner. Small groups then wandered back to their rooms to chat, play cards, or read. There was a feeling of loneliness throughout the camp, which fell silent earlier than usual. Like me, most of the residents were probably lost in their dreams of home. By morning, however, the men's mood became more upbeat. General Lu, head of the Chinese SACO soldiers, decided to throw a party for the men, and a water buffalo had been butchered for the feast. Around 2:00 in the afternoon, the Chinese contingent began to arrive, bringing gifts, special foods, wines, and good cheer. For the first time, some of them brought their wives and children with them.

An hour later, everyone crowded into the mess hall for the Christmas dinner party. A bright fire burned in the fireplace that had been installed at one end of the room. Father Shannon started the affair by saying grace and then Doc Coleman pounded out Christmas carols on an old piano our allies had brought to Camp Six. The Americans sang *White Christmas*, *Silent Night*, and a number of other Christmas songs. Several of the Chinese joined in, and then introduced us to a song they had written titled, *The*

Americans Are Our Friends. The tune was great and, even though we didn't know very many of the words, we recognized "Megwa Wo Su Pon Yo" (the Americans are our friends). After dinner we drifted into the toasting ceremonies, which guaranteed that more than a few would leave the party carried out by their buddies. Of course, we had to toast the leaders of our two countries, the local Chinese and U.S. commands, and our friendship. I had only been in China for a few months, but I had already learned the trick of pretending to drink the clear rice wine by holding it in my mouth and then spitting it into my drinking water. That way one could endure the party and leave with only a slight buzz.

With the onset of winter, it became increasingly cold, but we were not high enough up for snow. When it was time for my evening watch, I would dress warmly, put on a rain slicker, strap on my .45, and head up the long rise to the RDF shack. Since the sun set early, it was dark by the time I set out. On rainy nights, it was pitch dark, and once outside the compound my small flashlight lit up the path only for a meager thirty or forty feet. Given rumors of a possible infiltration by Japanese intelligence, I kept a hand on my .45 pistol along the way, but left the defense to the guard at the gate once I was inside the RDF building.

One night, shortly after the New Year, I finished my watch, gathered up some records, picked up my flashlight, and started down the hill. I nodded to the guard at the gate, who said something to me in Chinese that I didn't understand. He then made a movement with his gun, pointing it down the hill. I assumed he was giving me some form of goodnight salutations. There was a slight mist falling, and the trail was as black as the proverbial ace of spades. About halfway down the hill, I heard a rustling in the brush off the trail. I swung the flashlight toward the noise and caught a glimpse of an individual running rapidly down the side of the path. He did not attempt to conceal himself. I pulled my pistol and shot in his general direction. However, he was already out of the flashlight's range, and the gunshot only sped up his flight.

When I got down to the camp, everyone was on the alert because the shot could be heard at both bases, ours and the Chinese. I was taken immediately to our new commander for debriefing, and a Chinese patrol was sent out to search. But whoever it was surely had long since left the area. Over the next several days, there was a lot of talk about the incident, and the CO might have thought that I had panicked. He asked me if I wanted an escort, and I said, "No, if the Japanese want to pick me off or blow up the RDF, one or two more men wouldn't make much difference." However, I did start carrying my carbine, which gave me more firepower and accuracy. This story took an interesting turn when the local Chinese military report was filed several days later. Apparently the guard at the gate had tried to warn me that he had seen a ghost down the trail. Local villagers confirmed that ghosts were frequently seen on the mountain and, according to them, came from the Chinese graveyard at the bottom of the hill. The ghost story remained in the report of the event, and the Chinese advised me several times that ghosts frequently were seen on the mountain. As for me, I never saw another individual or ghost on the hill, and the one I had seen sure made a lot of noise as he ran.

In early February, according to my recollection, the rooms at the far end of the compound were completed and the first-class petty officers were assigned, two to a room, to quarters in the new facility. I was bunked with a first-class boatsman's mate, a bearded, tough-looking man named Duncan. He was known for sometimes getting mean after a few drinks of chiew. I had chatted with him off and on and never had any trouble. I was pleased to get the new room with two single beds and no overhead bunks, regardless of who my roommate was. Dunc, as I called him, and I had the added privilege of being assigned our own houseboy, a thirteen- or fourteen-year-old lad named John. The room had lockers for each of us with small tables next to our beds. With several candles, it was possible to read after dark. Such luxury!

John took care of our laundry, cleaned the room, did tasks we assigned, and generally tried to make life easier for both of us. He spoke a little English but could understand more and, with the help of our rudimentary Chinese, we managed to communicate quite well. He was recruited from a country farm some fifteen miles from our camp and had been raised in a family that included his mother, father, and two younger sisters. He brought us each a bowl of hot water early every morning and, during the cold winter months, stoked up the charcoal stove. John had a great sense of humor and tried hard to improve our Chinese and teach us what he could of the local habits, culture, and survival in the hostile environment. Every evening he would spray the room for mosquitoes and check our tent meshes for holes.

In late February, the marines Daine and Rainnie took more than one hundred Chinese guerillas south toward Swatow, where they attacked a number of Japanese facilities, killing and wounding several dozen of the enemy with only minor injuries to their group. Rainnie, however, had returned exhausted and within several days he was down with pneumonia. Doc Coleman was swamped with both Americans and Chinese suffering from malaria, dysentery, dengue fever, scabies, and open leg ulcers. I had an ulcer on one arm and dysentery, while Duncan remained healthy and seemingly impervious to the local maladies. He would assure all of us who were sick that he would be there to "piss on our graves" if we didn't make it. Dunc made absolutely sure that there was no room for self-pity.

We added a few more men to our roster in early 1945, mainly gunners and boatsman's mates. After a few days at the camp, the boatsman's mates were sent to Changchow to join the local coast watchers. We also finally got a much-needed yeoman first class named O'Conner, who was from San Diego. It was natural that he and I soon became good friends. He was a bright man and had a great sense of humor and was billeted in the next room. Our supply of weapons increased along with the manpower as we trained more and more Chinese units. In my spare time, I helped in the armory, taking cosmolene off the weapons and breaking them down, reassembling them, and readying them for distribution. We continued to receive more Composition C, blasting caps, primer cord, etc. It became increasingly obvious that we were headed for a series of offensive actions. I continued to monitor the RDF from late afternoon until after midnight and was able to intercept a great variety of Japanese communications. Chunking, however, remained silent to my inquiries; I had dropped off the end of the world.

The training of Chinese recruits became more intense, and I spent more and more time helping with that effort. An ensign named Mattmiller was teaching a group of Chinese how to use explosives to destroy enemy ships. Part of the course included training candidates to swim and carry weights to increase their endurance. Because I was a good swimmer, Mattmiller asked that I assist him on this special mission. The plan required the Chinese trainees to swim across the River of Nine Dragons and back, carrying five- to ten-pound rocks lashed to their waists with a rope tied in a slipknot. If a swimmer got in difficulty, he could pull on the rope and jettison the rock. The trainers, including Matmiller and me, would swim next to about eight Chinese to assist anyone who had trouble. The river at the swim site was perhaps sixty yards across, and during the morning we had moved about forty trainees through the exercise without incident. None of the swimmers had to release their rocks.

In the afternoon, we were working with another group. Everything seemed to go well until suddenly we realized that one swimmer was missing. Neither the Americans nor his Chinese companions had seen the trainee go under. We initiated an immediate search, but the water was murky and the river was up to twenty feet deep. After several hours, the search was abandoned. Mattmiller and I both felt horrible about the loss. The body floated to the surface and was recovered several days later. We retained our training methods, but altered them to ensure that swimmers could easily release the rocks they carried.

I got a good work out running to and from the RDF facility, but Dunc and I decided that we needed to exercise to stay in shape. We requested Helprien's permission to have a weight set and a horizontal bar made available. Helprien gave the okay, so we asked the Chinese trade people in Huaan to make circular weights out of stones and install a horizontal bar behind the enlisted men's quarters. At first, a number of the men took advantage of the "gym" facilities, but the numbers dwindled and ultimately Dunc and I became the only ones working with the equipment.

The sport of fishing in the River of Nine Dragons using Comp C had become more attractive than doing chin-ups or lifting weights. The men would borrow or rent a small boat from the locals, arm themselves with a half-dozen loads of plastic explosives, and drift on the river near camp. As they passed over the deep ponds, they would light a fuse attached to a blasting cap and launch the explosive into the river. After it detonated, a variety of freshwater fish, some weighing up to four or five pounds, would float to the surface. The men would jump into the water to recover the fish using their arms as stringers to hold the prey. This fishing method was destructive to the local fish population, but it did diversify the camp menu!

By early March, the weather had turned miserable, with the temperature hovering around thirty-five degrees. It rained almost every day and was very windy. Our weatherman, Sinks, later told me that we had measured some rain for forty days in a row. The training of Chinese recruits slowed down, but the numbers of trained guerrillas had increased to between two thousand and three thousand. I continued to help on the pistol and rifle ranges, as well as in the armory. It had been almost six months since I initiated the RDF intercept activity and still I had not received a single communiqué from the radio-intelligence bases in Chunking or Kunming. Despite my concern about sharing my dilemma with a CO who was not familiar with my work, I went to talk with Commander Helprien. He listened, shook his head, and said, "There has got to be a snafu somewhere along the line. I'll try to straighten it out. In the meantime we will keep you busy." Although I didn't expect any relief in the near future, it was good to get the problem out in the open.

We had a change of pace when a navy dentist arrived late in the month to take care of anyone in need. He opened his business in one of the officer's rooms and began to examine each member of the camp. Duncan had a bad molar and had to have some drilling done. It wasn't good news, because the men who had already been to see the dentist had named his office the "House of Horrors." I accompanied Dunc and was amazed to find that the drill the dentist used was not mechanically driven. It was turned by a small wheel with a handle on either side, and was operated by a Chinese aide. If my recollection is correct, the doc worked without Novocain, so the whole process was pretty crude. I could tell Dunc was in a lot of pain, but he tried hard not to show his discomfort. After the dentist finished, Dunc returned to our room, lay on his bed, and moaned. Revenge was mine; I told him, "You better not die or everyone in camp will be there to piss on your grave." Thank god, I didn't need any dental work.

Near the end of March, our coast watchers in Changchow notified Camp Six that following an air raid of Amoy by a dozen U.S. Army B-25s, two planes had gotten into trouble. Their crews headed inland in hopes of getting out of Japanese-held territory and ditched their planes. SACO personnel located both crews and took them to our base at Changchow. Because they'd received a preflight briefing that suggested that the Japanese were in control of the land for about forty miles inland, they were amazed to find an American navy base within a short distance from Amoy. The crew was smuggled out of China and I assume returned to their base in the Philippines. We had also picked up a pilot from the 14th Air Force, whose P-51 had been downed in the same area in December and returned him to Kunming.

Camp Dentist and equipment, Camp Six

The news of the downed B-25s had no sooner cooled when we heard that a U.S. Navy PB4Y2 had been downed in Amoy Harbor. In this case, the men were not so lucky; six died when the plane plunged into the waters north of the airfield. By chance, a man named Tucker from our Changchow unit was en route to relieve another coast watcher and saw the plane fall. He moved quickly along the harbor trail to investigate and soon ran into seven Americans accompanied by two Chinese. We were told that he greeted them with, "Hi ya, fellows, who might you be?" They identified themselves as U.S. Navy and asked Tucker who he was. He answered, "I'm U.S. Navy, too." After the introductions, Tucker provided first aid and led the survivors, one seriously wounded, to Changchow. The PB4Y2 was based in Clark Field in the Philippines, so that was the second rescue of downed flyers from the Philippines that our group had made in one week.

The plane had been carrying a news correspondent, Bell, who worked for *Time* magazine. On his return to the States, he wrote, "Imagine our gasps of amazed delight when we were told that there was a U.S. Naval station just 27 miles away. Here we had been shot down less than a mile from a Japanese garrison, we had been shelled, we had been chased by motorboats and searched for by Jap planes two hours before, and here was a man telling us that we were within a few hours of safety. We met the Navy within 24 hours. When we saw Tucker swinging along with a Tommy-gun over one shoulder and a bag of iron (K ration) over the other —well, you can talk about a sailor's welcome but you haven't seen anything."

The uninjured flyers, like their army compatriots, were smuggled out of China and returned to the Philippines. Jimmy Warr, a radioman third class, was brought to Camp Six to be treated for a serious shoulder wound. On April 1, his four fellow crewmembers who had died in the crash were brought into camp. I was working in the RDF shack when the procession wound down the mountain trail that led from the coast. I could see a group of about twenty or more carriers bringing caskets slowly down the hill. It was almost an hour before they reached camp. The bodies were buried at the rear of the compound, where the Chinese had worked feverishly to build a respectable cemetery. Several hundred Chinese and all the camp personnel stood in silence as taps was played and a gun salute fired. On the same day, we received the sad news that President Roosevelt had died. The camp flag flew at half-mast for those who had given their lives in war and for a dead president.

A few days later we heard that Changting had fallen to the Japanese, and we had lost the remaining airfield that supplied the southeast China area. It meant that supplies coming via western China would be substantially slowed. We believed that a makeshift landing field would soon become operational, perhaps near Lungyen.

From January through April, the men at Camp Six were involved in a series of small raids on various Japanese facilities. With the rapid pace of training and preparations, everyone thought that we were going to become more deeply involved in the guerilla actions with our Chinese allies. Additional rocket launchers arrived at the camp along with several .50-caliber machine guns. Captain Daine took his commando group on another raid, and Ensign Mattmiller enlisted the swimming abilities of his trainees to sink a freighter in Amoy harbor. We also mined the river estuary. By then, the sense of uneasiness that some of the men had about the Chinese SACO contingent had melted away.

Sometime during this period, I received mail from home telling me that that my brother had fought in New Guinea and subsequently had been involved in an 11th Airborne drop into the Philippines. My father's ship was also engaged in the battles in the Philippine Sea. The thought that we were all in the same general theater of war made me feel closer to my family. Mom also told me that Oscar was fighting in Europe and Louis had managed to survive the B-17 air battles over Germany. Gene was training to become an army bomber pilot. I received several letters from Ruby, who was still working as a secretary for the main Sears store in Los Angeles. One of her letters enclosed a three- by five-inch color portrait that raised my morale 100 percent. The arrival of mail was always a time for the men to celebrate in the isolated mountains of China. Everyone waited for the next mail boat.

The U.S. Navy had kept the operation that functioned behind the Japanese lines in China a tight secret. But the return of the downed pilots from Camp Six and other SACO units had spread the word that there was a navy unit operating in the region bombing Japanese-held territory in China. The crew of the downed PB4Y2 had agreed to make airdrops of vital supplies needed by our group after they returned to their base. True to their word, in late May navy planes from Clark Field dropped a number of parachutes containing medical supplies—including the new wonder drug Penicillin—ammunition, and guns. Additional goodies included cigarettes and a hand-wound phonograph along with a selection of records such as *I'll be Seeing You*. I collected one of the chutes, cut away the white nylon from the seams, and put it into my sea bag.

I had been at Camp Six for a little over seven months, and the only things I had bought were several pairs of custom-made sandals in town. In late May, a peddler of Chinese jewelry, whose goods were said to be authentic, visited the camp. He went from room to room showing off his wares. When he came to our room, he laid out a variety of old silver bracelets, rings, and gold-threaded tapestry. He tried to sell me what he called a "tea diamond." It was a good-looking ring and seemed to be several carats in size. He asked the equivalent of $200 for the ring, which seemed far too cheap for a real diamond, so I turned it down. It was subsequently sold to one of the other men and turned out to be genuine. The buyer later sold the ring for $1,000! I did buy several silver bracelets with woven Chinese figures that reportedly told the story of a bride before her wedding; the peddler said they were from the Ching Dynasty. They were exquisite, and Ruby has kept them to this day. I also bought several pieces of white jade that I thought could be made into unique necklaces.

Outside the bamboo stake fence that surrounded our compound, spring reawakened activities that had structured the patterns of Chinese life on small farms for centuries. Farmers were hard at work in the rice paddies repairing the dikes surrounding the fields. They plowed the earth using a water buffalo and a single-blade plow that turned one furrow at a time. In the prepared paddies, men, women, and children were setting out rice plants. The importance of rice in China is reflected in the fact that the word for rice and food is the same. Rice not only serves as the major carbohydrate in the Chinese diet, but the husks are used to feed livestock and the straw (rice stocks) to make paper. Day after day, the peasants worked in the fields with hardly a glance at the nearby foreign encampment. They smiled when they passed us on the trails, but seemed little concerned with the war that had engulfed major portions of their country. Without radios, daily newspapers, or phones, the events of the Japanese conflict in the isolated mountain regions were but rumors and subjects for philosophical discussion among the old men drinking their tea. The real business of life—the rice cycle—had to continue. It had always been that way.

In late May, our houseboy John came to Dunc and me with his eyes filled with tears and told us that his father and mother were going to sell his sister. They had gotten into debt and apparently this was an accepted method of settling the matter. Dunc and I talked about the situation to our interpreter and one of the Chinese officers, who confirmed that such things were not uncommon. We then called John back to our room and asked him how much his parents owed. We then put together a sum of about 29,000 CNC, which was about $140 based on the exchange we got through Chunking, and gave the money to John with our blessing. The next day he proudly announced that the family no longer had to sell his sister. The family remained intact at least as long as we were in China.

Retrieving grain from rice grass

The Woo Su Affair

By late June more than half of the camp personnel were in the field with their Chinese colleagues, engaging Japanese forces at a number of coastal areas. I had just come down from the RDF shack when I heard that Commander Helprien wanted me to proceed downriver immediately and courier two .50-caliber machine guns to SACO forces who were about to attack a Japanese-held island just south of Amoy Harbor. The CO's radio message requested that I bring the guns and ammunition to Shima, a city about ten miles upriver from the large Japanese garrison at Amoy. I was to leave as soon as possible and deliver the guns within twenty-four hours. I didn't know what radio intelligence headquarters would think, since HQ did not want any of its staff near the front lines, but I didn't really care. Back in my room, I checked my weapons, laid out my clothes and helmet, pulled out my rubberized maps of Fujian Province, and then headed over to the sick bay to pick up some atibran pills. A contingent of Chinese would carry the guns over the coastal trail, past the treacherous rapids and falls that made boat traffic impossible below Huaan, for about five to seven miles downriver. At that point the guns would be loaded on a sampan and ferried to Shima.

Just after sunup the next morning I departed with my coolie carriers and two Chinese SACO soldiers. The trail to the coast ran across the rice paddies in the valley adjacent to Huaan and then climbed steeply for several miles up a mountain grade. When we finally reached the top of the pass, I could see why Japanese forces might be reluctant to attack the American mountain hideaway. The trail got steeper and steeper from the top of the pass down to the river and halfway down became difficult to traverse. We moved slowly and carefully down a path that frequently sloped at an angle exceeding forty-five degrees; in many places, we seemed to be going down the face of a cliff. As we neared the river, the trail descended through a small gorge of rocks that likely had been worn down by generations of travelers. The coolies carefully inched their way down the trail, but seemed to know exactly what they were doing.

I arrived at the river launch site about ten minutes before the rest of the entourage. At the foot of the trail, there was a small clump of buildings, several houseboats, and a dock. Tied to one side was a sampan, somewhat larger than the one I had used upriver. It had a mat-covered living area mid-ship, and in front was a charcoal stove built into a stone base. On board I could see the probable owner, his wife, and a small girl of five or six. When the carriers arrived, they immediately put the guns and ammunition on board and I jumped on deck. The boatman untied the sampan, and we were on our way. Standing on the bank, the two SACO men and several Chinese gave me thumbs up as we began our river trek. The rest was up to the sampan crew and me. If luck were with us, we reach Shima in about fifteen or sixteen hours. As it headed out to sea, the river was larger, deeper, and flowed at a slower pace. The mountains rose steeply on both sides of the riverbanks, as they did along the entire length of the River of Nine Dragons.

Top: Hat Sellers
Bottom: Transporting Family member

I sat just forward of the mat cabin with my carbine lying across my waist and watched the banks of the river slide past. The owner manned the stern oar and kept the sampan in the areas of strongest current. I could not see villages or people on the river bank, but several sampans and their crew passed us on their way up the river. Neither the boatman nor his wife tried to make any conversation, and their daughter, who sat quietly, watched me continuously. I smiled at her several times, but her face remained stoic and she seemed almost mesmerized. Shortly after noon, the boatman's wife put some rice on the charcoal stove along with a few greens. When it was cooked, she brought me a bowl along with some tea and a set of wooden chopsticks. She poured hot water over the chopsticks first, so I assumed that she had a sense of sanitation. She did not smile when she gave me the food, just gave a nod of the head when I thanked her.

Suddenly, I felt like little more than a piece of freight. It made me uneasy as I had always gotten along well with Chinese soldiers and civilians working and living around Huaan. I imagined that the boatman and his wife were not thrilled about having to take their boat so far down river in areas frequented by the Japanese. After all, Shima was not a Chinese name but was given to the town after the Japanese had occupied the region. There was also the possibility that whoever had arranged the trip had intimidated them or not paid them well. Over the next several hours, the young girl continued to stare at me without emotion or facial expressions while her parents tended the boat. I felt as though I had the plague. I sat quietly for most of the afternoon, occasionally checking my maps to gauge where we were. Just before sundown, the wife cooked some more rice, to which she added several chunks of fish. Again, she gave me a bowl and chopsticks and without expression moved on to take care of her family.

It became very dark after sunset with just enough stars and moonlight for the boat operator to navigate. Around 10:00 in the evening, I laid down on the mat in the cabin to rest, assuming that we would arrive in Shima before sunrise. The daughter was lying down near me, but on the other side of the sampan. She wasn't sleeping, just lying there watching my every move. My intentions were to take a short rest, but within a few minutes I fell into a deep sleep. It had been a long day, which started at 5:00 a.m.; we had hiked several hours and perhaps I had been a little tense about carrying vital cargo down the river to a town I had only seen on my rubberized map. It also was the first time I had to operate solo. Successful completion of my mission was in the hands of a boatman I didn't know.

Sometime after midnight, I woke up to the sounds of yelling from the boatman and his wife. To my surprise, I discovered that the boat had gone aground on a sandbar. It was so dark that I couldn't make out the banks of the river. I could see, however, that the boatman had gone over the side and was attempting to free the sampan. He had braced his back against the hull of the boat and was straining to move it into deeper water. After about fifteen minutes it became clear that he wasn't going to get the job done. I stripped down to my jockey shorts, put my clothes and guns under the covered area, and jumped into the water and onto the sandbar. The water was over my knees and shallow enough to get good leverage on the boat. Working with the boatman, I could feel the sampan move several inches at a time toward deeper water.

After five to ten minutes, the boat lurched free and began to swing sharply into the current. I had one hand firmly on the rail but with my body twisting, found it difficult to get in a position to pull myself on board. In the interim, the boat owner had scampered up on deck and I felt his hand reach down and under my arm to steady me until I could get my other hand on the boat railing. I soon was on deck, a little cold and almost naked. The wife motioned for me to come near the stove, to which she had added fuel, and at the same time handed me something resembling a towel. As soon as I had dried myself off, she gave her husband and me bowls of hot tea. They both looked at me for a few seconds and then began to laugh. It was infectious, so I joined them in the laughter. The ice had been broken; I was no longer just freight.

I took off my jockey shorts, put on my pants, dried the shorts next to the stove, and got dressed again. It was just before 3:00 a.m., and we were again heading to Shima, but I feared that we would arrive late. Less than two hours later, dawn was upon us. We were leaving the mountains, and another large river was flowing in from the right side. According to my rubberized map, we had arrived at the junction in the river where a branch from Changchow joined the River of the Nine Dragons. Shima appeared to be only several miles downstream. I looked at my watch. It was shortly after 5:00 a.m., and we were running late. I began to get nervous and in Chinese asked the boatman to hurry, but he just smiled and gave a gesture that I interpreted as "What can I do?"

About an hour later, we pulled into a relatively large dock at Shima. The town, made up of two-storied stucco or mud buildings, was built down to the river and ran several blocks inland. We tied the boat near a host of other sampans, but there was no one in sight to greet us. My choice was to stay on board and wait for someone to contact me or leave the guns with the boat owner and attempt to locate our SACO unit. During the almost eighteen-hour trip with the small boat family, we had bonded. I felt confident that they would stay put with my cargo. I got off the sampan and looked at the boatman and said, "Please wait, thanks" (Dunge dung, shi shi) in my best Chinese. He nodded, and I jumped up on the dock and ran into town where I expected to meet a Chinese SACO member. There were only a few people walking the streets, and all of them were strangers. I had been told that there was phone service into Changchow, but was still amazed when I saw a phone booth. Somehow, I managed to get hold of an operator and tried to get her to call the American base in Changchow, about thirty miles up stream, but my Chinese was just not adequate to make her understand.

I had just started back to the boat when a Chinese SACO soldier ran up and motioned for me to follow. We quickly returned to the boat, where we transferred the cargo to a smaller sampan. Just before departing, I took one of the USN pins off my shirt collar and pinned it on the little girl, shook hands with the boatman, and thanked his wife. As I jumped into the adjacent boat, both waved a Chinese goodbye (with the arm held out and the hand flat and still). They smiled, and at last, the daughter's face broke into a grin. I wasn't sure if it represented a friendly gesture to me or her happiness because the large white foreigner who had invaded her life was leaving. It didn't make any difference because the grin was instantly captured in my memory and has remained there for more than fifty years. It had been a good trip, and they had brought me safely to my destination. The guns' fate, and mine as well, had to rely on another unknown boatman and a fellow warrior.

We immediately started down the river, and I had no idea where we were heading. Looking at the map, it was obvious that we were moving into enemy-held territory. After about an hour of drifting, I could tell by examining the riverbanks that we were in an area of tidal influence. I began to think that this SACO soldier, someone I did not know, might be an enemy agent who was about to deliver the sampan, guns, and me to a Japanese patrol boat. After all, $100,000 was a lot of money. However, within minutes, the soldier asked the sampan boatman to pull over to the riverbank. There were a number of buildings about twenty yards from the river. He motioned for me to wait and ran quickly into the small town, where he disappeared. A few minutes later he returned, and pointed downstream. I was more than a little nervous, but my choices were limited. Trust the SACO soldier or abandon the effort to deliver the guns. We continued downriver and in less than five minutes, I spotted our navy unit standing several yards away from the riverbank. I was relieved and the men on shore were equally happy once the guns that were so vital to the success of the mission were in their hands. My confidence in my fellow Chinese had skyrocketed!

In the next few days, things moved so fast that it is difficult for me to reconstruct them with any accuracy. From the site near the estuary of the River of Nine Dragons, our unit moved toward the island of Woo Su—the target of the planned attack. In addition to taking the island, we intended to capture a set of codebooks from the Japanese radio station. After moving through two villages (called Gong-way and Wong Dung, I believe), the SACO commandos arrived at TauBe, a small town occupied primarily by fishermen on the mainland, close to Woo Su. We traveled to TauBe at night, and the steep and narrow trail was difficult to follow. Although all arrived without major injuries, a number of slips and falls left some of the men with bruises and sore muscles.

The attack started at sundown as several groups aboard sampans approached the island from different directions. West and several other SACO men were left on the mainland with one of the .50-caliber machine guns and another .50-caliber gun was taken to Wu An, a small island next to Woo Su. The machine guns were supposed to protect the landing party, but Murphy's Law was at work. Anticipating the attack, the Japanese had reinforced the island and were ready with patrol boats and artillery. When our sampans were about fifty to a hundred yards offshore, the Japanese shore-based defenses let loose, their patrol boats began to approach our boats, and they began to shell the area around TauBe. I'm not sure in what order all of this happened; all I know is that things did not go as the navy planned. With bullets flying through the sampan sails and sides, the boatmen fled back to the mainland with the American and Chinese commandos. Some were thrown into the water, but managed to swim ashore.

My role in the adventure had been minor, but I was glad the machine guns had been delivered. West and others used the guns to battle the enemy patrol vessels, which allowed our men to evacuate the scene. Helprien did not allow me to join the attack force, noting that I had been up for almost forty-five hours with little or no sleep. I was told to make my way back to Shima and then to Changchow. On one hand, the short battle had been a snafu, and we were unable to get any of our men on the island. On the other hand, we managed to fight off the Japanese patrol vessels with our forces largely intact. Despite the bullets and shells that filled the evening skies and some sunken junks, only a few were injured and we did not lose a single American or Chinese guerilla member.

While the men were regrouping on the mainland, I set off for the base camp. I was offered a local guide, but in the few days since I left Huaan, my confidence about making my way through the Chinese countryside had grown immensely. I reached Shima before sundown on the day after the raid. There, I stayed at a riverfront inn. The pace of the previous few days had caught up with me, and I went to bed without dinner and slept straight through until about 8:00 a.m. When I arose, my entire body was covered in large, swollen bites. I hadn't secured my mosquito tent properly and had been attacked by those parasites and also by a horde of bedbugs. I believe that they took forty or more bites out of my hide. It was another lesson in survival, and I was less concerned about the bites themselves than I was about the worse things that could come from them.

I dressed and went outdoors; the street was crowded with people going to work, moving produce through the streets, and shopping. I found a small restaurant and ordered some boiled eggs, rice, and hot tea. It was a good breakfast. Then I returned to the waterfront to secure transportation upriver. A young woman, who was between seventeen and twenty years old and knew English quite well, asked if she could help. When I told her that I needed to get to Changchow, she took me to an office, where I purchased a ticket for a river ferry that would leave in about an hour. I thanked her, and she smiled and moved off into the crowds on the street.

The boat was docked at the same location as the one that had brought me from Huaan. It was a diesel-powered riverboat about sixty-five feet in length, "beamy," and could carry about sixty to eighty passengers, including adults, children, and various livestock. Several Chinese soldiers were on board, but I could not tell if they were part of our SACO group since the local police dressed in a similar way. They seemed to be accompanying a couple of prisoners, but I wasn't sure. The trip upriver began well. Someone had given me a large tangerine, and I sat outside the main cabin eating the sweet and juicy fruit in the warm sun. The passengers seemed happy and were chatting among themselves.

I took out my map and looked at the route. We would go up the River of Nine Dragons for several miles and then take the branch that flowed southwest toward Changchow rather than proceeding up the main branch toward Huaan. The countryside surrounding Shima was mostly a flat coastal plain that ran from the ocean to the base of the mountains. The mountains dominated the countryside for hundreds of miles inland. Small farms growing rice and occasional fruit orchards occupied the land adjacent to the river.

Perhaps an hour and a half later, we arrived at the confluence of the two major river branches and started up the southern tributary. The river was smaller and the current faster. I could see the mountains rising to the north, but the valley here was much broader than the gorge that the river followed down from Huaan. We had been proceeding for about thirty minutes when a fight broke out between the Chinese soldiers and several of their prisoners. As there were about twelve to fifteen passengers between me and the scuffle, I couldn't make out what was going on, except that there was a lot of pushing and yelling. Suddenly one of the prisoners shoved a guard down, jumped up on the rail, and dove into the river. The two guards screamed at the escaped man who, with an excellent free-style stroke, headed toward the riverbank. The guards were helpless because they had only wooden clubs (guns and ammunition were scarce and the navy only issued them to individuals who had passed our training course). Soon they were pointing and yelling at me, and I presumed they wanted me to shoot the escaping prisoner. By the looks on the soldiers' faces, I could tell they were enraged, but I couldn't shoot the prisoner without some understanding of the situation. This infuriated the soldiers, who screamed louder "Shoot, shoot" in Chinese. When

I looked at the civilian passengers, however, it was obvious that they didn't want me to get involved, and an old woman sitting next to me kept shaking her head.

I put my gun down and sat back against the cabin. I had caused the soldiers to lose face, and my inaction had made them my enemies. Several hours later we arrived at Changchow, and the soldiers hurried ashore to report to their superiors. When I reached the American base, I told the officer-in-charge about the incident. He told me that I had done the right thing and not to worry, he would talk to the local SACO headquarters. Later I was told that the Chinese commander had said that while under their rules of war, the soldiers would have shot the prisoner, they shouldn't have expected me to intervene. Later I met the two guards and, through an interpreter, told them that I just didn't know what was going on and explained my responsibilities as a member of SACO. We parted friends, perhaps because I gave them an explanation. Still they probably would have been happier if I had shot the swimming escapee in the back. It was the local way of doing business.

The base in Changchow had hot water, indoor plumbing, and good food, including more fruits and vegetables than were available than in Huaan. I had been traveling in sampans and walking for the better part of a week. I cleaned up and crashed on my bunk, and when I awoke the next morning most of the raiding party from the Woo Su attack had returned. There were a thousand stories and at least a dozen versions of the raid, but no one pointed fingers. The attack didn't go well, but we had learned something about our enemy, our comrades, and of the need to plan for the things that are not supposed to happen.

In Changchow, I started to prepare to return to Huaan and Camp Six. Before I had completed any arrangements, Ensign Mattmiller told me, "Our intelligence people tell us that the Japanese are preparing to move a large contingent of their forces on Amoy down to Swatow [about one hundred miles to the south]. We are not sure of the route, roads, or trails they may take, but it is our intention to harass them all along the way with everything we have. We have worked out an integrated plan [with the Chinese] and the first thing we have to do is make sure they don't come upriver to Changchow and then move south. I want you, West, Birr, Cannon, Warner, and Tucker to be ready to leave to go downriver tomorrow.

We will set up an ambush in Shima. Take along extra ammunition because you are likely to be in the field for some time."

The luxuries of Changchow were short lived. Early the next morning, we left by powered ferry for Shima. I took along a few extra ammunition clips for my carbine, loaded up my ammunition belt, picked up a couple of hand grenades and K rations, and put on my pistol, canteen, and helmet. It was early July, and the weather was warm and humid. I sat outside with my back to the cabin, along with West and Mattmiller. Our first topic of discussion was about our chances of stopping any significant Japanese force. The Chinese intelligence had estimated that 3,500 Japanese would evacuate Amoy and move down the coast. We were armed with carbines, pistols, and grenades and had three World War I Lewis machine guns. Our contingent was made up of six Americans and a few hundred SACO Chinese guerrilla fighters. Our tactics in the past had been "hit and run." We had no mortars, artillery, land mines, or motorized vehicles at our disposal. Mattmiller sat quietly listening to our chatter. Then he broke in and said: "That's all true but we have surprise on our side and perhaps we can slow them up enough to get our SACO units in place in the mountains south of here. That's where we expect to do the real damage. Most all of the Chinese forces trained at Camp Six, and the remainder of our naval unit will be waiting for them."

A little later, I asked Mattmiller where we were going to spend the night. "The inn on the river front," he responded. I laughed and said, "Oh yeah, the Bed Bugs Plaza." West and the others then joined in. "We will be out of blood before the Japanese even get there," one said, while another suggested that perhaps we could recruit the bed bugs to fight for us. Mattmiller, lost in other thoughts, finally observed that we had better keep our mosquito tents tucked in because the so-called inn was an ex-brothel. Cannon, already unhappy about the situation, then remarked, "Shit, we have to sleep with bugs that probably have venereal diseases."

We arrived at the dock in Shima a little after noon and immediately began to set up our defenses. Mattmiller chose the sites for the machine guns and assigned two Americans and more than a dozen Chinese soldiers to each site. Each emplacement had a clear view of the river downstream and was situated just below the riverbank. Working with the Chinese, we sandbagged each defense position and discussed evacuation plans in case our positions were overrun. The other Chinese were positioned in strategic locations to prevent our machine-gun implements from being outflanked. After working for most of the afternoon, we returned to the waterfront inn, leaving the night watch to our Chinese SACO members. After dinner and sundown, an American aircraft bombing run began on the Amoy defenses. We could see tracers from the antiaircraft guns lighting up the sky and the occasional light from exploding bombs. It was the Fourth of July, and we were treated to an array of fireworks. West noted that this would be a good time for the Japanese to leave Amoy.

After the fireworks, the night in Shima was quiet with no evidence that the Japanese forces were planning to move upriver. Mattmiller, who was in touch with other SACO elements, felt that we needed to remain in Shima for at least twenty-four hours to ensure our mission was completed. In the morning, we returned to our machine-gun positions and spent the day watching downriver and attempting to make small talk with the Chinese contingent. We took advantage of the local supply of fresh fruit, purchasing watermelons, bananas, tangerines, and guavas. After midday, West spotted a dead SACO soldier floating down the river. As the corpse came close, its smell was so putrid that we had to put gauze or cloth over our faces. The Chinese officer in charge ordered West and me to recover the body so that it could be identified. We rowed out in a small skiff, but heaved before we were within twenty feet of the body. When we reached the corpse, we tied a rope to one of its arms a rope and pulled it to shore. From that point on, we left the identification process to the Chinese. The episode left me with the chilling thought that nothing can smell worse than a decaying human body.

By late that afternoon, we had received word that the Japanese were heading down the coast road, but at times were moving across the coastal range to the inland southern trails. They obviously knew that our forces were in the area and were playing a game of hide and seek. Mattmiller called our group together and laid out a course to join up with the SACO forces and intercept the Japanese marching to the south. We were to move out the next morning at dawn. That evening, several of us located a small restaurant and sat down to a meal of chopped chicken, rice, and fruit, including mangos. As we were about to leave, the girl who had helped me catch the river ferry to Changchow tapped me on the shoulder and said, "Thanks for what you are doing. The people in Shima have been alerted to the possibility of having to evacuate, and we all realize why you are here." She then introduced her husband and departed saying, "Good luck to all of you, and God bless you." It was a touch of Christian gratitude in a land with many gods. We all said thanks and then quietly returned to the inn.

We moved out the next morning at dawn and marched for five or six hours with only a couple of rest stops. The pace was relatively fast and, for most of the morning, the trail inclined on a gradual grade. Just before noon, we began to struggle up steps of stone that had probably been part of the five- to six foot-wide trail for centuries. When we finally reached the summit, we found a tea stand with a number of wooden tables covered with bowls of tea. Knowledge of our arrival had apparently preceded us. Later we were told that the Chinese had sent a messenger ahead the night before. I sat with my fellow Americans in the shade at the base of an old stone wall and drank my tea, nibbled on some K rations, and tried to get a little rest.

After lunch we got back on the trail; my legs had stiffened during our thirty- to forty-minute stop. Mattmiller continued at a pressing pace in an attempt to get us in front of the Japanese column. By nightfall, we had been on the road for sixteen hours and we were all tired, strung out, and hoping to find a place to rest for the night. We kept on the move until about 10:00 that evening. We finally were put up in a small Chinese farmhouse. The local owner helped us prepare some tea, and we turned to our K rations. The meal was not of any great importance because we were beat. The night watch was posted, and the rest of us crashed.

We were up again and on the trail early the next morning. A check of my survival compass told me that we were moving southeast toward the coast. The march was somewhat easier because the trail was winding down into a well-farmed valley. At the far side, I could see another range of mountains and suspected that they lay between us and our adversaries. Mattmiller kept the pressure on, and that evening we marched into a relatively large town located at the base of the western slope of the mountains. We were again tired, but most of us seemed to have gotten our second wind. We rounded up fruit, tea, and steamed rice and discussed our situation: we had made up some ground but needed to cover a lot of territory the next day. We slept on a small stone patio outside of one of the city buildings.

On the third day after we left Shima, we crossed the coastal range and headed south on the heels of the Japanese. As we descended to the coastal plain, we could see the Pacific stretching out to the east. Out there somewhere was the U.S. fleet of carriers, destroyers, and the *Honolulu*. To the south, my brother Frank was fighting in the Philippines. For those of us from the West Coast, home was some 7,000 to 8,000 miles to the east, and waves from the same Pacific Ocean were breaking on the beaches in San Diego. That morning, one of my SACO friends, Cannon, and I had taken the point along with three Chinese soldiers. We were perhaps a quarter of a mile or more out in front of the main column when we came to a fork in the trail. We took the trail that led to the coast, presuming the group would follow. We had been warned to keep an eye out for Japanese snipers, but made no contact with the enemy.

Within a half hour we entered a small coastal village and, much to my surprise, one of the stores was selling guava sherbet and cold watermelon. I was leery, but our Chinese companions implied that the food was okay. Something inside my head said no, but my stomach hormones triumphed. All of us feasted. We had just downed our sherbet when the local Chinese warned us that a large contingent of Japanese was coming toward town. They believed that it was too late to run and, to protect us, hid the five of us in a foul-smelling outhouse. We remained there for what seemed to be an eternity, and every minute the smell got worse and the flies thicker. After about two hours, we were informed the coast was clear. Escape from the outhouse came none too soon as the odor had just about done us in. Afterward our SACO Chinese colleagues interrogated the villagers. There seemed to be some confusion about whether the Japanese had searched the village or even entered it. I am inclined to believe that it was a false alarm that led to a stinking situation.

Because of our delay, we had to force-march for several hours before catching up with the remainder of our forces. The five of us who went to the coastal village all became nauseated. By late afternoon, we were vomiting and having trouble catching up to the rest of the column. Fortunately, we had almost reached that day's destination. When we arrived in the village chosen for the night, we collapsed. Doc Coleman set up cots with mosquito tents and gave us each several teaspoons of some type of medicine. We became so sick that we could barely lift our heads to vomit, and our temperatures were over 103 degrees. Our illness was recorded as food poisoning, and I have never felt as sick as I did that night. We all continued to vomit well into the early morning.

By sunup I had managed to keep some water down and get a few hours sleep. All of us refused to be left behind and, although very weak, we joined the march, though at the rear of the column. It was the fourth day of the chase, and we were moving back across the coastal range. We had been told that the Japanese had crossed back into the valley east of those mountains. It took us the better part of the morning and the early afternoon to cross them. We descended on a trail that hung to the side of a steep slope, expecting to encounter the Japanese at any moment. It became evident that we were on the right track when we heard artillery fire and shell bursts to the west. As we made our way down the foothills, we came to a position overlooking a large valley, with a village that was directly below our position. After scanning the village for a few minutes, we determined that it had been evacuated, as no one could be observed in the town or in the rice fields surrounding it. The Japanese were either in the town or nearby. From our position, we could see what appeared to be Japanese activity at the far end of the valley. Those with binoculars could see men on the top of a hill observing our movements.

Confirmation came quickly when a number of Japanese artillery shells began to howl and snap, passing over our heads and landing several hundred yards behind us. Our SACO forces immediately took cover behind the crest of the hill we were occupying. Fifty or so of our men were sent to an adjacent hill to ensure that our flank was covered. Without a weapon that was big enough to return fire, and given the limited range of our small arms, we hunkered down while several of our men with sniper rifles took aim and fired at the Japanese, who were probably a mile away. The men on the hill took cover and shelling of our position continued for an hour or more. Then the shelling stopped as quickly as it had started, and it became very quiet. Our defensive position was excellent as the hill dropped sharply below the crest, and there were a large number of extremely large boulders, ten to fifteen feet in height, strewn over the hillside. They provided shelter and limited the impact of the shell bursts. It was difficult for the Japanese to drop their shells behind the crest of the hill and only a few minor wounds—the result of flying pieces of rock—were inflicted on the SACO guerilla fighters.

The Japanese held their position until it was dark and moved out sometime around midnight. At dawn, we made our way across the valley, passed the hills that the Japanese had defended the night before, then moved down a long valley that ran southeast in the direction of the coast. Once again, we were warned about snipers, and the column moved close to the rice paddy dikes, which offered good protection. Still feeling washed out from the food poisoning, I fell behind our main force. The trail followed a relatively large but shallow river that drained the valley. By 8:00 a.m. we had been on foot for several hours. Suddenly several shots were fired from a clump of trees and bushes at the foot of a hill to our east, kicking up water in the rice paddies adjacent to the river. We took cover along the river bank and tried to see where the snipers were hidden. They appeared to be well over a quarter of a mile away, and no one was visible. We returned fire with our carbines, spraying the general area, but our chances of hitting anyone were slim.

The shooting continued, and their shots were getting closer. The snipers were probably in the trees, but we couldn't make them out. The enemy had us pinned down, and we seemed to have two choices: move out quickly and expose ourselves or make our way down the river using the bank as protection. I went down the bank and into the water and the Chinese followed suit. Although the current was relatively fast, it was only several feet deep. Staying close to the three-foot embankment gave us good protection, and we proceeded rapidly downriver. When we had moved several hundred yards, the river began to narrow, the current got swifter, and the water rose above my hips. We were being carried into rapids, and I was having a hard time keeping my weapon dry. I moved to shore and tried to make it up the bank that, by then, was about four to five feet high and almost vertical. My Chinese companions were having similar difficulty, and it was soon obvious that the three of us who had chosen this route would have to cross the rapids.

The current soon was moving so rapidly that we could only let our feet slip over the bottom, and the water was high on my chest. The Chinese soldiers, shorter than the Americans, were in a worse predicament. All of us were carried quickly downstream. At the end of the rapids, the river dropped into a calm area that was five or more feet deep. The Chinese had difficulty keeping their heads above water, but we all made it to a sandbar, where we wrung out our shirts, got the sand out of our shoes,

174

and cleaned our weapons. We climbed back up the bank, which was only a few feet high, and surveyed the area. We were more than a quarter of a mile downstream and apparently out of sight of the snipers, as the shooting did not recur. We got back on the trail, picked up our pace, and in a couple of hours caught up with the end of our column. The hot sun soon dried out our clothes, and I was relieved when part of our unit, including me, was asked to set up a defensive line behind two large rice paddy banks next to the main trail.

After spreading our forces out along about two hundred yards, we set up two .30-caliber machine guns and waited. During the lull, I had time to take off my socks and dry them, recheck my carbine and pistol, and reload the weapons with dry ammunition. We had barely secured our position when the Japanese began sniping again and, unfortunately, we did not have a sniper weapon among us to return fire. The shots, however, splashed harmlessly into the rice fields in front of us.

During our first encounter with the enemy, we had guessed that there were two to three snipers hidden in the trees, but there seemed to be only one firing at our position behind the levee. We only heard ten or so shots during the next thirty minutes. We had been in position for less than hour when a runner motioned for me to follow him. I moved a hundred yards or so along the paddy dike, keeping low, and then we stopped to talk with a Chinese officer through an interpreter. The officer said that we could not waste any more time and needed to move toward the coast where the main Japanese forces were concentrated. I agreed, noting that one or two Japanese snipers were holding down our whole unit. While we were talking, several more shots were fired. We decided to leave a few men behind the dikes to engage the Japanese if they tried to follow the unit. I returned to the machine gun, only to receive some stunning news. The man assigned as a loader had just been hit, almost dead center in his head. The wound appeared to be from a small gun, perhaps a .25-caliber. Had he been wearing a helmet, he might not have been killed, but few helmets available were issued to the Americans. It was a lucky shot from the standpoint of the sniper and a very unlucky one for the loader. His body was carried away to be buried, and the rest of us moved on toward the coast.

Our group soon joined with several other SACO units, and the column stretched the better part of mile down the trail. Looking backwards, I could see our forces coming down the long valley and the Chinese SACO members began singing *The Americans Are Our Friends*. The song had a catchy, marching tune, and I felt like I was watching a scene in a movie; all it needed was a Chinese John Wayne at the head of the column. The men stretched out along the river, moving like a winding serpent. I was lost in thoughts of home when the sound of a plane brought me back to reality. A large U.S. Navy PB4Y2 was coming up the valley, banking along the river to get a good view of the column. We weren't sure the plane crew would be able to differentiate us from the enemy, so everyone took cover.

Someone handed me a mirror, and I began feverishly sending out the message "USN down here," knowing that the navy units in the Philippines knew of our presence in the coastal area. The plane made one more circle over our group and then I heard the sound of a .50-caliber machine gun. At first, we thought they were attacking us, but I soon realized that one of the gunners on the plane had responded to my message by shooting out a crisp "USN" in Morse code. We immediately formed an arrow on the sand that pointed to the Japanese position. The plane then turned and headed back to the coast.

We moved on toward the city of Pan Toa, which apparently had been taken over by the Japanese garrison, approaching the city after sundown. It was midsummer, so it probably was about 9:30 p.m. when we reached the outskirts of town. We stopped for a few minutes to plan. Our scouts said that the city was silent—there was no sign of the Japanese and the inhabitants had left and not returned. We moved slowly, watching the top of every building. The town was an excellent place to ambush our incoming column, so we cautiously went from street to street with our guns readied and checking the best places for cover should we come under fire.

After about fifteen minutes, we entered the city center; there still was no sign of the enemy. We gathered on the steps of a large temple and then spread out to check the remainder of the city. The Japanese apparently had left minutes before our arrival and set up defenses in the hills just south of Pan Toa. Shortly after we secured the city, the Japanese began shelling the area. The explosions could be heard all over town and continued well into the night, but most were well out of our range.

Late in the evening, I accompanied one of our pharmacist mates to a building where the Chinese wounded had been taken. We found a hundred or more seriously wounded men crowded into a makeshift hospital. They were lying on cots and on the floor, and some were leaning against the walls. All had serious injuries, and it was difficult to decide where to start. The pharmacist mate began with those in the worst shape, which may have been a mistake because we lost two men in the first ten minutes. He then turned to the less seriously wounded, and we managed to patch up a number of men and provide plasma to those who had lost a lot of blood. Before all of the men could be attended to, we ran out of bandages and plasma. We tried to scavenge materials, but the ones we found were dirty and unsuitable. It wasn't my game, but I did what I could before we returned to the temple late that night.

In the morning, the battle moved south of the city and raged throughout much of the day; at times, the two sides were only fifty to a hundred yards apart. The losses on both sides had to be heavy as I periodically saw wounded men being carried off on stretchers. The fight broke off somewhere north of Choan, where the Japanese were expected to get reinforcements from the Swatow area. It was hard for me to assess the damage we had done. Communications between the American units (accomplished with hand-powered generators and Morse code) had been very difficult, and there seemed to be considerable confusion about the dispersion of our troops and those of the enemy. We later were told that the Chinese lost several hundred men. The Americans were lucky; only one man had been shot and that was through the arm. The Japanese reportedly lost almost a third of their forces. At any rate, Chinese authorities were very pleased, and a message from Chiang Kai Shek read, "SACO troops, combined with American assistance, have performed excellent gallantry during recent campaign in annihilating stubborn enemy. I am deeply pleased and wish to extend my congratulations."

After the battle, we returned to Pan Toa where several of the locals, who had returned town, marched through the streets with the heads of Japanese soldiers on long bamboo poles. It was a hideous sight, but the men watched quietly, seemingly immune. We returned to Changchow, and I was subsequently sent upriver to Huaan and Camp Six. It was the third week of July 1945, and I still had heard nothing from headquarters regarding the RDF problem. I returned to my nightly watches, taking as many bearings as possible but with a lot less enthusiasm. Sometime in late July, a man named Walker, a radioman first class, arrived at the camp with instructions to straighten out the problem, but he had no radio-maintenance training and soon found himself as stymied as I had been. He could get bearings out of the RDF, but the reciprocal device just wouldn't work. He reviewed the matter and then sent a message to headquarters asking for help. His message went into the same bin that mine had fallen in, and received no response.

In early August, I and other members of our group received embroidered scarves from the people of Pan Toa. It was their way of saying thanks for our efforts against the Japanese, a thoughtful and unexpected gesture. Later in the month, when the first atomic bomb was dropped, the camp became a beehive of talk. We could not fathom the extent of the bomb's damage, and everyone seemed to be caught off guard concerning the possibilities for peace. We waited anxiously for further news and had no concept of the changes the nuclear era would bring to our lives. We were ecstatic when we heard several days later that the second bomb had been dropped and the Japanese had surrendered. Our minds then turned to what would come next. Would we have to make our way back to Kunming, or would we move to the coast and be picked up by a navy ship or plane?

The answers came rather quickly. As soon as the formal declaration and instrument of peace was signed on the *Missouri*, we were told to leave Camp Six and move to Changchow, where we would remain until the navy worked out the logistics of getting us home. We immediately began to pack our gear. Excess or unneeded equipment was given to the Chinese, including all arms except our personal weapons. Within a few days, the men were ready for departure, and suddenly we realized that we were abandoning the camp that had bound us together. On the morning we were to leave, I rose early, checked my sea bag, wrapped and put the Flint doll away, and then walked outside the camp toward the village downstream. The farmers were already out checking their rice fields. I passed a young girl leading a water buffalo along the path; she smiled and went on her way. The crop of rice looked good, and life in the village continued with little sign that anything had changed. The rice would soon be harvested, and the seeds beaten from the straw. Huaan would settle in for the winter rains.

We moved out later that morning, about twenty-five of us heading down the river trail. The men were jubilant about taking their first step toward home but grew quiet as we headed over the mountain. Almost everyone in the group stopped periodically to look back over the valley that we had lived in for almost a year. Each one had his own personal memories of Camp Six and our Chinese colleagues. The farther up the mountain trail we got, the more I realized that a door was closing on a part of my life, an experience radically different from any I was likely to encounter in the future. I took a long last look and said my farewell to Camp Six and Huaan. Our houseboys traveled with us to Changchow, where they continued to work for us.

The coast watchers and men in the field poured into Changchow, quickly filling the school that had been taken over to provide quarters for the Americans. We were all anxious to get information on how and when we would depart, but time crawls when you're waiting for the word. Some of the men played softball in the afternoons to keep busy or just hung around the quarters reading. I joined in the games but also did a little shopping in town. It was rather interesting that goods that were impossible to get several weeks earlier seemed to have found their way onto the merchants' shelves. Suddenly, merchants were looking for the opportunity to make teak chests and shoes or to locate jewelry that had been scarce.

After about two weeks the main quarters in Changchow became over-crowded and a new facility across the river was set up to relieve the situation. I joined Duncan and several others at the "farm," which was a two-story mud building about two hundred feet from the riverbank. It had been a chicken farm, but the rooms were larger than those at the school and the food similar to what we had become accustomed to at Camp Six. We had been there only a day or two when news came that we were to proceed to Amoy and take over the Japanese base, plans that were interrupted by heavy rains that continued for several days. The river rose gradually, then flowed over its banks, and encroached on the farmhouse. By the morning of the third day, the water was more than a foot deep over the entire first floor of the building, so we abandoned the lower level and moved upstairs. The local Chinese farmers took several sampans and went to higher ground, but not before offering to take us to Changchow.

We had heard that the city, which was adjacent to the river, had been flooded several blocks inland and that the water was still rising. Commander Plank, who was in charge of our group, expected us to be evacuated by a small craft at any moment but it was slow to arrive. By the next morning, the water had risen another six to eight feet, until it was just a foot or so below the second floor level. We were all getting a little edgy. The river seemed to extend over the entire valley. The only things we could see were the occasional tops of other two-story buildings. Around noon we spotted several power boats moving toward the farmhouse. When they got close, I could see that the boats were operated by three men from Camp Six, who had come twenty-five miles upriver from Amoy to rescue us. They made a number of jokes about our situation, and then we loaded our gear into their boats. Powered by ten-horse engines and running downriver at flood stage, the boats made the trip to Amoy in about two hours.

When we arrived, we were taken to Goo Long Gu, an island about a quarter of a mile off the coast of Amoy. The foreign contingent and wealthy Chinese had inhabited the island before the war. It was a very attractive area with large, old trees and a road that ran to the top of the island, from where there was a beautiful view of Amoy and the large estuary of the River of the Nine Dragons. We were quartered in the Sea View Hotel, which had been taken over from the Japanese. It was high living in comparison with our recent experiences in China. Much to our delight, it even had telegraph service, and all of us sent a wire home. I wrote Ruby that I was "loafing near Amoy and hoped to be home soon." Other luxuries included potatoes, ketchup, showers, beef, and a great variety of fruits. We felt that we were back in the civilized world. It was mid-September.

By this time, the Japanese had surrendered to the SACO forces, and the island, airfield, and all of the Japanese military equipment had been transferred to the Chinese. I always wondered what the Japanese forces thought when they saw us, a somewhat ragtag band of Americans, many unshaven, wearing an assortment of clothing, and all heavily weighed down with weapons. Whatever they thought, they followed orders and treated us with great respect, standing at rigid attention along both sides of the road as we passed.

The city of Amoy was a short ferry ride across the narrow strait between the two islands. It was made up of a variety of two-story earth buildings that housed various merchants, but there was still very little in the way of merchandise. The merchants demonstrated their ingenuity by opening a bar, dining room, and dance hall to honor our arrival, which helped to lessen the burden of carrying around the local currency. Craftsmen also came to the hotel offering to construct teak chests, make wooden carvings, and sell a variety of jade and other jewelry.

We were in Amoy for two or three days when we spotted a group of four minesweepers clearing the outer harbor. We watched the vessels move from side to side in the main entrance to the harbor, slowly progressing toward the shore. Watching the vessels through binoculars, we eventually saw they were flying the New Zealand flag. Once they were about a half mile from Amoy Harbor, several of us signaled a welcome from the U.S. Navy. There was no response; the New Zealand sailors must have been shocked and wondering who on the beach was sending such a message. When they docked, at first they were sure that we had been prisoners of war, and it took some convincing to assure them that we had operated in the region for more than a year.

We had only been in Amoy/Goo Long Gu for about ten days when the navy made arrangements to take our group to Shanghai. Approximately twenty or so went north on the first plane, a C-47, the following day. With the exception of a few who were to remain and transfer war materials to Chiang Kai Shek's troops, the rest of us readied our gear for a flight out the next morning. It dawned on me then that our close-knit group would soon break up and that our China experience was coming to an end. It was lonely with half of the group gone, and memories of the war we were fighting just a month ago seemed unreal. All of us began to understand the friendship and bonds we had developed with our houseboys, who watched the first plane depart with tears in their eyes. Duncan and I put together the equivalent of almost $200 in CNC currency and gave it to John before Dunc left with the first group. John took the money and began to cry.

The next day, the second C-47 arrived and the rest of the group said farewell to the remaining Americans and our Chinese friends, boarded the plane, and flew off to Shanghai. I can't remember how long the trip took, but I do remember flying over continuous mountain ranges interspersed with the large rivers that served as the transportation links with the inland areas. Before we arrived at Shanghai, the mountains gave way to a large, flat coastal plain drained by numerous rivers. We set down at the Japanese-built airport outside of town and from there were trucked to the Shanghai American School, where SACO units from all over China were billeted. The schoolrooms had been converted to barracks, each room packed with bunks and lockers. We were back in the real navy with a mess hall, sick bay, and all. We retained our unique look, however: soldiers with USN pins on our khaki shirts. I was arranging my belongings when I heard someone holler my name. It was Duncan and O'Conner, who had flown up a day earlier. Both had already received assignments in two of the various naval facilities that were opening in Shanghai. They told me that the Camp Six group was spread throughout the school, and its members were being reassigned to various activities in town.

The next morning I was reassigned to a communications unit being established on the riverfront, or Bund. The Bund was the arc of the city along the Huangpu River at the foot of Bubbling Well Avenue. To get there, I would walk one block from the school entrance to Bubbling Well Avenue and then down a mile or so to where the road ended on the riverfront. The building sat just upriver from where a large slough joined the Huangpu. If it rained, I took a rickshaw to work. When we first arrived in Shanghai, the cost of a bicycle rickshaw ride from the school to the radio base was about a dollar. The city was exciting because it was rebounding from Japanese control. New restaurants, dance halls, bars, nightclubs, and shops with jewelry and local artifacts were springing up all over town to entertain and relieve the servicemen of their extra greenbacks.

After returning from work on the third day, Duncan noted that we could move off campus into an apartment of our own. I asked how long we were likely to be in China before we were sent home and discharged. He didn't seem to know and neither did the local CO. I decided to remain at the school, believing that we could ship out any time. Most of the SACO unit had adequate discharge points (which depended on time in service, time overseas, time in combat, awards, etc.), so we expected to be returned to the States as soon as transportation could be arranged. But it wasn't really that simple as replacements had to be on hand before we could be relieved.

Duncan and O'Conner found living quarters off campus. That evening the three of us went to a new restaurant called the Shanghai American Steak House. We arrived about 6:30 p.m., and there was a line waiting for tables. It seemed that every service person in Shanghai had the same idea. It amazed me how quickly the merchants had been able to respond to the changing control of the city and arrange for supplies and foodstuffs sought by the growing number of Americans in town. The restaurant had steaks of all types, along with hash browns, fries, and green salads. The food was good, but the management was running the clients through as fast as they could, and it looked like they were trying to make their fortune in one night.

The city was hosting an ever-increasing number of servicemen as elements of the Pacific fleet moved up into the Huangpu, and the navy, marines and air force swarmed into the city. The Shanghai we had known was rapidly transforming before our eyes, and prices began to skyrocket. Rickshaw rides that had been a dollar several weeks earlier increased tenfold, as did the price of food and most store-bought merchandise. What had been a buyer's paradise was suddenly a seller's market. I had intended to buy Ruby a set of pearls at the Palace Hotel down on the Bund, but waited too long to get a great deal. Nevertheless, I bargained with the store owner until I got a nice necklace for $100, a set that could have been purchased for just over $40 two weeks earlier. The crowds of servicemen began to create difficulties for the local police as well as the military police (MP) and shore patrol (SP). Servicemen were going into the tough parts of the city, getting into difficulty with the local inhabitants, and starting fights. The situation deteriorated until a number of Americans were found dead, drifting down the sloughs and into the Huangpu River. As a result, major sectors of Shanghai became off limits to U.S. personnel.

I was just finishing a week's duty at the communication center when I was asked to help an SP unit. The unit needed someone who could speak a little Chinese, particularly when during patrols of the off-limits areas of the city. I tried to explain that my Chinese was limited to essential survival language and that conversational Chinese was out of my league, but didn't sway anyone and was signed up for a Saturday shore patrol duty. My job was simply to aid discussions with local police or inhabitants when necessary, although my Chinese was totally inadequate, and what little I understood was mostly Fujian dialect. We operated out of a small truck, and the night started off quite calm but then became more interesting. Our work largely involved getting drunks off the street, stopping small fights, and intervening in disputes over how much someone really owed a merchant in CNC. We had not been called to the off-limit areas, however, and, as of midnight, the duty had been less challenging than I anticipated.

Sometime around 1:30 a.m., we heard that some sailors were in trouble in an off-limits area. We traveled quickly several miles from the city center, stopping at an alley entrance that the local Chinese had blocked. A policeman kept telling us not to enter and that the American sailors were "no good" (boo how). We presumed that someone was in real difficulty and the five of us, four SPs and I, proceeded down the alley to a door where we could hear a lot of yelling. We pushed our way through the door and found ourselves immediately surrounded by half a dozen Chinese men with bamboo poles sharpened to a razor point, which were directed at our stomachs and backs. We had entered a hornet's nest. There were about ten American servicemen, some without pants, pinned against the back wall. On the other side of the room, six to eight half-dressed prostitutes were screaming at the top of their lungs. I was scared, and I knew that if anyone started a fight, there would be a lot of blood, and some of it might be mine.

Trying to stay cool, I began talking to the woman who seemed to be the madam. She didn't speak English, so I asked in my primitive Chinese what the problem was. With gestures and a few chosen words, she explained that some of the Americans had cheated them by not paying and one of the young girls had been beaten. The Chinese men, whom I believe were the prostitutes' husbands, stood silent and stoic with their sharpened bamboo poles within inches of our group. The sailors looked like all the blood had drained from their faces.

The madam yelled for someone to come, and a girl walked in from a side room with blood gushing from her nose and a bruised lip. I asked who had done this, but the sailors stood silent. I told them that if we didn't settle this to the satisfaction of the Chinese, there would be a lot of bodies floating down the river in the morning. Finally, one of the men admitted that he'd hit the girl because she had charged him "way too much." One of the SPs said, "It's going to cost you a hell of a lot more to get out of this mess." I asked the madam how much money she wanted, which started a yelling match between the men and their girls or wives. I was finally able to get some words in, saying basically "the Americans are your friends and helped to fight the Japanese." It sounded corny, but it seemed to work. The madam then pointed out three of the men and told me they were "boo how" and said one English word—"jail." She asked for about $80, which was about five times the going rate for sex. But then they held all of the cards (and sharpened stakes), and the sailors had been way out of line. I discussed the matter with the SP in charge, and he made the group pony up the money and took the three men the madam had pointed out into custody. The Chinese men grudgingly put down their poles, and we retreated to safety. I think I was closer to death that night than any time during the war.

The behavior of the American troops in Shanghai seemed to be directly related to their numbers. As they became more numerous, there was an increased level of belligerence and a heightened tendency for them to demonstrate their masculinity. Shanghai was a great liberty town for those who had been at sea and were searching for fun and pleasure. There were a great number of White Russian, beautiful Chinese, and a sprinkling of French girls in town. Most of the men took advantage of the welcome mat that had been spread before them. A fair number, however, seemed to have forgotten that the Chinese had fought on our side and treated them with disdain and frequently beat them up. It was embarrassing and did nothing to endear the locals to the American way of life.

Sometime in November, I met Lee Hall in Shanghai. He had been in my radio class on Point Loma and had served at Imperial Beach. He had just come in from Chunking and was anxious to see the town. We made a date to have dinner, and I made reservations at a nice Russian restaurant several blocks up Bubbling Well Avenue from the Park Hotel. I had eaten there many times, and the food and atmosphere were excellent. We first walked down to the Bund, where I showed Lee our communication center, and then to the hotel, which faced the Huangpu River. The hotel had a touch of class and an excellent bar, which we decided to visit. As we were entering the hotel, a second-class boatsman's mate got out of a bicycle rickshaw and got into an argument with the rickshaw driver. The mate yelled at the driver and knocked him off the bike. He then barged into the Palace Bar and took a seat on one of the stools at the bar.

Lee Hall and I followed the sailor into the bar and, as there were no other seats, sat on two stools next to him. The mate told me that he was from the cruiser *Nashville*, which was in the Huangpu just off the Bund. The fellow was a little over six feet, about one hundred ninety pounds, and seemed friendly enough, so I thought I would try a little diplomacy. I asked if the rickshaw operator had tried to overcharge him, and he answered in the affirmative, swearing about the "damn little cheats." Most of the time the rickshaw drivers asked for sums that were out of line, I pointed out, but could be argued down without violence. His mood changed in an instant; he stood up and with one swing knocked me off my stool and onto the floor. I got up, took a boxer's stance, and began to defend myself. He came at me like a windmill, but didn't make much contact. On his second rush, I got a good left jab right on his nose that drew blood and made him ferocious. He began to rush me over and over and, although he landed only a few blows, he was strong and pushed me around. I got in another good left before the SPs came in and broke up the fight.

I was glad they had come because his nose was bleeding rather profusely, and he was madder than an enraged bull. They took us outside and said we could either go on our way with no more trouble or spend the night in the brig. It was the only real fistfight I had been involved in during my three and a half years in the service. We shook hands and were about ready to depart the scene when a Chinese policeman came up and talked to the SPs about the incident with the rickshaw driver. The mate ended up paying the driver, who was unhurt, an exorbitant amount.

I inquired about my return to the States several times, and the answer was always the same: "any day now." It was late in November, and I was hoping to get home for Christmas. The weather had turned wet and cold. I was at work at the Shanghai naval communication center when the CO asked me to come to his office. He said, "O'Conner took his .45 and shot himself in the head last night. He blew off half his head." I was shocked and didn't know what to say, but asked, "Did he leave a note or tell Duncan about his problem?" He answered, "No, no one seems to have a clue why he did it." He asked me if I could think of any reason for the sudden suicide. "No," I replied. "I had dinner with O'Conner and Duncan several weeks ago, and O'Conner seemed to be cheerful and enjoying the Shanghai scene." "Well," he said, "he apparently told no one about his problem. The brass are expecting you, Duncan, West, and [several other men, whose names I have forgotten] from Camp Six to serve as honor guards at the funeral."

Two days later, just before noon, O'Conner was buried in a Shanghai cemetery. Using our carbines, Dunc, four other Camp Six SACO members, and I fired the salute. It was a rainy and cold morning and, after a wrenching taps, played by a sailor from the *Nashville*, the casket was lowered into the earth. Everyone was caught off-guard by the O'Conner affair, and subsequent investigations revealed nothing about what had caused him to take his life. A few months later, I visited his family and fiancée in San Diego but had nothing of substance to tell them. He just picked up his .45 one night, put it to his head, and pulled the trigger. He shared his problem only with his God. After the funeral, the six pallbearers went to a bar and toasted our friend. It seemed rather futile, but what else could we do? The irony was that O'Conner was due to go home within a month.

About two weeks before Christmas, I was told that I could go home aboard a "can" (destroyer) that was going to San Diego. The catch was that I would have to serve as one of the radiomen during the nine-day trip. My problem was that I had copied only Japanese code over the past three years and not a great deal of that. After working as an RDF and fighting in the field with the SACO unit, I just was not up to the job of a first-class radioman aboard a ship. I declined to go, knowing it would blow my chances of returning home before Christmas. Later I found out that I would be shipped out on a transport shortly after New Year's Day, 1946.

Several days before Christmas, I was invited to a party given by Chiang Kai Shek and General Tai Li. The affair was for all the navy personnel who had served under their command during the war.

I had thought the SACO unit had been forgotten, but the Chinese were planning a regal party for their American friends. Held in a large hall several blocks off Bubbling Well, the extravaganza lasted the better part of two days. It featured acrobats, scarf dancers, singers, magicians, and an elegant band. There was an ample supply of food and alcoholic and soft drinks. It was one of the greatest parties that I have ever attended, and it helped stave off the loneliness caused by being away from home on Christmas Day.

The party started at 11:00 a.m., and I arrived early in the afternoon. The spacious room was crowded with our Chinese hosts, their guests, and several hundred American SACO members. I moved down a center walkway between rows of tables, looking for a place to sit but without success. Then, much to my surprise, a Chinese officer grabbed me by the arm and asked me to sit next to him. I didn't recognize him, but someone soon advised me that I was sitting next to General Tai Li, Chiang's feared head of the Chinese Intelligence. He had personally headed the SACO efforts in China.

Numerous stories related how he had savagely and brutally tortured his enemies. I suspect that some were true (but consistent with the rules of historical Chinese warfare) and that his enemies invented others. Regardless of his reputation, I sat next to him for the rest of the day and well into the next morning. He was a delightful dinner companion with a good sense of humor and one of the few military leaders who could listen as well as he could talk. Perhaps I was biased because he had rescued me, a lonely enlisted man looking for a seat, and escorted me to his table, which was occupied by Chinese brass and civilian dignitaries. After he asked numerous questions about where I had served, who my CO was, where I had been trained, and where I lived in the United States, he introduced me to the girl sitting to my right. I can't remember her name, but I shall call her Milee.

According to Chinese etiquette, you did not initiate a conversation with a lady until you had been properly introduced, so I said nothing to her while I was talking with the general. She turned out to be one of the Soongs (a prominent and powerful Chinese family), but her exact relationship to this famous family was not clear to me. Later that evening, she invited me to come to her house and have Christmas dinner with her family. I didn't know what to say, or whether the invitation constituted a date. I told her that I was engaged. She laughed and made it clear it was not a date; she was asking several servicemen to the dinner so they would have a good holiday. The general intervened and said, "It's okay, okay, and it will be properly chaperoned." Milee gave me a formal invitation with a map and an arrival time of 7:00 p.m. I slipped it into the pocket of my navy-blue blouse (we had been required to get rid of our SACO uniforms and dress in navy blues) and thanked her for the invitation.

Shortly after dinner, the Chinese began giving presents to all of the Americans at the party. They must have called names for about an hour when General Tai Li asked me to follow him. He walked to the front of the room, where the presents were stacked in a large pile, went around to the side, picked up a nicely wrapped package and presented it to me. At first, I thought that he had just taken a present at random, but then I saw it had my name on it. His intelligence network was as good as was reported. He seemed to know exactly where to find the gift. It was a beautiful bowl from the early Ting dynasty packed in a velvet-lined, wooden display box. It has occupied a special place in our dining room hutch for many years. I never met the general again—he died in an airplane accident the following month—but we had exchanged small talk and toasted each other, the end of the war, and to a lasting friendship between our two nations.

On Christmas afternoon, I hailed a bicycle rickshaw and gave the driver a copy of the map to Milee's house. It was a fairly long ride, and we ended up in an area with large homes surrounded by walls topped with barbed wire. The driver stopped in front of a house set well back from an attended gate. After a short argument with the driver about the price for the ride, which we had previously negotiated, I paid the original agreed-upon price and proceeded to the gate. The attendant looked at my invitation and called the house. After a few minutes, a servant came and escorted me inside. I was led up a circular stairwell and met Milee at the top. She introduced me to her parents, Chinese friends, and four naval officers—one ensign, one lieutenant junior grade (JG), one full lieutenant, and a commander JG. A quick glance around the house told me I was in "the high-rent district." I felt a little embarrassed, outranked, and out of place. I was the only non-commissioned officer at the party.

Milee introduced me as a close friend of General Tai Li. I thought it best not to set the record straight and tell them that my conversation with the general had been a fortuitous encounter. At the moment, that so-called relationship was the only thing that gave me any stature among the guests. I joined with the group in tête-à-tête conversations, drinking the hosts' fine wines, and looking at the Chinese art and artifacts displayed around the house. I spent most of the evening talking to Milee, her parents, and the other Chinese guests. They were all very kind and gracious and made the evening very pleasant. One of the lieutenants asked me about my relationship with Tai Li, and I laughed and told him that we had fought on the same side during the war. It seemed to satisfy him, because he then asked what my expectations were after the war. It was a difficult question to answer, as I hadn't given the slightest thought to my postwar life.

Dinner comprised a number of excellent dishes, including soup, fish, rice, shrimp, sweet and sour pork, and noodles with beef, etc. Milee's mother announced that the food would be served American family style. Instead of one course following another, as was normal at many banquets, the food—except the soup, which was served first—was placed on a large Lazy Susan. As the food circled by the guests sitting at the table, each took as much as he or she wanted. I felt a little smug when Milee's mother noted that I handled the chopsticks very well. After dinner, the chitchat continued and, everything considered, it was a good party. Before leaving I thanked Milee's parents for a great evening. Milee walked me to the door, said she was glad that I came, and wished me a good trip home and a wonderful life. I told her that I wasn't really a special friend of the general, only a party acquaintance. She said, "I know, but it sure gave the guests something to talk about." I then flagged another rickshaw and headed for the Shanghai American School. During the ride, I found myself wondering what Ruby, my parents, and my brother were doing.

Orders for me to proceed to San Pedro (Terminal 1) arrived in early January. I was anxious to leave. Many of my friends had departed during the last weeks of December, and others were scheduled to depart during the first two months of the new year. The American SACO unit was rapidly entering into the civilian sector. My transportation, a small troop ship, was scheduled to depart Shanghai on the ninth of January. I guessed that it would arrive at the Golden Gate late in the month.

I spent my last few days in China shopping for a few gifts, saying goodbye to the friends who were still in town, and writing down addresses for future reference. Dunc left on a rig heading for Seattle about a week before my departure, and there was word that Jim Murphy had been shipped home back over the Hump and was en route to Washington, D.C. As one might expect, I had my sea bag packed and ready to go several days before my departure. I placed the more expensive gifts and artifacts that I had collected during my China stay, including the Flint doll, near the bottom of my bag where they were less likely to be stolen.

My orders, passport and, other important papers were stuffed into my pea-jacket pocket. Those last few days in China seemed an eternity.

On the morning of January 9, 1946, I took my sea bag and other gear down to a small dock on the Bund where a shore boat picked me and others up. After traveling down the Huangpu River a short distance, it delivered us to the gangway of the ship that would carry us home. The vessel was much smaller than the *General Anderson*, but I got a good bunk in a compartment one level below the main deck. Several of my SACO companions were in the same hold. We had no sooner stored our gear when we were told that liberty was available to all hands. I had spent the past three months in Shanghai, enjoyed its culture, nightlife, and cuisine, and watched as it rebounded from Japanese occupancy. The message in my head was, "You've seen Shanghai, both its good and its bad side. You've managed to stay clear of trouble, and you're on your way home; don't chance getting into difficulty."

I went topside and for several minutes looked at the busy and vibrant life along the Shanghai Bund. China had been a wonderful experience, but it was over. After the shore boats departed, filled with sailors and marines looking for a last night of excitement, a number of sampans pulled alongside selling a variety of souvenirs. I spotted a vendor with a model sampan about twenty inches long. It kindled a number of memories, and I negotiated with the boatman for almost half an hour before purchasing it for less than half the original price. I went below to my compartment, disassembled the sampan, and stowed it in the top of my sea bag. Lying on the bunk, I began reading another one of Zane Gray's books, determined to be with the ship when it sailed in the morning.

Early the next day, the ship pulled anchor and, with the help of two small tugs, made its way slowly down the Huangpu. The waterway was crowded with U.S. Navy ships, foreign freighters, and myriad junks and sampans that seemed to move out of the ship's path at the last moment. As Shanghai faded from sight, we entered the broad waters of the Yangtze River. There, the ship traveled more slowly as it had to navigate through a mass of sunken ships that cluttered the river's bed. Their masts and rigging looked like a forest of deciduous trees. Once we cleared the Yangtze burial grounds for ships lost during the war, we picked up speed and China rapidly disappeared from the horizon. I had been gone almost twenty-one months from San Diego. With the wind in my face, a small swell causing a gentle rolling of the ship, and the bow plodding on to the east, my thoughts of home were burning strong.

Very little of the first part of the trip home remains in my memory, probably because there were few events that made any lasting impressions. I do remember sorting through my war years; they had started off quietly in San Diego, where I worked on an intelligence effort that set the stage for the defeat of the Japanese at Midway, a devastating attack on their submarine fleet, and the shooting down of the infamous Admiral Yamamoto. Things changed rapidly when I was transferred to the SACO unit and dispatched to China. There are many who will tell you that the American navy unit that fought within China, behind the Japanese lines, was subject to extreme danger and constant enemy harassment. Perhaps that was true for some, but not for most. The servicemen who spent one day on the beaches of France during the invasion of Europe were probably in much greater peril than we were during our entire stay behind the lines in China. On the other hand, no other unit fighting the war was exposed to such a bizarre, strange, and fascinating experience.

A group of young Americans, most of whom had never been more than several hundred miles from their homes, were placed under the command of the famous Tai Li, and lived and fought with a ragtag, under-armed Chinese guerrilla unit. The Americans were swept into, and expected to acclimate instantly to, a society whose civilization evolved centuries before our country was colonized and whose ways, behavior, and culture were as different as day and night from those that molded our character and beliefs. This was an experience unique to the few hundred navy men who volunteered for SACO and who, within months of their arrival, adapted to living with a price on their heads in isolated villages and cities throughout China. These were men who quickly learned to live within an environment inhabited by malaria, dengue fever, scabies, dysentery, hepatitis, yellow fever, and a host of other exotic diseases that were previously unknown to most of us.

The China we knew was made up of extensive mountain regions connected by occasional roads, trails, and vast river systems that were familiar only to the boat people. Our lives were intertwined with the loyal and brave Chinese soldiers, with whom we'd trained and fought. We watched and, at times, chatted with the peasants who tended the rice fields; small girls leading water buffaloes to plow the paddies; women planting rice, the grain of life; men plodding across mountain trails that were centuries old, with heavy loads at the end of their yo-yo sticks; families who spent their entire lives on sampans; and old men drinking tea and chattering children, who found us unusual and entertaining. SACO units lived with the mountain people of western and southeastern China and the plains and desert nomads of northern China. For the most part, until the war ended, we lived among the have-nots, who always retained a sense of humor, despite their poverty. When the war ended, many of us had the privilege of meeting and mingling with China's leaders. We witnessed the ebbing of the warlord era and the beginning of Chiang's defeat by the communists.

While I mused about all this, the ship sailed slowly across the expansive Pacific Ocean. Several hundred miles north of Midway, we made a detour and helped a ship that had lost power. As we came alongside the vessel, the crew stuck out their thumbs to hitch a ride; we secured a long, heavy manila rope to its bow and started again for home. Those onboard the rescued ship were elated, but on our vessel, most of us found little joy in the fact that our journey would be extended ten or more days. The two ships moved at about eight or nine knots, edging closer to San Francisco each day. Sometime in mid-February, around noon, the Farallon Islands came into view. A tug came and relieved us of our tow, and I realized that we were just a few hours from the "City by the Sea." From there, it was only several hundred miles to Ruby, who had written almost daily since my departure, waiting patiently for the war to end.

An hour or so later we sailed under the Golden Gate Bridge; my service life would soon be over. Cars, trucks, and buses were moving across the magnificent bridge. We sailed into San Francisco Bay, passed Alcatraz, and docked at one of the piers on the waterfront. As the returning sailors walked down the gangway, a band played military marches and the Red Cross handed out lemonade and coffee. We were loaded into a line of buses and driven to Treasure Island, which was being used as a transit base. The navy kept us there for two days, waiting on orders, and then shipped us by train to Terminal Island, California. There, navy personnel checked out our sanity, teeth, and physical health; gave us some parting shots; handed each of us an envelope filled with papers; and sewed a gold discharge eagle on our blue blouses. All that took three days and, when it was over, they briefed us on the veterans rights we were granted by the GI Bill, gave us several hundred dollars, and said, "You are civilians and on your own!"

I jumped onto the red trolley car in Long Beach and headed for Los Angeles. From there, I took a couple of streetcars to Maywood and set off for Ruby's house. From the street, I could see her working in the kitchen. I watched for several seconds, and then knocked on the back door, and we were reunited. I asked Ruby if she would come to San Diego with me. She went straight to her room, packed a bag, and, within an hour, we had jumped on the train. A little over two hours later, we were at the San Diego depot. From there, we grabbed a streetcar to Ocean Beach, climbed the hill that formed the backbone of Point Loma, and were home. I passed out a few gifts and showed everyone the famous Flint doll, which by then was covered with the names of all the cities to which she had traveled. My mother, father, and Frank sat up until early the next morning trading war stories. After that night, I don't think we ever mentioned those stories again. It was time to move on and build a new life.

Chapter VI
Looking for a Niche

Upon returning to San Diego, I moved into my parents' home on Point Loma. I had never really lived there, although I had visited it when I was stationed at Imperial Beach. Frank and I had a small bedroom furnished with the double bunks from our old house on Orange Avenue. One important component of the family was missing—Kamehameha. My mother told me he had died several months after I left for China and was buried in the back lot, next to a flower bed. Our only real pet, he had been at our sides when we were in Hawaii and tagged along with us at every opportunity after we moved to San Diego.

I spent the next several days talking on the phone to school acquaintances, the east San Diego crowd, and hanging out with my brother and a set of new friends he had met in the Point Loma area. Jim Fraser who had been with us in Wailupe, Honolulu, also lived on the point. It wasn't long, however, before everything we did seemed anti-climatic, and I began asking myself, "Where do I go from here?" Never one of those people who held tightly to a planned course, I had dreamed of being everything from a ship's captain to an explorer of new worlds. Had astronauts had been in vogue, my fantasies probably would have included that career. During the war years, Uncle Sam had done all my planning for me and added a little insurance for the future in the form of the GI Bill.

Frank had gone back to San Diego State College, as had Louie Black. Oscar had taken a job as a boat carpenter, and Gene had not yet returned from the Pacific, where he was serving in the Air Force. During the first week home, the only decision I made was that I would marry Ruby, so we set a date for June 16, 1946. Together, we had saved several thousand dollars during the war years, and we weren't at all worried about making a living after marriage. (We were in love and a little stupid!) Come to think of it, we didn't even discuss the issue of finances. We talked about where we would go on our honeymoon, where we would live, and how many children we wanted. We also agreed that we would both work until we had children, and after that Ruby would stay home. Being a full-time mother was still fashionable; you weren't likely to be called a "male chauvinist pig" for suggesting such a course of action.

I wasn't sure what the future held in store, but I wanted to bask a little in the past. It was a little early to hit the beach, so I decided to spend a week in Idyllwild. I thumbed my way up to Hemet, where Uncle Bob met me and drove me up to the mountain retreat. It was still winter and Idyllwild was quiet. I hiked down to Strawberry Creek and around the valley, but something was missing. Ruby came up on the weekend, but memories of the past seemed to haunt me. My grandfather and grandmother had died just before and during the war. Aunt Jimmy and Uncle Bob were operating a new photo souvenir shop that had been built just before the war began. I remember scanning some of the postcards in the shop and finding one of my brother holding up a string of trout that he had caught in the creek. It stirred up memories from the "never, never land" that we call our past, events that are recallable but not renewable, and that made me sad. After Ruby returned to Los Angeles, I thumbed my way back to San Diego, filled with nostalgia.

When I got back, my brother told me that there was to be a funeral for Eldon Crotts, the only one of our original six-man football team who hadn't made it through the war. His body had been shipped back from Europe to San Diego, and a military funeral was set for early March. Frank, Louis, and Oscar, all ex-army, were to be a part of the honor guard. On the day of the funeral, I put on my dress blues, including my war ribbons, and headed to a cemetery somewhere near the state college. Frank had left earlier, so I crossed town alone, taking a bus from Ocean Beach to downtown and transferring to another bus to the east part of town. The trip took longer than I had anticipated, and I was a few minutes late. When I arrived at the funeral parlor, there were several main doors, so I chose the one closest to me. I entered and found myself at the front of the parlor where the Crotts family was seated. I started to leave so I could enter through the back when Eldon's younger sister, Pat, whom I had dated frequently in high school, stood up, looked for a moment into my face, took my hand, and asked me to sit with the family. She was sitting next to a young man in a uniform, whom I presumed was her boyfriend, and joining them didn't seem appropriate. Feeling awkward, I went to the other door and took a seat in the rear of the hall.

After the ceremony, Eldon's casket was taken to the cemetery next door, the chaplain said a few words, and Frank and several other uniformed men fired a rifle salute. Frank looked good in his 11th Airborne uniform. Taps was played, and my mind slipped back to the graveyard in Shanghai where my good friend O'Conner was buried. By the time I snapped back to the present, one of my first childhood friends had been laid to rest. When it was over Frank, Oscar, Louie, and I gathered together, but said little because it was difficult to talk about Eldon without becoming emotional. We discussed having a reunion of our 35th Street group, but decided to wait until Gene came home.

Oscar had a car, so I asked him to drive me back to our old house on Orange Avenue. During the war, my mother and father had sold the "little old homestead," but I was curious to see how it looked and hoped to find some comforting memories there. Oscar stopped at the corner drugstore a block from our former homes, which had been one of our favorite hangouts. We went in, had a fountain coke for old times' sake, and talked about the fun we had growing up, the war, and Eldon. After a short chat, Oscar, who had gotten married during the war, said that he had to get home. He left, and I walked slowly down 35th toward Orange Street, passing Louie's home, then Pat and Evie's house—sisters who had been my close friends during high school—and then on to the little white house where Frank and I had been born and where we had lived through junior and senior high school. I didn't know who had purchased the house. It hadn't changed, but the neighborhood seemed quiet and lonely.

I walked on to the next street, turned back, passed Gene's home, talked a few minutes with his parents, and then caught a bus back to Point Loma. We had grown up together and left for war as young boys.

Close to four years had passed, and then we had returned as men, each one tainted by the war and each with a set of largely unshared memories. It was a long ride to Ocean Beach and up Colorado Street to a house that had never really been my home. The landscape passed by unnoticed as I tried to sort out what I wanted to do. I sifted through myriad of possibilities, but nothing seemed attractive.

The following day was sunny with a dark-blue sky over the ocean, and Frank asked if I wanted to walk down to Ocean Beach and have a shake. Our home was almost at the top of the hill that rose from Sunset Cliffs and formed the backbone of Point Loma. If we walked straight down the hill, we could follow the beach and cliffs into town about a mile away. We were both in relatively good shape, so we ran all of the way down, challenging each other with sudden bursts of speed until we reached Sunset Cliffs. With the breeze in our faces, we seemed to have captured the carefree past for a few fleeting moments. We then slowed down and walked up the beach to the city center. Ocean Beach had one main street running up from the ocean and a few stores on the adjacent side streets. At the drugstore, we slid onto two stools at the fountain bar and ordered a couple of chocolate shakes.

It didn't take long to figure out why Frank had selected the place. Two attractive young ladies worked at the counter, one brunette and the other blonde. They were sisters, and Frank seemed to know them quite well. After our shakes arrived, Frank asked me why I didn't sign up for classes at San Diego State. A new semester would start the following week. My grades in high school hadn't been all that great, and I doubted my application would be accepted. Frank pointed out that GIs and others who didn't have the required grade points were being allowed to take a qualification exam. "I hear they are being relatively lenient on vets returning from the war," he said. We finished our shakes and made our way back up the hill, racing each other for a block at a time and then stopping to look back over the ocean. It felt good to be back home, and I began to let go of the war.

That evening, I gave some thought to entering college. During the war I had done quite a bit of reading, and my studies for radioman first class had given me experience in the practical application of math and physics. Nevertheless, I was concerned about my ability to pass the entrance exam, which major to choose, and the fact that I still didn't have a clue about what I wanted to do. I asked Frank about his major, and he said that it wasn't necessary to select a major right away and that most of the first-year classes were standard requirements for everyone. These included math, English, history, and science courses. After my first year, I could choose a major, and if the school didn't have the I needed, I could transfer to another college or university.

After talking to Ruby, I decided to give college a try. There was little to lose, and the GI Bill would provide the necessary finances. Two days later, I went to San Diego State College; Frank introduced me to an assistant dean, and I made arrangements to take the test the following day. At home, I pulled out some of my old high-school math, English, and history books and started to cram. I quickly realized that there was no way I could learn enough in one afternoon and night to pass the test. Either I already knew enough, or I would fail. I closed the books, called Ruby and chatted for a few minutes, and then told Frank, "It's time for another shake." We scampered down the hill and over to the drugstore in Ocean Beach. Life was good.

The next day I went to campus to take the qualification test. I arrived in a classroom filled with other potential students—about twenty men and a couple of women. From their demeanor, ages, and appearance, most appeared to be other veterans. Some of the young men wore military clothing, fatigues, or navy dungarees and seemed to be in their twenties. The professor asked us to sit down, explained some details about the test, and said that any sign of cheating would disqualify a candidate. We were then told to pick up our pencils and proceed. I don't recall a great deal about the content of the test, but it involved math, vocabulary, reading skills, and what I called "sorting out illogical logic." I worked my way through the exam and turned in my papers several minutes before the professor called for the papers. He also told us that our scores could be obtained within forty-eight hours and gave us a telephone number to call.

As I rode home on the bus, I mentally reviewed the test and tried to put a pass or fail grade on each section of the examination. I didn't know whether to feel good or discouraged, only that I had been able to answer more questions than I had expected. Growing up in the Great Depression, I had never presumed that I would go to college. Knowing that my brother and Louie had enrolled, however, and that Gene and Jimmy Fraser were planning to do so, stirred up an intrinsic fear that

I would be left out of the pack. The forty-eight-hour wait seemed to stretch beyond its prescribed period. I probably was one of the first to call for results. The woman on the phone took my name and then said, "Congratulations, your score is above the passing level." I didn't want to know by how much, but the door had opened and my brother had shoved me through it.

The following week I joined the ranks of thousands of GIs who had laid down their arms, returned home, and picked up a set of books to get an education. Many were probably similar to me in that they were not sure what they wanted to do, but if there was ever another war, they didn't want to be the grunts. The campus at San Diego State was still quite small, but the post-war boom had pushed the attendance up to over six thousand. The incoming students were a mixture of officers and other men and some women who had been in the services. My class schedule included English, history, and a refresher math course. It was a little difficult at first for me to get into a disciplined study mode, but within a week or so it became routine. Weekends were more difficult because I wanted to visit Ruby or have her visit me in San Diego. Regardless, I found enough time to keep my grades at a B or above.

A week after school started, I got a call from Gene, who had been mustered out of the air force and returned home. We made arrangements for a reunion of our 35th Street six-man football team. Several days later, Frank, Gene, Oscar, Louie, and I met at a bar in East San Diego. It was both a happy and a sad gathering. We talked about the loss of Eldon, the great fun we had on our six-man football team and growing up together, and there was a sad but unspoken realization that it would never be the same. Oscar was married, Eldon was gone, Frank and I had moved out of the neighborhood, and soon we all would be scattered by the demands of our future lives. We finished off the evening at a small restaurant on El Cajon Boulevard. It grew quiet. It was the last meeting ever of the 35th Street gang.

As May came and brought warmer weather, Frank, Jim Fraser, and two new friends, Snide and Dwight, who lived in the Point Loma area, began meeting after school at the foot of Orchard Street, where the beach near Sunset Cliffs was great for body surfing. It was a great feeling to once again crash down the face of a breaking wave and slide out ahead of it for a long ride into the beach. When we weren't body surfing, we frequently drove out to an area at the south end of La Jolla and dove for abalone. In those days, there was a great abundance of green and red abalone; some were nine to twelve inches across the longer side of the shell. With a facemask and a dive down eight to twelve feet, it was relatively easy to harvest the limit established by the Fisheries Department.

Sometime in late May, we set out to go abalone diving. It was mid-afternoon when we arrived, the sea was calm, and we had the area to ourselves. The five of us jumped in; towed an inner tube, fitted with a net to hold our catch, out to about ten feet of water; and began our dives. The water was clear, and we seemed to have chosen the right time and place. Using our tire iron, we pried off one abalone after another, and in a matter of forty minutes had taken our limit—or so we thought. We towed the inner tube filled with giant snails back to shore, only to be met by a California fisheries warden. He asked us to show our take, which we did willingly. As Frank unloaded the abalone, I made a quick count and noted to Snide that we were two over the quantity limit set by the state. The warden had taken out his measuring device to ensure that the abalone were all of legal size, but hadn't yet counted the harvest. Snide quickly tossed two abalone over the cliff and into the water. The warden must have caught a glimpse of this and asked if we had gotten rid of some of the abalone. Snide just smiled and said, "I think a few escaped." The warden must have been in a good mood because he said, "Okay, guys, you can go but watch your limit next time you are out diving." It was a good thing because none of us could afford a ticket.

We divided up the catch between the five of us, which left Frank and me with ten nice abalone. We cleaned them and removed the meat from the shells. Then it was Mom's turn. She sliced them about three-eights of an inch thick, pounded them with a steak hammer, dipped them in an egg batter, and fried them in a skillet. They were best eaten hot along with a green salad and, on a warm evening, a glass of cool beer. That spring, body surfing and abalone diving proved an excellent post-war tonic for my east San Diego and Point Loma friends. Interestingly, none of us ever spoke about our war experiences.

I continued my education at San Diego State, accomplished a fair amount of swimming, and visited Ruby, or she would come down and spend the weekend at my parents' home on Point Loma. By mid-May, Ruby and I were immersed in planning for our wedding and subsequent life. We had found a nice little one-bedroom apartment on Santa Cruz Street, about four blocks down the hill from my folks. It was built over the top of a garage, had a nice living room, kitchen, bedroom, and small bath. It was only two blocks from Sunset Cliffs and a few blocks from our favorite body-surfing beach at the foot of Orchard. We secured the apartment in mid-May to ensure that we would have a home after our June wedding. Our income would come from my GI Bill education allowance, which was $125 per month, and from Ruby's job. She had transferred from the main Sears store in Los Angeles to the downtown store in San Diego. She made slightly over $100 a month and that was enough to keep me in school and give us a good life.

We both wanted a church wedding, but neither of us brought it up. Ruby's folks were not really flush at that time in their lives, and no one offered to pay the tab. We decided to save our nest egg for a home and expenses that might come up during my education. Thinking back, I should have pushed Ruby, because she really wanted a church wedding and renting the church would have cost only $15. I later realized that it was a mistake, but then my hindsight has always been substantially better than my foresight. We were married in Ruby's home, and her aunt sang "I Love You Truly" and played the piano. She was just a little bit better than horrible, but she was loud and seemed to enjoy her performance.

Chris Vesper and Jim Murphy both attended the affair along with the Point Loma group and part of the east San Diego gang. Relatives from both sides of our families were there and, with Ruby's girlfriends, there was a full house. The preacher quickly went through our vows, reminded us of the commitment we were making, and wished us good luck. He was the same preacher who once said in a sermon, "Don't date servicemen!" Cake and punch were served after the ceremony. Ruby's dad was "dryer" than a hot day in the desert, but somehow Frank and the San Diego gang managed to spike the punch. Her dad was furious, but most of the guests seemed to thrive on the vodka-spiked beverage. Other than the embarrassment of Ruby's aunt singing, the wedding went well. We had made plans to go up to Big Bear for a week, and Ruby's dad loaned us his '39 Chevy. We sneaked out the back door, got the car started, and were down the road before any of our friends spotted our departure.

Nevertheless, a half-dozen or more friends got in their cars and took pursuit. We headed down Atlantic Boulevard with horns honking behind us, and we were scared they would cut us off and add to the junk that Ruby's father had already attached to the car. Luckily we made it through a stoplight before they did. Two blocks later we turned into a side street and sat quietly as our pursuers whizzed by. We waited twenty minutes, removed all of the shoes that were tied to the car, wiped the writing off the chassis, and resumed our trip unmolested. Ruby had worn a turquoise dress her mother had sewn, and I had a rather bulky looking suit. We both wanted to change into something less dressy, so we pulled into a drive-in restaurant and changed clothes in the restrooms. Within a few hours, we started up the grade and passed the turnoff to Lake Arrowhead. We arrived at Big Bear around 4:30 p.m., found the inn where we had reservations, and were soon situated in a small cabin several hundred yards from the lake. It was a beautiful day, and a great beginning to a wonderful honeymoon.

In the evening we went down to the lodge for dinner and dancing. By the time we left, the moon had risen and the water shimmered across the lake. We walked to the cabin slowly and drank in the pervasive scent of the pines. I had given Ruby some white nylon that I had cut from a supply parachute in China. She and her mother had made a gorgeous nightgown from it, which Ruby wore to bed on our wedding night. (She also made a blouse out of the nylon and wore both items throughout her life.) The next morning we scouted the area's entertainment options. There was local shuffleboard—ugh! —hiking, swimming in a cold lake, horseback riding, and fishing. We flipped a coin to decide between fishing and riding and ended up on horseback by some trails near the lake. I don't remember much about the ride, except that both of us had sore behinds the rest of the day.

The next day, we rented a small boat so we could try our luck at fishing. We drove along the lake toward the dam before setting out our gear. There were a half-dozen boats near by, so we assumed it was a good spot. We started to fish, using eggs and worms as bait and fishing on the bottom in about twenty feet of water. A warm morning sun made it quite comfortable. We were there about thirty minutes when Ruby got a bite and landed a beautiful fifteen-inch rainbow. We carefully placed the fish in our reed creel. We must have fished another hour before we had any action, then I got a good bite that landed another rainbow. This one was a little larger, and we were pleased with our results, two bites and two fish. We continued fishing for another couple of hours without a bite before we got bored.

Back at the boat landing, several people saw our fish and suggested that we have them cooked for us at the inn. That seemed like a good idea, so Ruby and I went to the lodge and made arrangements to have them made for dinner. We then returned to our cabin for a rest. I had gotten burned on the face even though I had worn a hat. The inn did an excellent job of cooking the trout and served them with small red potatoes and creamed peas. We were pleased with the feast, but we weren't prepared for the check. It was several dollars higher than if we had ordered the trout off the menu. We didn't say anything, but it took the charm out of the evening. We went fishing one more time during the week and caught a couple of somewhat smaller fish, but fixed them ourselves at our cabin.

We also hiked a bit, danced most evenings, explored the Big Bear area, and for the most part had a great time. The week seemed to fly by, and we weren't ready to leave, but our allotted honeymoon funds had been spent. We packed up and headed back to Los Angeles and Maywood. After making a few turns down the grade, I put my foot on the brake and felt the bottom drop out of my stomach—the pedal went to the floor of the car! We managed to get around the curve and I tried the brakes again, but to no avail. Then I noticed that the hand brake was on and quickly released it as I shifted the car into second gear, which slowed us somewhat. I suspected the brakes had overheated but, still scared, managed to maneuver around the next two curves, although we crossed over the dividing line. I tried the brake again and noticed a little improvement, and after a few more minutes of driving, the brakes slowly returned to normal.

When it was safe, we pulled over to the side of the road and let the darn brakes cool. We kissed each other and decided that it was an event that was better left untold. We returned to Ruby's parents' home without further problems. They had gone to New Mexico, so we had their house for the evening. We washed our clothes, went to a restaurant on Atlantic Boulevard for dinner, and turned in early.

Ruby's father and mother got home around noon the next day. My cousin Florence May was getting married the following afternoon, so we stayed over to help them celebrate. About an hour before the wedding,

Ruby and I went over to my aunt's house. Frank and my parents had already arrived, along with Katherine Jo's husband, Bill Ming. After a little discussion, Frank, Bill, and I decided to find Florence's getaway car and decorate it appropriately. One of the parents disclosed its location, and we quickly ran to where the car was supposedly parked. Much to our disappointment, it had been locked in the neighbor's garage, but that only slowed us down for a few minutes. We unscrewed the garage door hinges, laid them down on the lawn, and decorated the car and windows with soap and colored paste. That just whetted our appetites. We were delighted to find the keys in the car along with a jug filled with lemonade and a picnic lunch. We salted the lemonade and then drove the car and parked it in front of my cousin's house, blasting on the horn, unaware that the wedding ceremony was underway.

It took a long time before we heard the rest of the story. My cousin and her new husband, Watson, were heading across the desert bound for Arizona when the car ran out of water. They used the salted lemonade to tide them over to the next gas station, but the car overheated and died. It was a few years before we were considered acceptable company; Ruby was greatly embarrassed.

The next afternoon we caught the train to San Diego, arriving home about 5:30 in the afternoon. When we entered our apartment, we found that every piece of furniture in the house had been moved to another room, and it was a mess. There was a gift in the middle of the floor with a note that said, "When you get through sorting out the house, you may then go to bed." Bed was underlined. It took Ruby and me the better part of two hours to straighten the furniture out, re-hang some pictures, and put away the pots, pans, and china that had been scattered throughout the house. When we finished it was almost dark, so we walked down to the beach and watched the sun sink into the ocean. When we returned, Ruby made us a snack. We listened to several radio serial shows (*The Lone Ranger, The Shadow, Zorro*, etc.) and turned in about 10:00 p.m. When Ruby sat on the bed and began to undress, we heard the tinkling of small bells. We checked around the room and didn't find anything, but when she sat down again we heard the bells. We looked under the bed, and there must have been a hundred tiny bells tied to the springs. Ruby got out the scissors, and we diligently cut off and saved each bell, which took almost an hour because they were tied in areas that were difficult to reach. We went to bed quietly suspecting that there were other surprises we had missed. Frank and his cohorts had been busy while we were gone. They were an unruly bunch!

After settling into our apartment on Santa Cruz, we discussed our financial situation. We still had a nice nest egg in the bank, collecting a handsome 3 percent, but unless I got a job for the summer, we would have to skim off our savings to live. I had been talking to some of the local Slavic lads, whose fathers were part of the expanding tuna seine fleet that fished out of San Diego. They urged me to try the High Seas Tuna plant on the bay side of Point Loma, which had been looking for help. I put on a pair of Levi's and a clean T-shirt and went down to the plant to apply for a job. I was interviewed for a few minutes by one of the plant's foremen and told to start the next morning.

I can't remember the hourly rate, but those who worked full-time, with a little bit of overtime, made close to $350 a month, which was a fortune to both of us. Together, Ruby and I were bringing in over $400 a month. The work involved various manual labor jobs, including stacking the boxes of canned tuna; helping to take the cooked tuna, stored in large steel-wheeled baskets, in and out of the cookers; helping on the canning line; and loading tuna on trucks and freight cars. The plant, which had a number of young men about my age and younger working at various jobs, seemed like a pleasant working environment. The workers were mostly of Slavic origin and were generally muscular, well built, and with dark brown hair.

I started on the 8:00 a.m. shift the next morning. The supervisor told me to report to Joe, "the old Slav" stacking tuna cases in the warehouse. Joe, who looked well into his sixties, told me to take the boxes, which weighed a little more than twenty pounds each, off the conveyor line and pass them up to him. He was standing on a moveable ladder about ten feet off the ground. He would then stack them on top of the lower rows as we worked our way toward the ceiling. The work seemed easy, as I was in relatively good shape and the boxes were not that heavy. After several hours, however, I became a little tired. We stopped for a ten-minute break and I was glad to rest and sip a cup of coffee. By noon, I was rapidly becoming exhausted and began to wonder how Joe, who was maybe 140 pounds and wiry, managed to keep going at a steady pace.

At lunch, I sat and chatted with some of the young Slavs, who seemed to be keeping an eye on my ability to keep up with the old man. I decided that somehow I would make it through the day no matter how tired and sore my muscles got. After lunch, I was back on the job, moving box after box up to Joe. He seemed to move with ease, while I felt like I was on the verge of collapse. Nevertheless, I stayed with it without falling behind and savored the afternoon coffee break. We completed our work just before 5:00 p.m. I was sweaty, sore, could hardly move my arms, and my legs were stiff from climbing up and down the ladder. I looked at Joe, and he hadn't even broken into a sweat. I realized that Joe served as the testing ground for new recruits. As I left the lot and headed for home, one of the lads put his hand on my shoulder and said, "Good job, not too many can stay with Joe." I never revealed my exhaustion.

The work was never as difficult as it was that first day, and within a week my body had responded to the new tasks and could handle whatever was thrown at me. A couple of weeks later, I moved to the ramp where they loaded outgoing tuna cases on trucks, took them down to east San Diego, and unloaded them onto freight cars. There was a certain competitiveness among everyone working the loading shift, and each of us attempted to move the cases from the conveyer and stack the freight car as rapidly as we could. Working with two well-built Slavs and a six-foot-six Norwegian, I became part of a well-oiled team that could stack a freight car in under an hour.

It was a wonderful summer for a couple of newlyweds. In the afternoon after work, Ruby and I routinely walked down to the foot of Orchard and spread out our blanket on the sand in the late afternoon sun. We rode a few waves and chatted with other beach lovers, which frequently included my brother, Jim Fraser, Dwight Young, Ralph (Snide) Malloy, and others. It was the good life. From time to time, Ruby and I would invite the group up to the apartment for dinner or a snack. Sometimes, if class got out early, the guys would purchase the makings for spaghetti and leave them with a note to Ruby that said we were swimming but would return for dinner. Sometimes I would have dinner ready when she got off the bus. Because Jimmy was a race-car buff, we began going down to the Balboa Park Stadium to watch the midget cars race around the oval track. We always sat at the closed end because that is where most of the crashes took place. Ruby wasn't thrilled with the location because we were frequently splattered by mud.

We also spent some evenings going to the famous drive-in at the foot of Point Loma, a gathering point for many of the San Diego State crowd. That was back when scantily clad young ladies brought your hamburger, coke, and fries to the car on a tray, while the customers interacted with friends and acquaintances in the neighboring cars. To fill our calendar, Ruby and I made occasional trips via train to see her parents and my cousins in Los Angeles. When we got tired of the routine, we packed a picnic lunch and headed for the hills just above Scripps Institute. There, we sat and watched the waves roll in on the rocks below and the sun sink into the sea. We had our share of spats, but we were happy.

The summer passed quickly, as time does when you are busy and enjoying yourself. Just before the fall semester started, Ruby and I took a few days off and visited Idyllwild. When we returned, we sat down and discussed our future, but after several hours we had decided nothing. I wasn't very keen on engineering even though I liked math and physics, and a major in business administration seemed almost revolting. Teaching was a possibility but it didn't really excite me. Ruby thought that I would be an eternal student. I reminded her that I only had one semester under my belt and had considerable time to choose a major that would determine my career for the rest of my life. Without locking up my future, we decided that I should emphasize science and math courses and take whatever was required for graduation.

Throughout the spring of that year, I bummed rides to school with Snide and Jim Fraser or rode the bus, which took a big bite out of my time—almost an hour and a half. During the summer, Ruby and I decided to buy a car, but there was a waiting list for every type of new car. During the war, all of the plants had been converted to defense construction, and now a lot of people were looking for transportation but no cars were available. It was a seller's market, and manufacturers were taking orders. If you were clever, you ordered a car from every company and then sold your spot to an anxious buyer for several hundred dollars. If you played the game well, you could earn enough money to pay for your new car.

But we didn't play the game. We signed up with Hudson Terraplane, a make of car that was well known in the pre-World War II era. We expected to receive the car before the fall quarter started, but by September it was clear that it wouldn't be delivered until late in the school year. So Frank and I continued to get rides from our friends when we could and use the bus as a last resort.

As the fall quarter began, I went back to school on the GI Bill, and Ruby continued to work as a secretary at Sears. I had signed up for zoology, algebra, and English. The math teacher was all right, and I particularly liked the zoology professor. The English teacher was probably no more than a couple of years older than her students, mostly GIs, and seemed to have a chip on her shoulder, particularly with respect to older men and servicemen and women. We had mutual respect for one another, but not very much.

As you might expect, English turned out to be my most difficult subject and it seemed that the harder I tried, the more difficult it became. The class mostly involved writing and, as I recall, we had to submit two typed themes a week. The professor had a grading procedure that was somewhat unique. First, she gave a grade for the quality of the story and how it was told. Then she added a weighted grade that accounted for originality, grammar, and spelling, in that order. Your final grade was a culmination of the process. That is, whatever you started off with was the best you could do, so the grade either remained the same or went down. Mine always seemed to start with an A or A-, but by the time the other elements were taken into account, I was lucky to get a C.

From my perspective, she had an arrogant way of putting down people who asked questions. I sensed, perhaps incorrectly, that she had an intrinsic dislike for veterans and delight in criticizing those who challenged her commentary. With two weeks left, it seemed impossible for me to get anything but a C in her class, so I decided to write something in which she could really sink her teeth. I worked hard on a story about a vet who had won a number of medals, but after his discharge he became a boozer with complete disdain for those who had not seen action. He wrote off both men and women if he believed they didn't understand what he had sacrificed for his country. He had a particular dislike for politicians and educators, who both indulged in rhetoric to enhance their own feelings of superiority and power. He eventually mellowed with time and married a woman who wanted to be the mother of his children.

The central character was a classic male chauvinist pig, but one who finally recognized that people who had not been directly involved in the war were still respectable. I took extra care with the story and Ruby helped with the spelling and grammar. It was certainly original. After submitting the paper, I waited with some anticipation to see how the professor would respond. It was the first, but not the last time in my academic career that I would deliberately write something to irritate a teacher. When she returned my paper, she said, curtly, "You don't like me very much do you?" I answered, "Well, I just don't like your teaching methods." She had given me a B+. At the bottom of the paper she had written, "The story isn't very good but you have given greater attention to your spelling and grammar." Three cheers for Ruby! For the remaining two weeks the professor seemed to be at peace with the class, or perhaps I was with her.

The GIs with whom I interacted in school challenged their professors from time to time and took their education seriously. Generally, they were less intimidated by the lecturing scholars. Frank, Jim, Dwight, and I were all in the zoology class, and we liked the professor. He seemed to enjoy exchanges with the class and the opportunity to expand on certain points. He took the class on a field trip to La Jolla, where we gathered a variety of invertebrate creatures from the intertidal zone. Those of us who could dive collected a number of abalone for the class, which allowed us to show off to those less acclimated to the surf and ocean. We obviously also made a hit with the professor, who was anxious to get his hands on the giant sea snails and shells.

Sometime during the outing, the Point Loma clan decided that we should have some spirited fun with our teacher, so we saved the entrails of the abalone and, before the morning class, sneaked into the room and hid the smelly guts under the podium. The professor mentioned something about an odd odor, but the guts were not ripe enough to cause much disturbance. When we arrived at class the next day, however, the odor was repugnant and four members of the janitorial staff were trying to locate the source of the stink. We watched for about ten minutes before the professor told everyone that class was dismissed for the day. We recognized by then that the prank had outrun its intent, but we were too concerned about getting thrown out of school to own up. The bad guys were never found.

I was never asked to join any fraternities and had no special interest in them. Had my marital status been different, I might have had a different view, but I doubt it. Transferring information into a permanent place in my brain never came easily, so a good deal of my time was dedicated to poring over my books and lecture notes and preparing myself for tests. I remember Ruby giving me the test questions on psychology on Thursday nights before the Friday tests. I would always come up with the wrong answers until I figured out that most things that a person presumed to be logical would be incorrect according to the psychology book. From that point on the course became a snap. In the multiple-choice physics test, I never chose the answers that seemed logical. I got an A in the course.

We were surprised in late April when we got a call telling us our car would be available in a week at a dealer in Pasadena. Ruby and I were elated. Having wheels greatly changed our lifestyle. Frank and I could drive to school, and Ruby and I could use the car to visit her folks, drive to Idyllwild, attend football games, and visit friends in east San Diego. Our new Hudson was comfortable, well built, and a treasured addition to our property. In fact, with the exception of the clothes on our backs, wedding rings, and cheap furniture, it was our first large investment. It set us back close to $2,000, including tax and license. We were like children with a new toy, washing and waxing it every weekend, and ensuring it got the maintenance suggested in the manual.

The school year went quickly and, by the spring of 1947, I still had not selected a major. However, my brother once again entered the scene. He had been reviewing various college catalogs looking for a major. He was in the same boat as I was, except that he was about to finish his second year at San Diego State, and he needed to make a move. Sometime in May, he pointed out that the University of Washington offered a major in fisheries with a focus on marine and/or freshwater fisheries biology. The course work included relatively heavy emphases on mathematics, zoology, and selected courses in fisheries, taxonomy, anatomy, life history, behavior and population dynamics.

The major question in my mind was, what did marine and freshwater biologists do? The two of us discussed the matter for some time and decided that the main employers would include government agencies associated with wildlife/fisheries management and running hatcheries and academic programs associated with aquatic life sciences. I had a vision of living next to some beautiful stream, river, or lake and running a fish hatchery. It all seemed romantic and isolated from the hectic life of private enterprise. As a result of our conversation, I decided to make a trip to Seattle in June and check out the university's School of Fisheries. Ruby felt the career choice was good, although we recognized that the salary for such work was unlikely to make one wealthy. Yet, it could lead to a life of quality, and she encouraged me to explore the opportunity.

When school let out in June, I headed north with Bob Wedgewood, a member of the Point Loma gang. I had also asked Frank to accompany me, but he already had a summer job. We picked up a couple of maps at the gas station—at no charge in those days—and plotted a course north. We decided to take Interstate 99 through Bakersfield, then cut over to Hollister, south of San Francisco, and drive north to Oakland. From there we would follow 101 up through the redwood country to Crescent City. To avoid the long windy road along the Oregon coast, we would swing up the Smith River and over the coast range and down into Grants Pass, Oregon. Our route selection had been influenced by the fact that we knew absolutely nothing about the highways north of Los Angeles. I also had gotten in touch with Lee Hall, a buddy from Shanghai, and he asked us to stop by his home in Hollister.

Ruby packed us some snacks and lemonade for the first few days of our journey and bid us farewell. She couldn't get time off from work to join me and, since the trip involved three days up, three days in Seattle and two days home, it really didn't make much sense for her to come. The first day went well except that it was very hot from Bakersfield to Hollister. When we arrived at Hall's home, I was surprised to see that it was a large house, with four bedrooms, a pool, and a guesthouse. It turned out that his father was a medical doctor who had prospered, even in the Depression. We had a great dinner with the family, and Lee and I chatted about China and his current plans. He had entered into a local college and, as I remember, was planning to follow in his father's footsteps. When we retired sometime after 11:00 p.m., he took us out to the guesthouse, which included a large bedroom, a bathroom, and a small kitchen with a refrigerator filled with soft drinks. It far outclassed any other place we stayed during the trip.

The next morning we got an early start and headed north to San Francisco, crossed the Golden Gate Bridge, and followed the Russian

River into the redwoods. I had never been in big tree country and was impressed by the beauty of the region, particularly as we followed the Eel River toward Eureka. We traveled on a two-lane highway that wound through the coastal mountains and small lumber towns. The road was not crowded with tourists or cars, but we seemed to be constantly behind logging trucks hauling large segments of freshly cut redwoods to mills situated along the highway. From time to time, Wedgewood and I would check out redwood souvenir shops along the highway. Buying redwood burls was apparently popular, and we saw many being carried off to the tourists' cars.

We stopped and spent the night in a small town named Alderpoint right in the middle of the coastal redwood country. We hadn't made great progress because of the heavy truck traffic and our frequent stops to see various sights, including what was billed as the tallest tree in the world. During the day we had passed through grove after grove of the giant trees, which were fueling the economy from Santa Rosa to the north. Trees were cut, raw logs were hauled to plants where they were turned into finished lumber, and the lumber was then transported to markets in the south on large trucks, tractors, and trailers.

After breakfast, we started north again and made it to Eureka, where we finally saw the coast. We slowed along the bay to check out the salmon trollers and crab and bottom-fishing vessels tied to the piers or anchored close to shore. We had noticed a number of recreational/sport fishing camps and lodges as we moved down the Eel River, clearly a salmon and steelhead sports fishing Mecca for many summer tourists.

It was late when we stopped for a bite to eat, but we continued on north, as there was still a lot of sun in the sky. We wound up and down seemingly endless forested hills before we reached Crescent City. Wedgewood wanted to stop there for the night, but we were a day behind schedule, so I decided to try to get east over the mountains and down to Grants Pass. I thought it would be an easy day from there to Seattle. We climbed up the Smith River Gorge, winding along serene stretches of transparent green water making its way seaward. The two-lane road to the top of the pass, however, wasn't easy driving. Once behind a log truck, we stayed there until we hit a passing lane, and there were darn few of them. Just past the road to Cave Junction at the top of the pass, we spotted a nice little motel lodge and turned in for the night.

We went down the grade into Grants Pass early the next morning, turning north on Highway 99 toward Seattle. I thought the road from there north would be relatively fast, but that hope soon dimmed. We were in the heart of logging country and the Oregon timber industry. The road was a winding two-lane highway studded with small and large lumber mills being supplied by a horde of logging trucks, each hauling four or five large Douglas Firs. We found ourselves at the mercy of the local commerce, struggling between loads of logs. When we had the opportunity to pass, I always wondered whether the chains over the logs would hold or suddenly break and squash us flat.

At that time, the road went right down the middle of every logging town, including Roseburg, Cottage Grove, Eugene, Albany, etc., so the time we lost chasing trucks was balanced by the time we spent crossing through every town on 99 in Oregon and Washington. It was a long day, but it was summer in the Pacific Northwest and the light lasted until about 10:00 p.m. We pulled into the south end of Seattle at sundown and put up at the Old Richardson Hotel, which was just across from the train station. The next morning we drove through town, on Aurora Boulevard and east to the university. I was delighted to see that the city was bound by water, with Puget Sound to the west and the ship canal through to Lake Washington to the east.

I went to the Administration Building, where they sent me down to the fisheries department to see a Dr. Delacy. He was a very kind man who informed me that returning veterans from Washington and other local residents had priority over "out-of-staters," but I could get in if I wrote a petition and it was accepted. I decided to give it a try. After a day of looking at the town and possible housing, we headed home. We went straight down Highway 99, over the mountains in southern Oregon, down to Redding, and then through the hot and dry Sacramento–San Joaquin Valley to Bakersfield. From there it was up and over the Grapevine, down into Los Angeles, and south to San Diego.

When I got home, Ruby and I talked the situation over and, with hardly a second thought, decided to head north right after my brother's wedding, which was scheduled for early July. Ruby said, "If we don't give it a try, we will never know if you could have gotten in." Within a week or two, we began putting our meager belongings together and closing out our San Diego business. This meant selling the furniture that we

214

had purchased for the apartment just a year earlier. We withdrew the money we had from the local Ocean Beach branch bank, checked out with the landlord, and picked up all my school transcripts at San Diego State. We were ready for the northern trek that we presumed would last until I graduated.

In early July 1947, my brother married Meche, whom he had been dating for the better part of a year, in a church on Point Loma and we held a reception for them in our home above Ocean Beach. We sold the newlyweds our furniture and helped them make arrangements to rent our small apartment. Both Ruby and I were sad about leaving our first home. We had had some of the greatest days of our lives there, and we would soon be living a thousand or so miles to the north in a city where we knew absolutely no one.

A few days after Frank left on his honeymoon, Ruby and I packed our Hudson with everything we owned, drove to her folks' home, and said our goodbyes. The next morning we pulled out of Maywood at 5:00 a.m. and headed over the Grapevine, the two-lane mountain road between Los Angeles and Bakersfield, on the first leg of our trip to Seattle. We cleared Los Angeles before the heavy traffic was on the road and continued until we got to Gorman near the top of the grade. We had breakfast at a large restaurant on the right-hand side of the road, where the Greyhound buses made a rest stop, and then headed north over the pass and down the steep incline into Bakersfield. We arrived at Sequoia National Park early in the afternoon, drove through the big trees, and then on into Yosemite for the night.

The next day we drove into Sacramento and then west along the river delta over the coast foothills into Oakland. There we had dinner with a former SACO colleague Danny West and his wife, whom he had married just after the war. We had a great time, and Danny and I did a little reminiscing about China and exchanged addresses of other SACO members. The next morning we headed up through the coastal redwoods to Crescent City, then crossed over the mountains into Grants Pass, Oregon. On the third day, we completed our trip, winding north up the two-lane highway that made its way through every little logging town in Oregon and Washington. We drove out to the north side of Seattle just west of the university and found a motel near Green Lake.

That evening we walked around the lake and began to acquaint ourselves with the city.

Ruby and I went out to the university and talked to the administration personnel and later to Dr. Delacy about getting accepted at the U.W. Nothing had changed; the only possibility of entrance was to petition the school. I wrote an essay about how I had lived in many places, including the Washington state, that I had already moved to town, and my career would be put on hold if my application was denied. My petition was successful, and I signed up for the only summer courses that were still available—child psychology and developmental behavior. I cannot remember anything about either course, but they probably added a few words to my vocabulary.

After registering and getting my GI Bill education privileges certified, we went to a housing office and received a list of potential living sites. We first tried two housing projects relatively near the school, but neither one would have space available for a month or two. The last recommendation on our list was in a little town at the south end of Lake Washington called Renton. It was a World War II low-income housing facility built on a spacious flat area next to the Cedar River, which ran into the lake about a mile down from the housing project. A group of about ten multi-unit apartment buildings was scattered over an attractive grassy area with a few large hardwood trees. We were pleasantly surprised to find that there was a one large room and bath-furnished unit available. The manager told us that in several months we would be eligible to rent one of the larger units, which consisted of a living room, kitchen, bedroom, and bath. Rent at the small unit was only $19 a month, including heat and electricity. We took it without further questions.

We carried our few possessions into our new home and spent the rest of the afternoon in Cedar River Park. It had a large central building with a gymnasium and snack shop. The river ran down the south side of the housing project. On the other side of the river were railroad tracks that headed east and over the mountains and seemed to have a fair amount of freight and passenger traffic. What we hadn't seen when we checked in was a railroad spur that ran just to the north of our building. It was an active line, and twice that afternoon we watched as engines pulling several cars up the spur belched smoke that settled around our apartment. If our door was open, the smoke scattered soot on the hot plate, bed, and the few other pieces of furniture in the room. We also found out that the central heating in all of the apartments was fueled by a large coal-burning furnace that also tainted the air. I guess it all came with the low-rent package.

Over the next several weeks, I attended summer school while we settled into our new surroundings. It was apparent that we had been a little naïve about the furnishings of our flat. There was no refrigerator, just an old-fashioned icebox that required a twenty-five pound block of ice every few days. We then assumed that the icebox was designed so that the melting ice would flow into a drain, but that wasn't the case. We had been there about a week when our neighbor came and asked us if we were emptying the drain pan under the icebox. We didn't even know there was one. It had been overflowing, but we hadn't noticed because the excess water flowed under the baseboard and into the next apartment. We were humiliated by our ignorance, but that didn't help our integration into the neighborhood clan.

Ruby and I felt a little bit lost. When my homework was done, we would explore Renton, drive around Lake Washington and visit its parks, buy a bag of popcorn, and watch the ducks and geese feeding along the shoreline. About a week after we moved in, I met a young veteran, Chuck Ortloff, who had come down with tuberculosis in the service. He was on a recovery program that called for lots of fresh air and not too much exercise. He had taken up fly-fishing and worked the Cedar River each day for a few hours. He offered to teach me the art of using a fly pole and gave me a handful of flies that he had tied. Within a week or two, I had mastered the techniques enough to drop my line and fly reasonably close to my target area. I soon became an addict and fished the river each day for an hour or two, depending on my schoolwork.

The river had a run of Kokanee or silver trout (land-locked sockeye salmon) that made their way up from Lake Washington each fall, cutthroat and rainbow trout, young king salmon, and several species of white fish. In the late fall and winter, a run of steelhead trout traveled up to the spawning grounds in the upper watershed. The river wasn't that easy to fish, but if I was tenacious and didn't get bored. I always managed catch two to four fish that Ruby made into a meal. I wasn't picky and took anything of legal size home, but they weren't exactly trophy fish.

By the time summer school ended in late August, Ruby and I had become good friends with Chuck and his family, which included his wife Fern and a young son, who was about five years old. Chuck helped us explore the Cascade Mountains, which run from the southern part of the state to Canada and have a number of peaks over ten thousand feet. We soon became entranced with Mount Rainier and the number of beautiful parks, streams, and rivers that cascade down its eastern slopes. Ruby and I explored some of the small streams, which formed the upper drainage system of the Cowlitz River, and found that we could almost always take a good catch of cutthroat or rainbow after a few hours of fishing.

Late August brought a pleasant surprise. Wild blackberries were abundant and accessible throughout the countryside. Within a half-hour, we could pick enough for several pies. Ruby was an excellent pie maker, although I never understood how she managed on a two-burner hotplate with a metal box that served as an oven. Pies served as an excellent introduction to our neighbors and helped to extend our circle of friends in the housing project. Ruby had adapted well to our new financial limits, which I called the "$110 GI scholarship." She would pay the rent, then set aside four envelopes with $20 each that constituted our weekly budget for food, gasoline, etc. We generally made it through the month without having to make a draw on our savings, but at times it meant giving up our weekly movie, which cost fifty cents. The wild blackberry pies were an added bonus to a toned-down social life. It certainly was a change from the carefree beach life we had led in Ocean Beach.

Just before the fall quarter began, we had a surprise. Snide and his friend Bob Oversmith, another friend from Point Loma, who had a new bride, showed up in Seattle with the intent of entering the U.W. This was a great lift to Ruby and me, but Oversmith and his wife were turned off by the rain and left within a week. Snide, on the other hand, registered in the engineering school and took up residence in one of the campus dorms. He became a frequent visitor to our home in Cedar River Park and seemed to adapt to the Pacific Northwest. He even became a fan of the Huskies but, like us, was anxious to return to San Diego as soon as the opportunity presented itself.

In late September 1947, I entered the U.W. School of Fisheries and began taking courses in fisheries systematics, morphology, osteology, and so on. With the coming of fall came the notorious Seattle rains, which seemed endless and quickly soured our attitudes toward the Pacific Northwest. Ruby and I would go to the football games and cheer for the California teams, which usually won. The U.W. wasn't a football powerhouse in those days. Its single claim to athletic fame was rowing, and its teams frequently won national championships. The only nice thing about the fall was that I became acquainted with a number of the fisheries students. Almost all were GIs, and a number of them were married. The group of post-war GIs seemed to bond quickly, and soon Ruby and I had a number of friends, which lessened our sense of isolation.

In the middle of the fall, we moved from the one-room flat to a larger apartment upstairs. We had moved up to the "high life," but our rent increased to $28 a month, forcing Ruby to join the ranks of the employed. She got a job as a secretary in the local school district and made over $100 a month. Suddenly, we were rich! We even ate dinner out once in a while and took in a show almost every week. It was good to be among the upper class. The school years seemed to fly by, and Ruby and I, along with Snide, would look forward to each vacation, making plans to head south just as soon as the last final was over. We took numerous trips to California, and every one seemed to be a unique adventure.

Our first trip south occurred during the 1947 Christmas break. We decided not to take our car because, for $38, we could get a round-trip bus ticket to Los Angeles. We grabbed the Greyhound Special that left Seattle in the morning and headed south on Highway 99. It stopped several times between Seattle and Portland, then went to Eugene. I thought it would follow 99 all the way, but just south of Eugene it swung east into the mountains. The bus pulled steadily up the grade toward Oakridge, Oregon. We had barely started up the mountain when we ran into a silver frost. I had heard of the phenomenon, but it was still a surreal winter wonderland. The firs and other trees were covered with frost and ice crystals, and under a clear moon they sparkled like lit Christmas trees, particularly when the lights of the bus caught them in their beam. As we got higher, the snow level began to increase, and it was piled high along the road. We pulled into Oakridge, let off a passenger, and then continued the trip, going past Lake Odell, and stopping at the Crescent Mountain Inn. There were several feet of snow on the ground and the forest looked like frosted Christmas trees. The bus ran south through the night passing Klamath Falls, Shasta, and down the grade into Redding, California.

We arrived in Los Angeles late in the afternoon the next day and were met by Ruby's parents. We stayed there a few days and then took a train down to San Diego to be with my folks. We always had two Christmas dinners. We would visit Ruby's friends in Los Angeles and mine in San Diego. We had a continuous schedule of visits to see our relatives and friends, and the two weeks flashed past in what seemed more like minutes than days. When we got on the bus to head back north, we both felt as though we had been through a washing-machine ringer. We had only been on the road for four or five hours when we both came down with the flu, a variety they called Virus X. We were sick and vomiting, not a good situation on a bus. In Portland, we had to change buses, and we ended up having to sit on the bus station floor because all of the waiting-room seats were filled. I suspect most of the passengers thought that we had had too much to drink and were suffering from hangovers. Our neighbor picked us up at the Seattle station, and we stayed in bed for a couple of days before the winter quarter began.

The next time we took the bus, the south leg of the trip went smoothly, but the trip home was full of problems. When we hit the Grapevine just out of Los Angeles, we ran into a snowstorm. Cars were off the road all the way to the top of the pass. We were delayed several hours, but then we dropped down into the valley and ran into fog, which made the trip through the San Joaquin and Sacramento valleys slow and dreary. Still, Ruby and I felt a little smug because her folks had packed us a large bag of food, including a half-dozen ham sandwiches, cookies, and pickles. Even though we were starving, we planned to wait until we headed up the grade to Mount Shasta before we took out our emergency rations. However, just past Lake Shasta, we ran into sleet and before we got to Dunsmuir it was snowing heavily. Somewhere between Dunsmuir and Mount Shasta, the bus ground to a halt, with the traffic ahead sliding off, or stalled in the middle, or turned sideways on the road. We knew we were in for a long wait.

While we waited for the traffic and the snow to be cleared, we turned to the goody bag. As Ruby carefully unwrapped our ham sandwiches, I noticed that almost everyone around us was watching. Suddenly we felt very guilty. Ruby then started passing out half a ham sandwich to all those within reasonable proximity until our supplies dwindled to cookies. She then proceeded to pass those around too. Everyone seemed willing to help us annihilate what originally had seemed like more food than we could ever use. Several hours later, the bus broke free of the backup. We arrived in Seattle about five hours late, but without further incident.

Our social life in Seattle greatly improved during the summer of 1948 when we received a surprise call from Vern Benedict. He had been browsing in the phone directory when he came across Dayton L. Alverson. He said he couldn't believe there was more than one Dayton L., so he gave me a call. He had married a Seattle girl, Jackie, and moved from Oklahoma to Washington. "Benny" had become a butcher and lived on Mercer Island, which is situated in the middle of Lake Washington, just east of the city. Although I had kept in contact with Jim Murphy, Benny was the first SACO member I had met in Seattle. We immediately began visiting each other and traded dinners for many years. Ruby also had the good fortune of finding that her girlfriend Jane had moved to Seattle. She was married and living in a small house on the west side of Lake Washington; she and her husband provided Ruby with a California connection.

In October 1948, Ruby delivered our first offspring, an eight-pound baby boy whom we named Robert Dayton. Ruby had given up her job a month or so before the event, and we were back on a rigid budget. However, she later babysat other people's kids and this added money made us feel rich once again. We had moved to a ground-level apartment in Cedar River Park, and our son's birth was somewhat of a community affair as most of the newlyweds there were still without children. Bob was a local celebrity until the event became contagious, and a host of young toddlers came on the scene. We had been in Cedar River for a little over seventeen months, and it had become a delightful place to live. Most of the inhabitants were World War II veterans struggling to make a living, many at Boeing or Paccar, while others were furthering their education. They were low income, for the most part, but none of them felt poor.

We were on an upswing, and life was pretty good. Around this time, one of our neighbors got a seven-inch television. Four or five families would crowd into the Balcoms' home several days a week to watch wrestling, a local performer named Stan Borison, or other popular programs. T.V., even a seven-inch set, was a novelty and served as a social magnet. We would watch the magic of electronics, drink coffee, and exchange views on politics, the economy, and our expectations for the future. Everyone who came brought a different snack, so it was difficult to become bored, even if we had heard the rhetoric before.

Many of the men in the complex gathered on the large lawn and played football or baseball, weather permitting, and in the winter went to the gym to play basketball. Sports brought us close to the people across the park, because each year we managed to break their window playing softball, and each year the players would chip in and replace it.

During the spring quarter of 1949, a number of summer jobs were made available to the fisheries students. Most were positions assisting salmon biologists working in Alaska, Washington, and Oregon. A few of the jobs listed were in northern California, but there was nothing south of San Francisco. I mulled over the list and chatted with the other students to see what they were planning. Most found the Alaska jobs the most exciting and adventurous. After discussing the options with Ruby and several of my friends who were heading for Oregon, I opted for a job at the Oregon Department of Fish and Wildlife. The research headquarters were just outside of Portland in a small town called Clackamas. The work involved field studies on the Willamette River, a tributary of the Columbia River that ran south along the east side of Oregon's coastal mountains.

Plans were made to have Ruby and the baby ride south with Snide to Los Angeles just before I headed to Clackamas. Thus, in mid-June, Ruby and Bob headed south in our car to spend the summer with her parents, and I headed south with our car for my first job as a fledging biologist. When I arrived at the research headquarters, I accompanied a group of biologists to the McKenzie River and helped on a project marking young (fingerling) king salmon. The group drove south and then east up the river, where we were put up in a nice lodge just across the street from the river. It seemed like the ideal situation. We spent the day clipping off the adipose fins (the fatty fins on the back near the fish's tail) on thousands of fingerlings that averaged three to five inches in length. The clipped fins helped identify the origin of different stocks and cohorts of salmon when they were caught in the Columbia and at sea. In the evenings, we could hike along the river, fish, or just sit and look out the lodge window at the beautiful river and the occasional drift boat passing by. I was enjoying this life, but my budding romance with salmon biology was short lived.

After five days of fin clipping, we had marked more than a hundred thousand fingerlings and returned to Clackamas on a new assignment. Bob Schoning, one of the chief scientists at the research center, told me, "We are going to have to transfer you to Astoria. The summer worker we have assigned there as an assistant marine biologist is chronically sick when out on a boat, so we're bringing him here and you are to be his replacement." The next day I packed up and took off for Astoria, driving past Portland and then westward on what was called the Sunset Highway. I arrived in mid-afternoon and was delighted to see one of my classmates, Sig Westerhiem, at the lab. The commission was undertaking a study on the migrations and behavior of several flounder species, which involved a tagging program on three species of flounder: petrale, English, and Dover sole. Sig filled me in on the routine.

I was to contact any number of trawl fishermen working out of Astoria, request permission to accompany them on one of their trips, and select viable fish for tagging from the catch. The vessel had to have a saltwater running hose to fill a watertight wooden holding tank, where the retrieved fish were placed. After filling the holding box with sole, we selected one fish at a time, measured it, attempted to identify the sex, and then forced a long pin with a numbered disk through the upper back of the fish. When the pin and disk were secured on one side, a disk was placed on the pin on the opposite side and the end of the pin was twisted with electrician's pliers into a figure-eight knot. The whole process took twenty to thirty seconds.

The day after my arrival, Sig took me down to several of the fish plants where the catch was landed. There, we measured a number of fish to obtain data on age and length, and he showed me how to tag the flounders. Later he took me to the docks where the fishermen tied up their boats and introduced me to several skippers. Most were friendly, seemed interested in our work, and asked us to join them for a cup of coffee. The coastal trawlers ranged from forty-five to about seventy feet. The wheelhouse was forward and the area aft of the pilothouse contained two spool-like winches that were wound tight with half-inch cable that was used to pull the trawl and the wooden doors that opened the mouth of the net on the seabed. The decks were strewn with trawl netting, floats, and cables, and crewmembers were busy mending and tending the nets.

The next morning, I returned to the waterfront alone and measured several hundred petrale sole and extracted their ear bones (otoliths) for aging. After lunch, I went upriver to the boat marina just above the New England Fish Plant to see if I could arrange a tagging trip. There was a relatively large trawler called the *Harold A.* tied to the dock. When I asked two men working on a trawl net lying on the dock if I could speak to the boat's skipper, they told me that he was on board. I jumped on the gunnel, then down to the working deck, walked forward along the wheelhouse, and saw the skipper inside working on some electronics. I introduced myself and told him what I wanted to do. He looked at me, apparently sizing me up, and said, "Well, I don't know how many petrale or English sole we are likely to catch because I'm going to be fishing rather deep, but there will damn sure be a lot of 'shit sole' (Dover or slime sole) to tag." I said, "That's great," and he said, "Throw your gear up in the fo'c's'le [forward storage and bunk area] before sundown because we will be pulling out at sunup."

Around 8:30 p.m., while the sun was dropping below the horizon, I brought a tagging box to the boat with Sig's help and secured it on the port side, aft of the winches, in a place I assumed would be out of the way. I could see that the cable leading from the port winch to the davit (A-frame) would pass just outside of the tagging box and, when slack, would lie across the box. Regardless, it looked like the only reasonable location to stay out of the way of the crew working on deck. I checked my gear to be sure I had the required Peterson disk tags (a button-like tag with an identification number), electrician's pliers, clipboard, and recording forms, then said goodnight to Sig. I stowed my belongings and foul-weather gear on an empty bunk forward. The bunk was just over six feet long, about two and a half feet wide and had a mattress that was about two inches thick. It looked like it had been well used. The fo'c's'le was empty except for me. I supposed the crew lived in town and would board in the morning. It was still early, but I knew it would be a long day, so I climbed into my bunk and tried to get some sleep. The smell of fuel oil was strong and permeated the mattress. Sleep did not come until sometime after midnight.

I was in a deep sleep when I heard the main engine roar into action, and the smell of diesel smoke entered the sleeping quarters. The crew were on deck clearing the lines, and the vessel maneuvered out of the mooring area. I fell back to sleep for a short time and then awoke feeling the swells of the Pacific moving the *Harold A.* gently up and down. It wasn't long, however, before the bow began rising sharply with the oncoming swells and then plunging down the backside of the swells. Within twenty minutes, the boat was tossed about in violent motions. The sea had obviously turned nasty. The bow would rise and then the vessel would heave sharply to one side. It seemed like it hung there forever, then struggled to right itself. Before the boat was level, another wave would crash against the hull. I had heard about the reputation of the Columbia Bar and how rapidly the sea conditions could shift with a change of tide. We were in trouble, and I knew it.

I got out of the bunk, grabbed my clothes, dressed, and held on to the side of the bunk to keep from being thrown to the deck. I didn't bother to put on my foul weather gear; my life depended on getting topside before the vessel capsized. I quickly opened the hatch door between the fo'c's'le and the engine room. Holding on to whatever was available, I made my way aft to where the ladder went up into the galley, pulled myself up, and managed to push the hatch open. I suddenly felt like a little boy running to get out of the dark.

One of the crewmembers was sitting at the galley table taking a drink from a bottle of Canadian Club. As he removed the bottle from his mouth, tobacco juice ran down its side. He had been chewing "snooze." A man with a hook attached to his upper right arm was busy cooking breakfast, and eggs were sliding back and forth in a half-inch of grease. I looked out of the galley window and saw that we were in a cross sea with good-sized swells tossing the boat about, but it was obvious from the crew's behavior that it was just a normal crossing. I was embarrassed, but the crew was unaware of the fear that had catapulted me topside. I sat down and the crewman at the table passed me the bottle of whiskey without saying a word. He would test the mettle of this young biologist. I picked up the bottle, wiped off the tobacco juice with my hand and took a drink although it didn't go down easily. We were underway!

As we moved on across the Bar and passed the outer buoy, the sea began to settle down, with rather long swells moving in from the southeast. The man at the stove introduced himself as Jackie Ray and the other fellow as Ben, who was from Holland and didn't speak very much English. Before breakfast was served, the skipper, Buddy, came back, sat down, and helped himself to a shot of whiskey. He then filled his coffee cup and returned to the wheelhouse. I looked around the room, thinking to myself "There's Ben, about six foot one, with a deep scar on one side of his face, Jackie with a hook on the end of his arm, and the skipper who was about five-ten, stocky, and smoking a pipe." He could have been a Popeye look-a-like. I began to wonder when the skull and crossbones would be hoisted. Although I was somewhat of a unique guest on their boat, I suspected that they would not miss a chance to have a little fun with me. After all, marine biologists were a curious lot.

The vessel kept on a southwest course for several hours before the skipper yelled, "We are going to set." Jackie and Ben went onto the aft work deck and, at the skipper's signal, tossed the trawl net over the stern. The crew then let out the cable until a shackle arrangement and G hook were along the trawl davits and doors. The trawl cables were linked to the inside of the trawl doors, and the weight of the net shifted to the back of the door. The trawl cable then was let out simultaneously from each winch as the trawl doors and net sank to the seabed. We were in about ninety fathoms of water (540 feet), and it took five to six minutes to let the gear out. When the trawl doors reached the bottom, they sheared outward from the water force, spreading the mouth of the net open. Floats on the top and a weighted footrope gave the net a vertical opening. The result was a large mouth net, which tapered back to the end, where the fish were collected in a heavy twined bag or cod-end.

The fishing gear was towed along the seabed for almost two hours before the skipper gave the signal to haul in the net. When the doors were next to the trawl stanchions, the doors were freed from the cables and the net was then pulled into the stern. I stood watching the process and then noticed a sudden surfacing of an island of fish behind the *Harold A.* Jackie must have seen the surprised look on my face and told me, "We have a large catch of rockfish [referred to as snappers] and their air bladders literally exploded and extended through their mouths as a result of being brought up quickly from the deep water." The net, floating on the sea thirty to forty yards behind the stern, was bulging with brilliant red, rose, orange, and black rockfish. The crew brought the net along the starboard (right side) of the boat's stern and then, choking off part of the fish, began to hoist and drop the catch into "checker-board" holding areas on the deck. The skipper said it looked like they had caught ten thousand pounds. It took four or five lifts (splits) to empty the net on the stern, which formed a mountain of fish over three feet high. I didn't see any flounder come on board until the last couple of splits. Their density was greater than the rockfish, partly because they had air bladders that could adjust rapidly to the changing water depth; thus they sank to the bottom of the net and were the last to come on board.

As soon as the catch was on deck, the skipper told me to get my fish. I immediately began to pick up lively looking Dover sole, which were the most abundant, and carry them to the tagging box, which I had filled with salt water. The crew also threw some fish into the box. After retrieving about forty flounders, mostly Dover but also a few petrale, I began to tag. After about fifteen minutes, I was done. Then I turned my attention to the mass of rockfish on deck. The crew was sorting out the larger marketable red rockfish, sole, and lingcod. I scanned the catch several times and couldn't believe the diversity of rockfish species. There had to be at least twenty or more species in the checkers, ranging in length from ten to twenty-six inches. The skipper, who was helping to sort fish, said to me, "Lee, what the hell species are these? I am fishing deeper than we normally fish and we don't usually catch these species of rockfish. I was told by some of the fishermen down at Newport, Oregon that they had been picking up a small species of red snapper in the deep water and they were being sold as Pacific Ocean perch."

I had studied the taxonomy of rockfishes, but everything I had learned had come out of a museum bottle. Furthermore, it was a difficult group to identify because there were so many species with different colors but the same general morphology. The stacks of red, orange, and tinges of yellow rockfish were completely unfamiliar, and I was embarrassed. But I responded by admitting my educational deficiency and asking if I could set a group aside and attempt to identify them after the catch was put on ice. It took almost an hour to sort out the marketable species based on size. In the end, I had put close to forty rockfish of all colors and shapes into a separate pile.

As we started running to make another set, I went down to my bunk and got out a copy of Clemens and Wilby's *Fishes of the West Coast of Canada*, a book that describes the marine fishes of the Pacific Northwest and Canada. The differentiation of the rockfish species was largely based on skull shape, spine placement, and color patterns. It took me more than two hours to make my preliminary identification. I was amazed by the diversity of this one family; there are more than eighty different species in the northeast Pacific Ocean. We were well into the second tow when I told Buddy, "The most common rockfish species we have caught is called the long-jawed rockfish. A variety of other species are among the catch, almost all of which are noted in the book as being uncommon or rare. You sure as hell have changed that scientific view. I think it must be the same species the fishermen down in Newport call Pacific Ocean perch."

When the second tow was hauled in, the trawl net surged to the surface and there was even a larger island of fish floating off the stern. Buddy looked back and said, "I moved out a little deeper, 105 fathoms, and the fish are thicker than flies around dead meat. Christ, the bottom must be crawling with long-jawed rockfish." He then looked at the crew and laughed, saying, "There has got to be at least eight tons of fish in the bag." The snouts of the rockfish were sticking out of the mesh all over the net. When the catch was brought alongside, I could see there was a lot more fish than there had been on the first tow. Jackie watched as albatross and seagulls swooped down and pulled at the entangled fish and said, "A whole net full of damn good fish and we will have to toss more than half back, what a god-damned waste." I was puzzled and asked why. The answer was that the market only accepted the larger red rockfish, and there was a limit on the amount of snappers that could be brought in on any one trip.

The catch was brought on board, split after split, until every checker was filled and the deck was loaded down with rockfish, all with their air bladders pushing out of their mouths. There weren't a lot of flounders in the catch, and I only recovered a couple of dozen viable fish. I again surveyed the catch and set aside a variety of rockfish that looked different from those taken on the first tow. After two tows, I had recorded twenty-seven different species. About six remained unknown. According to my taxonomic bible, authored by two Canadian scientists, most were considered uncommon or rare. With the skipper's consent, I set aside a number of the species that were reported as rare to take back to the lab at the U.W. School of Fisheries.

While Ben was working on deck, Jackie walked up behind him and poured hot coffee down the inside of his boot. Ben let out a howl that could be heard back in Astoria. He then yelled something in what I thought was Dutch, cursed a little in English, raised his fish pue, a stick several feet long, with a curved steel rod that is pointed at the end, and pointed it at Jackie. But Jackie just stood grinning like an impish elf and raised his pue in defiance. Buddy then came aft and told them, "Let's get the catch down. You can settle your argument when we get on the beach." Both Ben and Jackie went back to sorting fish, taking only the larger snappers that weighed more than several pounds. More than half were thrown back in the water, providing a picnic for the gulls and albatross. It was well into the evening before the catch was sorted out and the keepers were put on ice. In the meantime, I was having a field day sorting through and identifying different marine species, including dogfish sharks, a variety of skates, sturgeon, hake, smelts, an occasional codfish, sculpins, black cod, sea poachers, etc. During my first trip at sea, and in only two tows, I had collected a number of species that were seldom seen in the university's collection.

That evening as we sat having dinner, the skipper said that we needed to fish where we could catch a higher percentage of large marketable rockfish and a few flounders for the biologist. He looked thoughtful for a few minutes and then announced, "We will run south to Hecate Bank just south of Newport. There are a lot of large snappers down there." He then went forward and set a course down the Oregon coast. Just before sundown, he returned to the galley where the crew was drinking coffee and chatting, and said, "Lee, can you take the first wheel watch?" "Sure," I responded. I filled my coffee cup and went to the wheelhouse with Buddy. He pointed to a flashing Sub-Sig depth sounder and told me, "Just keep us around the sixty-fathom contour." As he turned to leave, he paused and said, "Give me a call at 1:00 a.m."

After Buddy left I noticed that the boat was on automatic steering and that the wheel could not be turned unless the "iron mike" was taken off line. It was pitch dark, and I couldn't see how to uncouple the wheel from the automatic steering device. I sure as hell didn't want to go back to the galley and tell everyone my dilemma, so I continued to search for the light switch, but with no luck. In a little more than an hour, we had gradually slipped inside the contour to a depth of about fifty-two fathoms. I was quite sure that there were no rocks or shoals in the area, but I was getting increasingly nervous. I finally found a flashlight and then the iron-mike lever. Once the wheel was free, I nosed the *Harold A.* off shore and was soon back on the sixty-fathom contour. The moon was high in the southwest sky, the sea was calm with only a long low swell rolling, and I could see the lights along the coastline to the east. I put the wheel on the iron mike, walked back to the galley, and poured a cup of coffee, as if I knew just what I was doing.

The rest of the night went smoothly, with the sea calm and the light of the moon reflected across the oncoming swells. There was almost no traffic, and I saw only a couple of boats running up the coast. At 1:00 a.m. I woke up the skipper. After looking at the course, the fathometer, and the coastal lights for a few seconds, he said, "We're south of Cannon Beach." We talked a few minutes about my research, and he gave his view on the migration of flounders. He told me that flounders moved into deeper water during the winter months and inshore in the spring. The deep water species move up on the outer flats of the continental shelf in the summer, and the shallow water species move over the edge onto the slope in the late fall. I asked on what he based his opinion, and he said, "I fish all year around the clock, and if a fisherman can't follow the fish, he can't make a living." That was good enough for me; I wrote down his observation in my field notebook.

Around 1:30 a.m., I climbed into my bunk, too tired to undress. Within minutes I'd fallen asleep.

The sound of the winches spooling off the cable woke me, and I knew the trawl was being set. It was just a few minutes before 5:00 a.m. The crew had finished breakfast and was on deck, setting out the net. After a cup of coffee, I went to the wheelhouse where Buddy greeted me. "I decided to make a set here," he said, pointing to the chart. "Sometimes we catch a fair quantity of English sole in this area." I looked at the fathometer, which showed that we were in about fifty-three fathoms of water, and asked Buddy how he knew where we were. He said, "By the fathometer, the coastal mountain range, and by the magic of this loran. It's surplus military equipment, which if properly operated gives you digital readout that can be checked against numbered lines that are superimposed on the sailing charts." He pointed to the lines on his chart and said, "It sure makes fishing a lot easier." I observed for a few minutes and noticed that he was watching the fathometer and holding his course as close as he could to the fifty-two-fathom depth contour.

When I returned to the galley, Jackie and Ben had returned from the work deck. Jackie asked me if I wanted some breakfast, but the sea had come up somewhat and we were fishing almost in the swell trough. The vessel had a strange roll as it pulled against the fishing gear. I was a little queasy, so I opted for toast. I got out my Clemens and Wilby and studied the descriptions of the rockfish group. It wasn't long, however, before the crew began hauling the gear back. This time there was no floater, and the net hung almost straight down. When the catch of about three tons was dumped into the checkers, there was an abundance of flounders along with a mixture of dogfish, sculpins, hake, and several species of shallow-water rockfish, most of which were shades of brown and black. There were also a few large Dungeness crab, which Jackie threw into a bucket for later consumption. I quickly filled my tagging box with at least a hundred or more English sole and a few petrale. The fish here were in much better shape. They had been taken in shallower water, and they didn't have spiny rockfish rubbing against them in the net.

I tagged perhaps fifty or more fish and then threw them back; most swam off immediately, heads down making for the bottom. I thought their chances of survival were high. I pulled up a petrale sole, somewhat larger than the English sole, from the tagging box. By the shape of the extended gonad, I could see that it was a female and, by the feel, could tell it was in good shape. I tagged it and, as it went over the side, my peripheral vision caught a glimpse of something large being tossed into the tagging box. A wolf eel about four feet long was pushing its head into one corner of the box. It was somewhat stunned and not thrashing about the tank. I had studied the skulls of wolf eels and knew they had a set of sharp, strong canine and molar teeth. I instantly suspected that Jackie Ray had tossed the wolf eel into the tank. I watched it with my back turned to the crew, and it seemed content to stay with its head in the aft corner of the box. I reached down slowly, moved my hand forward until it was just above the gills, grabbed the back of its head, stuck my thumbs in its gills and threw it over my shoulder in the direction of Jackie. Catching trout at night in Strawberry Creek as a boy had taught me something useful.

I turned and looked at Jackie and Ben. Both had a smile on their faces, but said nothing. I turned back and continued to tag the remaining fish. Before the first tow was done, there were more than a hundred fish swimming around about three hundred feet below with Peterson disks. When I returned to the galley, we were into the next set and Ben and Jackie were sitting at the table drinking coffee. From the smell, I suspected that it was spiked with Canadian Club. I poured myself a cup and sat down at the table. Neither said anything about the wolf eel. I think I had passed some sort of test, however, because from then on Jackie was very friendly and went out of his way to help me with my tagging chores. By late afternoon I had tagged several hundred fish, mostly English sole.

Late in the afternoon, after finishing up a tow, I went forward to the wheelhouse to chat with Buddy. I had become comfortable talking with him. He had fished all his life, catching sardines off California, salmon in Puget Sound, and dogfish during the war when the species had become valuable for vitamin A. He was using the echo sounder to check the water depth and bottom type when someone called him on the radio. Buddy exchanged some information on fishing results and told the other skipper that he was going to move south to Hecate Bank the next day. He also passed the word along that he had a "plankton picker" (a name often given to fishery biologists) on board who was tagging sole. The talk continued for several minutes, and then the other skipper said, "Maybe we will see you at the blackberry patch tonight." Then they both laughed and ended the radio talk. I thought it rather odd that the two macho fishermen would go ashore to pick blackberries, but didn't ask any questions.

Just before dark, Buddy headed the boat into Newport Harbor. It had been a good day, I had tagged a lot of fish, and the thought of sleeping in a sheltered bay buoyed my spirits. Several hours later we arrived at the harbor entrance. The channel ran inside a sand spit and then opened into an extended bay. We tied up on the south side of the harbor just after 10:00 p.m. The crew and Buddy asked me to join them for an excursion to the Abey Hotel, where they were going to have a few beers. I was tired but wanted to fit in, so I joined the shore party.

The Abey Hotel sat on Front Street, which ran parallel to the waterfront location of a number of the fish plants. We sat down at a round table with several other fishermen known to the crew. The beer came, the talk grew lively, and everyone seemed to be having a good time. Jackie made his way around the large room, chatting with friends and bringing each of them to the table to introduce the fledging biologist. Each one had his own pet saying or joke about "plankton pickers." Regardless, they were friendly and I think Jackie's prodding encouraged each of them to buy me a bottle of beer. Within an hour, at least twenty bottles surrounded me and the crew of the *Harold A.* was greatly amused by the sight. There was no way I could even begin to down what was in front of me, but I could join in the fun. I was laughing and conversing with those at the table when someone tapped me on the shoulder. I turned and there was John Garret, assistant director of research for Oregon Fish and Game, looking down at me. He asked what I was doing in Newport. I explained that I was accompanying the *Harold A.* on a tagging trip and that we had come in for an evening break. Then I added, "I am attempting to improve the working relationship between fishermen and biologists." He looked at all the beer bottles, shook his head, and left. I thought to myself, what a great way to kick off a new career.

Sometime after midnight, Jackie got into an argument with one of the local fishermen, and things began to turn ugly. Buddy intervened and, with some difficulty, got Ben and Jackie to leave the hotel on the understanding that we would go to an after-hours bar, the Blackberry Patch. The radio conversation that I had heard on the boat now started to make sense, but there was more to the story. When we arrived at the Blackberry Patch, it quickly became apparent that it was more than an after-hours joint. A number of scantily dressed girls were working the bar and tables. I asked Buddy, "This is the local house of prostitution, isn't it?" He laughed and said, "Yes, but don't worry, we're just going to have a few drinks and leave."

Before I could say anything more, the fisherman who had been arguing with Jackie at the Abey Hotel came in the door with a couple of friends, and he wasn't looking for sex. He was on the war path and immediately started for our table, with Jackie and anyone else who was in the way as his target. Within seconds a fight had started, but the bouncer and several other visitors quickly got the mess under control, telling both sides that they had already called the cops. That was enough for Buddy, who grabbed Jackie and said, "We're going back to the boat." It was only after we were a block from the Berry Patch that Buddy revealed that Jackie had just been released from jail in Astoria for punching out a number of parking meters with his steel hook. Nevertheless, Buddy was

233

feeling his oats and, in a show of strength, wrapped his arms around a stop sign at the corner and struggled for a few minutes before pulling it out of the ground. I thought, this is all I need, to get arrested and thrown in the local Newport jail. We did make it back to the boat, however, without further incident.

It seemed that I had just fallen off to sleep when I heard the sound of the main engine starting. It was just after 4:30 a.m., and we were heading out to the fishing grounds. I began to understand that fishing was a tough way to make a living. It was well into the morning before the gear was set. Buddy told me, "The bottom is tough here but if we can tow through, we should have a good catch of snappers." I looked at the fathometer, which gave a reading of about eighty-five fathoms. We fished for about two hours when the skipper came back and yelled to the crew, "We are hung up solid."

They began to haul in the gear, the winches were smoking, and the boat was backing down with water breaking over the stern. I was watching the activities on deck when the port (left side) cable parted and went slack. "It must have turned loose just in front of the door," commented Buddy. They pulled in the slack port cable and then turned to the starboard side. Jackie yelled out, "We still have gear, and it's free of the bottom." The crew and the skipper worked well into the afternoon before the left door came up to the davit. Then the net exploded to the surface with a giant island of snappers floating behind the vessel. Buddy said, "We are going to salvage something out of this tow." Working until after sundown and then under the deck lights, the crew of the *Harold A.* struggled to bring the catch on board. When they were done, more than 20,000 pounds of snapper had been iced down.

With wire and the net strewn over the aft work area, Buddy turned the vessel north and headed for Astoria. I took one more wheel watch that evening and let the crew get some sleep, turning in about 1:00 a.m. We arrived in Astoria the next day after being gone for almost five days, with a catch of about 42,000 pounds, which seemed like a lot to me. Unfortunately, bottom fish prices were not very good. The snappers brought in four cents per pound, and the flounders about five cents. The gross was just over $2,000 and, since the boat's take was about 37 percent, that left only $1,300 for the crew to share after expenses (including food and fuel).

I left the boat feeling that I had made some friends in the fishing community. Even Jackie, who had a reputation for being mean at times, treated me with respect. It had been an experience that would live long in my memory. More than anything, the trip on the *Harold A.* changed my perspective on what I wanted to do with my life. I had learned more about fisheries and fish biology in a few days at sea than I had in the classroom over the past two years. I no longer had my sights set on being a salmon biologist, and adjusted my heading toward becoming a marine biologist. I had found my love and an ecological niche.

During the summer, I went out with a number of other Astoria-based draggers and one boat out of Yaquina Bay that was fishing for mink food. Each trip yielded special memories and put my taxonomic education to a severe test. But none provided the same sorts of challenges and experiences of the *Harold A.* By summer's end, I was anxious to head south and pick up Ruby and our son. I left on the Friday that my job ended and drove straight through to Los Angeles, arriving late the next afternoon. After a day or two of rest, Ruby and I headed back up Highway 99 for Renton. Our son was then ten months old and, during my absence, had started to walk like a trooper. He had changed from a baby to a little boy, and I had missed the transition. But he seemed to have fared well with his mother and grandparents and was a happy young child who could already talk up a storm.

By early September, we were back in our small apartment in Cedar River Park. Over the two years that we had been there, it had become home, and the rain and the absence of nearby relatives no longer intimidated us. We still wanted to return to Southern California, but our feet were becoming webbed. We had become U.W. Husky fans, and we were comfortable with our lives. Still, we were ecstatic when my brother enrolled at the U.W. to get his master's degree in fisheries. We made arrangements with the local housing authority so his family could get an apartment two doors down from ours. Frank and Meche had a son, Franklin, Jr., who was three months younger than our son. Bobby would have his cousin living next door. With Frank coming to Seattle, and Snide Malloy still living on the campus, we had our own California ex-patriot group.

Both Frank and I registered at the School of Fisheries on the same day. Frank chose the set of graduate courses of interest to him, and I realigned my courses to conform to my interest in becoming a marine biologist. We were heading down the same trail. After classes, I went to Dr. Welander, who taught taxonomy, and asked if I could do an undergraduate study on Pacific Northwest rockfishes. He agreed and noted that the first thing I needed to do was to identify the samples of rockfish I had sent up to the school during the summer. I already had a fair idea as to the identification of about twenty of them, but about a quarter of the total had me stumped. I finally sent off samples of the unidentified rockfish to Dr. Hubbs, a specialist at the Scripps Institute of Oceanography.

Late that fall I received a letter with the requested information. Most of the fish were recorded as rare species, but I suspected that was because there had been sparse sampling of the deeper waters beyond the shelf areas adjacent to the northwest coast of the Pacific. Yaquina Bay Fish Company was stimulating exploration of the deep-water fish inhabiting the waters of Oregon. The company owners were attempting to market the long-jawed rockfish as Pacific Ocean perch, taking advantage of the large market in the Northeast for Atlantic ocean perch. As the fishermen moved out into the deeper waters, a wealth of new information became available on the latent resources of the continental slope region.

The fall season came early in 1949, along with the traditional rains. I knew Frank and Meche were suffering, as we had several years before, from the constant overcast and drizzle. It wasn't an easy transition for anyone raised in southern California. Frank and I left for school every morning about 7:00 a.m and, as fall moved along, it was dark when we left Cedar River Park and dark when we returned. On Saturdays, Frank and I headed for the Husky Stadium to take in the football games, but the Dogs were no football powerhouse and sitting out in the rain, watching our team lose, didn't do much for our spirits. In the evenings, if we were not studying, we could watch little Frank and Bobby play together or, unless it was raining, fish the river. By Thanksgiving, Ruby and Meche were already planning a Christmas trip to San Diego.

The final exams for the fall quarter took place just before the holiday break, and Frank and I were both done by Wednesday, the week before Christmas. We made plans to caravan down with Ralph Malloy. We planned to drive straight through, stopping for only food, gas, and potty breaks. Around 4:00 in the afternoon, our two families crowded into our

Chevrolet (we had traded in the Hudson), joined up with Ralph and his three passengers, and headed south. The road to Eugene was good, and there seemed to be a respite between winter storms. We had decided to go east to Oakridge and then south on 97 through Klamath Falls and rejoin 99 at Weed. If the weather was good, the road south along the high plateau east of the mountains would be faster.

The grade up to the pass southeast of Oakridge went well, although there was a lot of snow on the road and the temperature had dropped to below twenty degrees. Even with the car heater on high, ice built up on all the windows, except right in front of the driver's seat where the defroster kept the glass clear. Ruby and Meche bundled the children in their snowsuits and covered themselves with blankets. Frank and I were a little better off because we were closer to the heater. As we moved south toward Klamath Falls, a light dry snow began to fall, making the road slippery and more difficult to negotiate. When we arrived at a long grade just north of Klamath Lake, we could see a number of cars off the road and others attempting to move slowly up the grade. It looked like trouble.

We had good tread on the tires, but no snow tires, and we weren't looking forward to putting on chains given the low outside temperature.

Ralph and I had made the trip a number of times and had a good deal of experience driving in snow. We realized that if we lost too much speed, we could lose traction and get stuck like the other unfortunate travelers. Ralph managed to move up the grade, staying just to the left of the cars pulled off to the side, but when he had to move around a car that was sideways in the middle of the road, he slowed down and lost traction. He was several hundred yards ahead of us, and I saw two of his passengers get out and begin to push the car. He finally got past the stalled car, picked up speed, and got to the top of the grade about a mile ahead. His passengers, however, were left behind to walk in the bitter cold. As we approached the stalled car, I yelled at one of Ralph's passengers to jump on the running board, forgetting that, unlike the Hudson, the new Chevy had no running boards. He made an effort that landed him on the side of the road. His expression of surprise is still etched in my memory, as is the vision of him flying into the snow bank. Somehow we got past the stalled car and made it to the top of the grade. There, we waited until Ralph's two frozen, unhappy passengers struggled up the hill and rejoined the caravan.

We arrived in Redding before sunup and made a food stop at an all-night café just inside of town. Our weary troop of ten sat at two tables, one with the Alverson clan and the other with Ralph and his passengers. It was apparent from the conversation at Ralph's table that he had a group of disgruntled passengers. Not only had two of them suffered through the climb up the grade, but Ralph's heater had broken, making the trip extremely uncomfortable. Two of them were talking about taking a bus the rest of the way, but the weather had improved, the sun was about to rise, and the worst was over. In the meantime, we had ordered for our table and the show had just begun. Little Frank and Bobby were at the grabbing age. They managed to spill two glasses of milk, drop the salt and pepper shakers on the floor, and spread their oatmeal all over the table. The waitress cleaned up the mess, brought more milk—which the boys managed to spill again—and somehow kept a smile on her face. We tipped her well and, when we stopped there a year later during the next Christmas break, she said, "I remember you folks from last year!" We had made an impression, of sorts.

We had a good vacation visiting our folks. On the return trip, I attempted to pass a truck and trailer north of Sacramento and had to exit off the left side of the road to keep from running head on into a Greyhound bus. It was the last time that we drove through without stopping. The wear and tear on the drivers and the family was just too much.

Over the course of our first years in Seattle, Ruby and I became acclimated to the rain and found ourselves strangely uncomfortable when we returned to California. I can't remember when the transition occurred. We had a great group of neighbors and were making new friends, who also liked to fish, camp, and hike. Ruby and I fell in love with the country east of Mount Rainier, where we pitched a small tent and spent many of our free weekends. Somehow we had been transformed and become captives of our new environment. If they only had surfing beaches with survivable temperatures, life would have been complete.

No life history is complete without a great storm story. Ruby and I endured several, but I will recount only one. In the winter of 1949–50, the worst snowstorm that I can recall slammed into the Seattle region. I was at a lab class for fish and invertebrate anatomy just after lunch when a cold wind moved in from the north. Within an hour, dry snow was blowing almost horizontally and was beginning to pile up on the lawns and to stick on the roads. I collected Frank and the friends who rode to school with us each day and started for Renton about 3:00 p.m. The snow had already reached a depth of almost six inches and the roads were getting treacherous. I decided to stay off the route we normally took and follow the perimeter road around Lake Washington. The main challenge was getting from the U.W. through the arboretum and down to the road that ran along the lake.

We made it to the lake, but then slipped off the side of the road as we rounded a turn. It was getting colder and, despite my riders' efforts, we could not get back on the road. Fortunately, within a few minutes a young U.W. student in a pickup truck pulled alongside, uncoiled a towrope, and yanked us out of the ditch. As we continued toward Renton, the snow became heavier and deeper, and few vehicles were left on the road, which is probably why we reached home without further difficulty. When we arrived at our apartment, the snow was about twenty inches deep on our front porch. Ruby and Meche had not thought about removing the snow from the entrance to our flats, and we could not open the doors. The excitement of the storm and the beauty of the snow had mesmerized the southern California ladies. They had watched as the snow accumulated and later turned into a block of ice. Frank and I had to enter through the windows. The snow lasted several weeks, and temperatures plunged into the low twenties. It was the only time we saw two- to six-foot icicles hanging from the roof or ice skaters on Green Lake.

In the spring of 1949, I started looking for permanent work since I needed only three additional credits to graduate in June. Rather than return to school the following fall, I signed up for twenty-one credits. Only eighteen hours could be recorded each quarter, so I wouldn't graduate officially until the fall quarter. Ruby and I were racing toward a crossroad in our lives. There were several job possibilities, one in central California, one in Oregon, and one in Seattle. If we were going to move back to California, there was no use wasting time. On the other hand, the job that I found most intriguing was right in our backyard. Don Powell, a good friend and classmate, had taken a job with a newly formed group with the U.S. Bureau of Commercial Fisheries (BCF), and he urged me to consider an opening that was coming up in the fall. The group conducted explorations of the marine waters off the West Coast and identified and quantified the region's fish resources. The Oregon job involved studies on the coastal salmon rivers. The California job was located at Stanford University and made studies of the California sardine and other pelagic fishes. None of the opportunities constituted a manifestation of our earlier dream—working in a hatchery and living in a small white house alongside a beautiful river in the mountains.

Don told me that in early 1950 the BCF would launch the first high-seas fishery research vessel in the region. It was to be named the *John N. Cobb* after a scientist who had worked for the old Bureau of Fisheries. The more I thought about exploratory work, the more I liked the idea. The work fit well with the interest I had developed during my summer in Astoria. Ruby and I decided that we wanted to live and develop our careers in the Pacific Northwest, and I applied for the BCF job in Seattle. The offices were at the Montlake Fisheries Center just across the Lake Washington Canal from the university.

When summer arrived, I was eager to start my new job, but couldn't qualify until my extra three credits were awarded. Even though I had completed the work, I had to register for the fall quarter to receive my degree. I spent much of the fall quarter working on my rockfish project and preparing a research paper, or key, that identified the fish. I was hoping to publish the key when it was completed. My marine studies were enhanced that fall by the opportunity to study with Dr. Harry, a graduate of Stanford, who was an eminent taxonomist and had spent much of his life in the Philippines working on tropical fishes. Almost fifteen years after leaving the Hawaiian Islands, I was putting scientific names to many of the fishes I had caught or observed as a young boy. In late October, I was notified that I could start work at BCF in late December and that the new research vessel would be launched in early February. I was twenty-five years old, had a bachelor's of science degree from the U.W., and was ready to start down a new path. My enthusiasm was high, and my goal to become a marine biologist had been given a jumpstart. The salary was only $240 a month, about $100 a month less than I made working at the fish cannery in San Diego four years earlier. Not a lot of progress for four years of study, but I was told that I was better educated!

CHAPTER VII
EXPLORATION OF THE SEA

By early 1950, I was on the job with Don Powell at the Montlake laboratory, where the Exploratory Fishing and Gear Research Base was located. The head of the group, Joe Ellerman, had worked for some years in the salmon industry in Alaska. He was anxious for our new research vessel to be launched so we could get our program underway. Don and I immediately began to put together a short- and long-term exploration program and to acquire various fishing gear and oceanographic instruments that were essential for documenting the results of our explorations. We were limited by our budget, but we purchased several trawl nets, a small dredge, a Dietz-Lafond bottom sampler, a bathythermograph (a device to measure the vertical temperature structure of the water column), Nansen water-sampling devices, and a variety of albacore troll-fishing lures.

On February 19, the *John N. Cobb* was christened, and it slid down the Western Boat shipyard launch in Tacoma, Washington. The weather was cold, and there were several inches of snow on the ground. I watched as the ninety-three-foot research vessel moved down the ramp and into the slew. It looked somewhat heavy forward, like a duck paddling through the water. There was still a lot of work left, including finishing the hotel facilities, installing the navigation equipment, and fitting the vessel with winches and fishing equipment. Still, we believed there was time to get the vessel ready for a shakedown cruise in the late winter.

Commissioned in early March, the *Cobb* had a short trial off Cape Flattery, Washington. During the trip, we tried out the various sampling nets and familiarized ourselves with the oceanographic instruments. While testing out the depth-temperature device (bathythermograph or BT), Don and I were chatting as we retrieved it, unaware that it was only a few fathoms below the surface. The BT rose rapidly to the winch, ran up to the end of the boom, snapped off the cable, and fell to the ocean floor. Both Don and I were embarrassed; obviously, the gear on the vessel wasn't the only thing that needed testing. The boss wasn't thrilled by the loss, because he had to buy another new BT before the maiden voyage.

Still, the trial run went well, and the first official cruise was scheduled for late March, when our survey of the shrimp populations inhabiting the inside waters of southeast Alaska would kick off. During the remainder of the year, we had planned an albacore survey and continuation of the shrimp studies in southeast Alaska. It was a demanding schedule for a three-person group, but I was anxious for the work to begin and was looking forward in particular to the deep-water exploration planned for mid-1951 off the Washington coast. Ever since my summer work in Oregon, I had read everything available regarding the fishes in the slope that extended from the continental shelf to the deep ocean seabed off the Pacific Northwest.

In the interim, I started collecting information from members of the Oregon trawl industry on developments in the expanding Pacific Ocean perch fishery. During my last year at the U.W., Dr. Welander and I had helped the Yaquina Fish Company in its effort to market the long-jawed rockfish, which was abundant in the deeper waters off Oregon, as Pacific Ocean perch. (By the way, I earned $25 a month for that work/fun at the U.W.) A similar species living in the Atlantic Ocean was the subject of a large-scale trawl fishery off the New England coast. Landings of Atlantic Ocean perch had risen to well over 150 million pounds and, along with cod and haddock, perch was one of the most important food fish caught by U.S. fishermen.

As mentioned earlier, in early 1950 I had worked on a systematic key to the rockfishes of the Pacific Northwest, with guidance from Dr. Welander. A letter from each of us to the state court led to the eventual approval of the common and market name—Pacific Ocean perch (POP)—for the long-jawed rockfish. Landings of the POP had increased each year since 1947, and the annual catch exceeded one million pounds in 1949. It was a start, but I believed that the species was even more abundant than was known and had the potential to be one of the major bottom fish caught on the West Coast.

Looking for Shrimp

In mid-March 1950, the *Cobb* sailed to Alaska on her maiden voyage. Its mission was to investigate the shrimp and other invertebrate populations in the many inlets, bays, and straits of the southeastern part of the state. The crew included experienced fishermen selected from the commercial fleet operating along the Pacific coast. The mate was an ex-halibut skipper, and the remainder of the crew was experienced in trawl, seine, troll, and other fishing methods. Only the young scientists on board were novices. We had boned up on the taxonomy of the marine fishes and invertebrates of the region, but it was a great leap from identifying a colorless fish in a museum bottle to dealing with a brightly colored marine organism flopping on the deck of a ship. Our lack of experience did not dim our enthusiasm or our commitment to document our findings properly. We just didn't realize the diversity of marine life that we would encounter.

The explorations of the *John N. Cobb* constituted the first significant investigations of the marine life inhabiting the ocean off the Pacific Northwest since the studies of the *Albatross*, the old Bureau of Fisheries research vessel. Conducted between 1889 and 1921, those studies estimated the magnitude of the region's fish and shellfish populations and their potential to supply marine protein to an expanding world population. They provided much information on the distribution of many fish and shellfish species, important taxonomic descriptions, and their general bathymetric distributions (depth).

Our work started just outside Ketchikan, Alaska, in a number of the bays and inlets joining Clarence Strait. At the start we deployed the eight-foot dredge, which, although functional, did not catch any significant quantities of shrimp. It did, however, capture a smattering of several species of fish and some invertebrates, and it was obvious that populations of pink, side-striped, spot, and coon-striped shrimp inhabited the region of exploration. Northern shrimp had a strange life history, developing as males over the first few years of their lives, and then changing to females for the last years of their life. Thus, all of the small shrimp were males and the larger ones were females. Fisheries developed for the northern shrimp in several areas of the world were based largely on harvests of the females using a beam trawl—a sled with a net between two runners that slid along the bottom. Members of the genus of northern shrimp (*Pandalus*) ranged from small pink shrimp, which ran thirty to seventy to the pound, to the large spots that could go several to the pound. The latter were normally found in rocky regions.

After testing the gear in Clarence Strait and other bays and inlets in the area, we moved down to Coral Inlet, just south of Ketchikan. Local fishermen had told us that the inlet had good quantities of shrimp and that it probably was a good place to test our gear. We ran up the inlet, sounding out the bottom. There were ample soft-bottom areas, but there also seemed to be a reef or rocks in the middle part of the inlet. To begin our test, we chose a section about a mile long that appeared to have a soft sand or mud bottom. The gear shot out clean, and the net dropped to the bottom, which was around 250 to 400 feet below. After towing for about a half hour, we brought the net to the surface and were pleasantly surprised with a catch of several thousand pounds of mixed shrimp and bottom fish.

I put on my boots and waded into the pile, sorting the fish and shrimp into different bins so that we could identify the species and estimate their catch. I tried to put my boot on the deck to secure my footing but made the mistake of stepping on the tail of a hundred-pound halibut that, with little trouble, tossed me onto the deck like an unwelcome guest. After we cleared the pile, we estimated that we had caught a ton of mixed side-stripe and pink shrimp and several hundred pounds of flounder, smelt, blennies, sculpins, and other fish. I spent the rest of the afternoon attempting to classify what we had caught and rubbing my arm, which had been bruised by the halibut's toss. I was more careful following that event but was surprised, nevertheless, several days later when a giant octopus (about twelve feet across) pushed his way to the top of the fish and shrimp pile, and moved quickly toward the ship's scuppers. Several of us grabbed at his arms, but he seemed to have little trouble freeing himself and squeezed his twelve-inch head, followed by his legs, through the scuppers and slid back into the sea.

Over the next several weeks, we moved through Beam Canal down to Burrows Inlet without any significant results. In early May, we pulled anchor and headed south for Seattle. Running down the Inside Passage was beautiful and, for the most part, calm. Through much of British Columbia, the mountains dropped steeply into a network of inlets and passages. Evergreens hugged the mountainsides right down to the high-tide level and the lower branches formed a canopy over the water's edge. The passage was largely protected from the open ocean, although a few spots were exposed to the offshore weather. It was a little rough from south of Ketchikan and the Queen Charlotte Islands in Hecate Strait until we sailed in behind Vancouver Island. From there to Seattle, we were protected.

Although it rained or snowed almost every day, the scenery was spectacular. High cliffs dropped hundreds of feet straight into the inlets, and the *Cobb* could maneuver within several yards of the shoreline. Long, narrow fjords penetrated well inland toward the mountains of Canada, and there were numerous small bays. The mist and fog frequently dropped to a hundred feet or less, hiding all but the base of the mountains that plunged into the water. At such times, it seemed like an enchanted and lonely world that extended only from the water to the low clouds or fog.

It was an interesting trip, and we learned a great deal about crimping, exploring, data logging, etc. As a result, we were better prepared for the next survey. We had hired a shrimp fisherman to help us with our gear rigging, acquire a beam trawl, and re-rig our otter trawls so that they would hug the bottom more closely. During the winter-spring exploration, we had not found any great concentration of shrimp in the waters around Ketchikan, with the exception of some good pot takes of large spot shrimp. We had caught, however, several species of fish that appeared to indicate new areas of distribution, but I could not be sure until we returned to Seattle. Once there, I would check the literature and confirm my identifications at the Fisheries school.

On the way south, I went over all my records and stored the samples of fish that I would identify with the help of the U.W. faculty. We checked the oceanographic equipment that we had used, played a good deal of pinochle, and scavenged the boat for reading material. It took about two days to run to Seattle and early on the third day we were on the south side of Whidbey Island. It wasn't long before we entered the locks and made our way to our berth, which was at the foot of Stone Way on the north side of the canal.

A Place to Live

It was good to be home and back with Ruby and Bob. My son had changed during my two-month absence. He could walk holding onto the chairs and couch. For the most part, he was a very happy child who loved his teeter-babe, a chair that he could make jump up and down. He worked it so hard that we had to brace it with a twenty-pound dictionary so he wouldn't go over backwards. It had taken me two years to pay off that dictionary at a rate of $2 a month. It looked impressive in our small apartment. My brother was finishing up his work at the U.W., and his family would head back to San Diego in June. Ruby and I had hoped they would stay in Seattle, but then we remembered how much we had hated the region for the first couple of years. Frank would receive his master's degree and probably work for the Tuna Commission in San Diego or the

U.S. Bureau of Commercial Fisheries, which had a big lab on the Scripps campus in La Jolla.

After Frank and his family left, Ruby and I felt rather lonely because we had enjoyed their presence in Seattle. Furthermore, Ralph Malloy had completed his engineering studies and accepted a job in Los Angeles. Suddenly, our family and friends had departed, leaving us alone in the great northwest. In addition, a number of friends we had made in Cedar River Park began moving out to new homes as they completed school or their economic status improved. Ruby and I took stock and decided that it was time to make plans for the future. We began searching for a lot where we could build a home.

We first looked around Renton, but we weren't thrilled with the housing and schools in the area, so we searched the north end of Seattle, where a number of my classmates had moved. We saw several lots and homes we liked, but everything that looked good was out of our economic range. Don Powell suggested we look in the Seahurst and Normandy Park area in the south end, noting that he knew a builder who owned some lots there. I knew the area because Don's brother lived in Normandy Park, which gave us access to the private beach, and we had gone there to dig clams.

Ruby and I set off on a Saturday afternoon to explore the area. We went west to Burien, about twelve miles south of Seattle, and then to Seahurst, which was on the hills west of Burien and near to Puget Sound. We found a nice lot several blocks off a main road that had madroñas and fir trees and several gray squirrels jumping from tree to tree. There were pleasant homes along the street, and it was only a few blocks to the crest of the ridge that overlooked the Sound. Our hearts had been set on finding a lot with a view, but they were all out of our financial reach. We settled on a lot that was a mile or so west of Burien in a wooded area near schools and the bus lines.

We returned to Cedar River Park excited about our decision, planning to return the next day and ask a neighbor about the asking price. On Saturday, the lot had seemed so easy to get to and within easy walking distance of Burien. But one day later, we couldn't find it! We went up and down every street in the area, but it was like the earth had opened up and swallowed our prize. After about three hours, we drove back to Burien and stopped at a drugstore for a shake and sandwich. We were dumbfounded. We went back for one more search but could not find the lot. For some reason, that lot was not part of our future.

Several weeks later, just before I was scheduled to leave on the *Cobb* for our summer exploration, we looked at the Normandy Park area, a mile or two south of Burien. Don had gotten in touch with his builder friend, Art Leonard, who said that he would show us some lots in the area. At that time, one paved road and several graveled roads served Normandy Park. There were a few nice large brick homes along the waterfront near the private beach. A few relatively new homes sat above the Sound. The neighborhood was attractive because there were many large, older fir, cedar, dogwood and hemlock trees. Several small creeks tumbled through the area, and there was an abundant supply of wild blackberry bushes. Normandy Park residents also had rights to a beautiful beach on the Sound—our little bit of California.

Several of the lots that Art showed us would have made excellent home sites. However, just like in Seahurst, the ones that had a good view of Puget Sound were too costly, ranging between $700 and $1,200. We decided on a large lot near the corner of Normandy Park Road and Brittany Drive. It cost us $400, which we felt was a stretch considering the cash we would need to build our dream home. Ruby and I sat down with Art and showed him some preliminary plans that we had sketched out. He pulled out several plans he had been using, one of which was very similar to ours. We negotiated for several days on the price and changes we wanted, and then committed to the construction of a three-bedroom, one-bath home in the fall. Because we were children of the Great Depression, when one- and two-bedroom houses were the mainstay of the lower-middle class, our planned home seemed unnecessarily large. But the partial view of the Sound and the beach access just two blocks away overrode our economic concerns.

Chasing the Albacore

As soon as I returned to the office, I started a report of the Alaska venture. Don and I put together a long-term plan for:

1. the exploration of the shrimp stocks in Alaska's inside waters and offshore;
2. ground fish trawling from Oregon to the Bering Sea; and

3. investigations of the open ocean (pelagic) fishes off Oregon, Washington, British Columbia, and Alaska.

Scheduled for mid-June, the first pelagic exploration was designed to investigate the migrations of albacore tuna. The species was distributed over significant portions of the major oceans of the world, but its migration patterns were largely unknown. Stocks of albacore traditionally showed up off the coasts of Oregon, Washington, and British Columbia during the months of July, August, and September. The rest of the year, they were in some other part of the Pacific. Thus, the fishing season in the Pacific Northwest was relatively short and frequently occurred seventy-five to two hundred miles offshore.

There was a general belief that the albacore migrated into the northwest Pacific from California, moving north as the surface waters warmed during summer months. Don and I scanned the weather and sea surface temperature patterns in the north Pacific region in previous years. We found that the warming of surface waters appeared to occur along a temperature (isotherm) front moving toward the coast in the summer. Our conclusion was that the waters warmed from east to west and pushed into coastal areas, first in California, and later into the more northern waters. If the albacore were moving in behind warm water, they would be found along an extensive area in a central stock inhabiting at least the north Pacific, not just in a line from north to south.

After surveying the scientific literature related to albacore in the north Pacific, we formulated the goals and objectives of the first major offshore pelagic fish investigation. They included a study of the distribution of albacore off the Pacific Northwest and their behavior in relation to ocean temperatures, salinity, ocean currents, and available food. The background papers we read suggested that surface water temperature was the ecological factor that most likely influenced the availability of tuna in the area under survey. As our plan evolved, we decided to start the survey off the southern Oregon coast, where the warmer water was between four hundred and five hundred miles to the west. The *Cobb* would move offshore and intercept the warmer waters, and then move back and forth across the 58-degree line (isotherm).

For years, fishermen had referred to the clear, deep-blue warm water, above 58 degrees, as "tuna water," but this was based on experience and not on demonstrated scientific fact. If the hypothesis was correct, we expected to find the albacore across the entire "warm water" front. We plotted a cruise schedule that would take the *Cobb* from Cape Flattery to several hundred miles west of Cape Blanco, Oregon. Once tuna waters were encountered, the vessel track line would be adjusted to the north as the surface waters warmed along the Oregon, Washington, and British Columbia coasts. The cruise was set to start in mid-June and continue to late September. Don was the chief scientist for the early part of the trip, and I would remain at the office and relieve him in mid-August. The plan called for a daily radio schedule and Don was to report to the main office at Montlake in Seattle at 11:00 each morning.

To support the cruise, we bought oceanographic equipment from a Seattle supply house, including accurate thermometers to measure surface temperature and BTs to measure the vertical temperature in the ocean. A small piece of smoked glass was removed from the BT after being cast to a depth of about a hundred meters and subsequently read against a temperature grid. The marked smoked glass provided a measure of the thickness of the warm surface water. Below the surface layer at the thermocline, the waters became colder and deeper. The vessel was fitted with large beams, or outrigger poles, that were tapered on the upper ends. The poles were secured to the base of the boat, one on each side, and rigged so that they could be dropped from a vertical position to a one that was level to the guardrail, adjacent to the side of the ship. The outer ends of the poles tilted upward so that they would not dip under the water as the vessel rolled. Three tuna lines were tied to each outrigger pole, and two were designed to fish off the stern, one on each side. In addition to the trolling method of sampling for albacore, we also would use several sections of monofilament nylon drift nets to harvest albacore and other pelagic fauna of the region.

The *Cobb* sailed from its berth in the Lake Washington Ship Canal on June 12. From Cape Flattery, it ran in a southwesterly direction toward a position well off the southern Oregon coast, where we expected to find favorable ocean temperatures. The surface temperatures remained cold (52 to 54 degrees) from Cape Flattery to south of the Columbia River. Don noted that life on the little ninety-three-foot *Cobb* was rather miserable. The ship spent most of the time bucking into southeasterly seas and, at night, the vessel rolled continuously, making sleep a real challenge. Once we were south of the Columbia, the water gradually began to warm. On the morning of June 17, Don noted that the surface temperature was close to 57 degrees, the warmer surface water extended down some fifty feet, and the weather had moderated somewhat.

Elliot, the head of operations, and I were both anxious for the next morning's report. Was our hypothesis correct? We both knew that temperature was no guarantee that the albacore would be present. The fast-swimming tunas might live in the relatively warm (for the Pacific Northwest) water, but they would be continuously on the prowl and searching for food. We would be lucky if both food and the right temperature were found at our designated point of interception. The next morning I turned on the radio receiver and transmitter about fifteen minutes ahead of schedule in the event the *Cobb* called in early. But for twenty minutes all I heard was the crackling of static. Around 11:05 a.m., we could hear Don, but the signal was weak and the transmissions difficult to understand.

After a minute or so, however, Don's signal became clearer. He spoke with excitement in his voice. "We couldn't have planned it better," he said. "We ran into 58-degree waters just before dark yesterday. This morning [June 18, 1950], just after dawn, we took our first albacore. We were 480 miles offshore and only an hour or so into 'tuna waters.' The fish hit the outrigger lines generally in pairs. They ranged from twelve to eighteen pounds." Of course, we were ecstatic, but Don would have to follow the warm front inshore and to the north, sampling first in the warm water and then in the cold water, to prove our theory. Over the next several weeks, the vessel followed the warm water as it moved inshore and up the Oregon coast. Small clusters of albacore were taken at a number of stations along the 58-degree front. It was becoming evident that the albacore inhabited the warm water across a broad sector of the north Pacific and migrated seasonally into the waters off Oregon and Washington.

As the vessel moved north, the crew made BT casts and took surface temperature readings and salinity samples throughout the day. Stomach samples offered evidence of the foraging opportunities available to the albacore, and net tows showed what types and quantities of zooplankton they consumed. At night, depending on the weather, the scientists on board could collect samples by -dipping a net under a night-light held on a beam over the water. The monofilament gill nets were set several times, but did not yield any albacore. There were always a number of blue sharks up to eight-feet long, a few pomfret, and an occasional ocean sunfish in the nets.

As we discarded the catch from the nets, albatross would surround the boat looking for a free meal, frequently pecking at the floating discards. If there was a large ocean sunfish in the area, it would swim under the albatross, which would peck at its sides. Interestingly, sunfish swim on their side with one eye looking toward heaven and the other down into the depths of the sea. The albatross and sunfish have developed a symbiotic relationship. The sunfish present themselves to the flock of albatross on the water and the albatross make meals out of the parasites on the skin of the sunfish.

A Surprise in the Sea

Throughout July, the *Cobb* sailed along the Oregon coast and by late in the month was north of the Columbia River. By then it was evident that the albacore followed the warm oceanic surface water shoreward and to the north, as the plume of warm water wedged in along the coast. However, they generally stayed clear of the continental shelf, which was cooled by the upwelling of deeper cold water. Heavy feeding seemed to occur along the front between the cooler coastal waters and the offshore oceanic regime. On August 1, while the *Cobb* was investigating the distribution and movements of a large albacore school off the southern Washington coast, the "flashing" echo sounder began showing a bottom at a depth of 860 fathoms, almost a mile down. The problem was that the navigation chart showed a depth of 1,200 fathoms, or 7,200 feet.

The vessel was maintained on a 353-degree magnetic course to the north, and by the time Don contacted the office for the 11:00 a.m. check in, the crew was recording depths on the paper-recording echo sounder. They had sailed up the slope of the seafloor to a depth of only twenty-two fathoms (132 feet). After reaching the minimum depth, the vessel reversed its track and retraced the echogram and then took readings from a course of about 270 degrees magnetic. Don asked me to examine several of the overlapping charts to check the bottom topography of the region. After about twenty minutes, we called the *Cobb* and told Don that "everything we have to look at shows the depth at about 1,200 fathoms, but there are a number of sea mountains to the north of you that run up into the gulf of Alaska." Don laughed, and answered, "Well, we have just discovered another one." Then he added, "The peak is shallow enough that the large swells feel the bottom and get steeper around the mountain peak."

The discovery of the *Cobb* seamount was fortuitous, but it was one of the great moments in the vessel's explorations. Pure luck, and maintenance of the vessel's operating instrumentation, had led to the finding of the seamount, which rose some 7,200 feet from the ocean floor almost to the surface. As the paper-recording echo sounder was limited to a depth of 400 fathoms (2,400 feet), it was difficult to obtain details about the seamount's bottom characteristics. Don estimated that there were close to thirty-five square miles of bottom above forty fathoms. "The sides of the peak are relatively steep down to 110 fathoms [660 feet]," he noted, "and then there appears to be some terracing."

Although it took away from the cruise mission, we (decisions of this type were made by the chief scientist and the skipper) decided to take some bottom samples and attempt a single set with long-line gear to get an idea of the types of bottom fish that might inhabit the seamount. It was difficult to get anything in the bottom grab, and most attempts came up empty, but several successful grabs picked up coarse calcareous sand. The top of the seamount was assumed to be largely rocky, with a few bottom sediments collected along the ledges. The single set of long-line gear (106 hooks hung from a heavier ground line) resulted in a catch of sixty-five red snappers (a species of rockfish unlike the snappers in the Gulf of Mexico), averaging about fifteen pounds.

I was in luck because we decided to continue the albacore investigation and set aside a period during the last half of the cruise for a further study of the mountain's fauna. We planned to check the sea life using bottom long-line gear and trawls, if possible. We would have a better idea of what sampling gear to use when we did a more detailed survey of the seamount topography. The underwater mountain, which was 273 miles southwest from Cape Flattery, Washington, and 290 miles north of Astoria, Oregon, was later named the Cobb Seamount after the research vessel.

In early August, I joined the *Cobb*. By that time, the warm water was moving toward Vancouver, and it was difficult to monitor and keep abreast of the advancing tuna water. By mid-August, water warmer than 58 degrees extended along the Oregon and Washington coasts, fifty to eighty miles offshore. Within a week, a wedge of warm water extended north to Dixon Entrance, off the north end of Vancouver Island. Over the next several days, we continued moving north along the Queen Charlotte Islands, catching a smattering of albacore on the troll lines. The weather was excellent, and the eastern Pacific looked like a giant lake, with only small swells moving under a flat sea.

It seemed like a good opportunity to watch the action from the crow's nest some forty feet above the deck. The surface temperature was just above 58 degrees and the warm surface water was close to sixty feet deep. It was late afternoon, and we all hoped for a good evening bite. We soon had hits on our two stern lines and, within minutes, had a half dozen fish at the same time. I could see a school of albacore off the stern. When a fish decided to strike, it shot out of the school like an arrow from a bow and hit the bait while moving at high speed. It was difficult to judge the swimming speed, but I believe it exceeded twenty-five miles an hour. When it had taken the lure, the fish immediately swerved off to one side and headed for deeper water. After a hit, the trolling lines would pull taut, and the rubber shocks, secured into each trolling line, would stretch to their limit. We then hauled in the fish by hand and brought them on deck.

Almost all of the albacore would strike the lures from behind or from the side, but on occasion one would run up past the lures, turn and strike the lure head on, like a fighter pilot attacking a bomber. Twice during the survey, a fish hit the lure so hard that it tore out its jawbones, which remained on the lure and were brought aboard by the fisherman hauling in the trolling line. The speed of the boat combined with the speed of the fish accomplished what once seemed unbelievable. A true fish story!

Before sunset we had caught and measured a number of fish, and also examined their stomachs, many of which were filled with juvenile rockfish. The weather remained excellent, which gave us the chance to set out our monofilament gill nets. Earlier in the trip, the weather limited the number of gillnet sets and we caught only the occasional albacore. The most common catch were blue sharks, which seemed to be ubiquitous inhabitants of the offshore waters. Just before sunset, we set the nets west of Tasu Sound. To our delight, when we hauled in the nets at sunup the next morning, we had hit the proverbial jackpot. We pulled in more than 150 albacore along with a surprising catch of about a hundred pomfret and a few blue sharks. As far as I know, it was the first substantial catch of albacore taken with gill nets in the northeast Pacific ever recorded.

I spent the better part of the day processing the fish, taking length measures, weights, and stomach samples and checking the sex and maturity of the fish. All of the albacore we caught were immature; the larger, sexually mature animals must have lived and spawned in some other region of the Pacific. The coastal fisheries off Oregon, Washington, and British Columbia all depended on the migration of hungry "youngsters" searching for productive zooplankton, small fishes, squid, etc. These food items were relatively abundant in the offshore waters of the Pacific Northwest during the summer months.

During the day, we traveled north, hoping to find other schools of albacore in the warmer plume, which by late August had formed a wedge along the coast up to southeast Alaska. We took a few albacore in the afternoon, then the winds began to pick up, and by late evening the sea had turned nasty. The skipper decided to give the crew a break and headed back south along Gram Island (the northern island of the Queen Charlottes) and then east into Tasu Sound. The vessel headed straight into the coastline, where the steep slopes of the mountains cascaded into the sea. As we drew near, we could see a small opening running between towering cliffs on each side. The entrance to the Sound looked only several hundred yards across, but beyond that it opened into an expansive inland waterway lined by large firs.

During the night, the storm blew over, and we headed back to sea at sunup, moving offshore to find tuna water. But the storm apparently had mixed the warm surface waters with the underlying colder water, and the surface temperatures had dropped to about 56 degrees. We found no sign of albacore, which had been so abundant the day before. They were just like "snow birds"—the first sign of cold weather and off they went to California or Florida. We surmised that the albacore schools were heading south and west, searching for a more comfortable environment.

In September, we moved back to the south and continued plotting the surface temperatures. Even when we found the warm water edge, however, we caught only a few albacore. They seemed to have spread over a broad area, and the mechanisms that had led to their high concentrations in early summer were less prevalent. Thus, in mid-September we headed back to the seamount that the *Cobb* had discovered in early August. The fishermen hauled out the long-line gear and readied it for deployment. As we ran offshore toward the seamount's recorded position, I wondered about the accuracy of our Loran A (a long-range navigational device developed during World War II) recording and if we would have any difficulty relocating the mountain. But on September 14 we arrived at the expected area and immediately saw that the bottom was getting shallower. We slowly approached the peak, and the recording fathometer showed the bottom rising steeply toward the surface. The peak of the seamount was only eleven fathoms below us, even shallower than we had first thought.

Over the next two days we continued to chart the mountain's bottom topography. We considered the bottom far too irregular to set down a trawl, so we explored the area's sea life by setting out several fifty-fathom sections (also called skates) of long-line gear. The gear set out between forty and seventy fathoms (240 to 420 feet) yielded spectacular catches of several species of rockfish, as well as sole, skates, and an occasional halibut. Of course, we also caught the ocean nomads—blue sharks. One long line section had a rockfish on almost every one of its 106 hooks. Two species dominated the catch: local red snappers, which were a deep orange-red, and vermilion rockfish, which were a bright red color. We sampled the grounds for several more days at depths between 40 and 110 fathoms. Fish life was abundant on the mountain peak several hundred miles offshore, but the bottom traces showed that the ocean floor there was not level and mostly hard, suggesting that commercial fishing, if it were to develop, would be limited to hook-and-line gear.

We returned from the seamount to Seattle in the third week of September loaded with samples of fish that needed further identification. We also brought samples of sea water to be evaluated for salinity, thousands of records about the sea surface, and BT slides of the vertical temperature patterns. It took several of us almost two years to collate and evaluate the data. The findings, which sharply altered our perception of the behavior of young albacore migrating into the Pacific Northwest, were published in 1952, along with a description of the seamount and its sea life. It was a good beginning.

Building a Home

Once I was back in Seattle, Ruby and I decided to complete the plans for our new home and start the construction. Ruby was pregnant, and the baby was expected in April or May. We hoped to move into our new home in spring 1951 before Ruby gave birth. Art Leonard, the builder, persuaded us to use his drawings rather than our plans because it would save us some money. The total cost of the lot and the house came to $12,400, which would place us on a very lean budget until I got a raise or two. We put up our $4,000 in cash and took out an $8,000 loan with a $64 a month payment. This covered our interest, payment on the principal, home insurance, and property taxes. Today, our total outlay would be less than the cost of a low-priced new car, but we felt that we had shackled ourselves to a debt that would take most of our productive lives to pay off. With an income of $240 a month, one child to raise and one on the way, and no furniture of significance, we were not sure how we would keep our heads above water.

Construction on our new home in Normandy Park began in early November. It was raining a good deal as the lot was cleared for the foundation, and the site looked rather messy. Most of the firs had been taken down, but a large madroña and an alder tree remained. We also managed to save one small fir, about six feet tall, which would be just behind the planned house. We checked on the construction each weekend during the late fall and winter. As the house took form, we became more enthusiastic and impatient for its completion. Bobby referred to the house as his "dream home."

My visits to the new home were cut short by a cruise to southeast Alaska. In the late fall and winter of 1950–51, we continued our survey of shrimp without any significant results. During my absence, Ruby moved into our new home with help from my father, who was working in Bremerton at the time, and the wonderful Renton folks. She and Bobby established our new residence at 17916 Brittany Dr. S.W., Seattle, Washington. We had little extra money for furniture and drapes, so Ruby made temporary drapes and curtains out of bed sheets and made do with almost no front-room furniture. We used a large apple box for an end table and ate lots of macaroni and cheese and spaghetti. We managed to pay our bills and still have a Friday night out with hamburgers and fries at Dan's, a precursor of McDonald's, for less than a buck and a half.

Bob thought it was great because there was a miniature fair with a merry-go-round and other rides across the street. That cost us an additional twenty cents. Life was good.

Joining the Published

Between trips, I worked hard in Seattle identifying the unknown species of fish taken in Alaska and finishing my work on a key to the rockfishes inhabiting the Washington and Oregon coasts. With Dr. Welander's help, I completed the study and submitted it for publication in early June. I was thrilled when a letter arrived from the editors of *Copeia*, a scientific journal focused on fisheries and invertebrate zoology, accepting my paper. It would constitute my first contribution to scientific literature. The paper had been stimulated by my experiences working off the Oregon coast the previous summer.

The trips off the Washington coast allowed me to continue to investigate the rockfish species. Since more than fifty different species inhabited the waters of the Pacific Northwest, the rockfish were difficult to identify. Many species were considered rare. In reality, most were relatively abundant, but there had been little scientific investigation of their distribution and abundance, particularly in depths greater than one hundred fathoms (six hundred feet). They were known, however, to a number of the more adventurous fishermen, who gave them a variety of popular names.

Trawl Surveys

Although I enjoyed the shrimp-survey work, I looked forward to the summer trawl survey scheduled for the Washington coast. Ever since my Oregon experiences, I had wanted to return to the continental slope waters of the north Pacific. To prepare, Don and I poured over the navigational charts. We were particularly interested in the submarine topography off the west coast of Washington and the southwest coast of Vancouver, Canada; this was the region we selected for the first trawl survey. Washington trawl fishermen fished heavily in the shelf region from south of Cape Flattery to Destruction Island (the bread line), and Swiftsure Bank from north of Cape Flattery to southwest of Vancouver Island. They avoided, however, the rocky bottom area at the south end of the large submarine peninsula that extended southwest from Vancouver

Island (La Peruse Bank), as well as the deeper waters off its western slope. Thus, the bank seemed like a logical point of departure for our expedition. It was avoided by trawl fishers, yet close to many Puget Sound ports—an area that could be easily accessed by fishermen if the results were good.

We discussed our plan with several fishermen from Oregon and Einar Peterson, Joe Chriscole, Johnny Courage, and Art Angle, trawl skippers from Puget Sound. They all agreed that the area selected was ripe for exploration, but Einar noted that it was a "rock pile" and that we'd need a lot of trawl nets on board. "You're sure to shred or lose a lot of gear," he added. We began to finalize a plan to investigate the demersal (bottom-dwelling) fish resources along the outer edge of La Peruse Bank. The survey would include the deep waters in the submarine canyon that extended southwest from the mouth of the Strait of Juan De Fuca, cutting across and interrupting the continental shelf between Washington and the south end of Vancouver Island.

We set sail in late June 1951, proceeding up the Sound to Port Angeles, swinging west out of the strait, past Sekiu, and then to Neah Bay, where we anchored for the night. As we passed the Coast Guard base at the harbor entrance, early memories of my life on Tatoosh Island surfaced, and I was anxious to see the island, which was just a couple of miles to the west. It was cold and windy, but I went ashore with several members of the crew to visit the village. There was an old movie house, several small restaurants, and sport fishing tackle-and-bait buildings along the bay side. At the upper end of the bay, a dock and fish-processing plant serviced the fleet of salmon trollers, draggers, and other small boats that fished off the northern Washington coast. I was surprised to find that Warshaws Trading Post was still in the village. It had been in existence when we lived on the island in the late 1920s, but had obviously been rebuilt. We purchased a few candy bars and magazines and looked over the store's Indian art. Then we went to the movie house and saw a John Wayne movie, which featured heroic battles against the Indians. Smack in the middle of the cowboy and Indian war, a mouse scampered across the stage and got the loudest cheer of the evening. The movie was poor choice for the Makah reservation. We returned to the boat early and played a few hands of cribbage before settling in for the night.

In the morning, we pulled anchor around 5:30 a.m. and moved slowly out of Neah Bay. There had been a good number of trollers and small trawlers in the bay, but most normally left just before or at sunup. As we rounded Wadda Island and headed out of the strait, I could see the circular flash from the Tatoosh Island. As we passed the island, I used binoculars to scan my old home. Most of the buildings were in ruins and had fallen to the ground. The radio station and lighthouse had been abandoned after the light was automated and better navigational equipment became available to ships. Still, the island looked much the same, and the bird populations on the lower rock outcroppings were in abundance. Two large California sea lions sunbathed on one of the exposed reefs, and two small boys ran along the top of the island, searching for mushrooms, as free as the birds and sea lions on the rocks. They faded as my mind switched to the job at hand.

We made our way out to the submarine canyon that ran southwest from the strait. We had laid out a survey plan, which called for exploration of the deeper canyons and depressions west and south of the canyon. We began our explorations in a submarine valley that was about 650 to 700 feet deep, which we located by using the echo sounder and taking radar bearings off Cape Flattery. After finding the valley, we "shot" the trawl. I marked the start bearings on the chart and logged information on position, depth, starting time, and other relevant information on a tow form that we had designed. We took a sample of the seabed sediment and salinity and temperature readings at the surface and on the seabed. We also began an echo-sounder paper recording of the bottom topography. The bottom appeared to be mud and sand, and there was no evidence of rock outcroppings. We started the tow at the edge of the depression and headed to the north and down the slope of the valley. We set the length of the tow for one hour, which would allow us to sample most of the valley bottom area.

I was somewhat excited, anticipating that there might be large quantities of deepwater rockfish, as there were off the submarine canyon off the Columbia River. When we had completed the one-hour tow, the skipper gave the signal to haul the gear. I watched from the galley door, hoping to see the trawl explode to the surface filled with various species of rockfish. After several minutes the trawl doors were secured to the aft davits, and the trawl came to the surface forty or so yards behind the vessel. However, it contained a small island of fish; I guessed that it was not more than a ton or two. When the cod end of the net was brought on board, however, there were more fish than I had expected. We had taken a mixture of flounders, black cod, some lingcod, and perhaps a ton of mixed rockfish. In total, we brought aboard one lift weighing about three tons, which contained a diversity of species. Most of the flounder species and the round fish other than the rockfish, I identified and recorded on the tow form in a few minutes. The rockfish required considerable examination and checking against taxonomic keys. As I recall, the tow included about twelve rockfish species, including perhaps a thousand pounds of Pacific Ocean perch. It was an interesting start, but did not meet my expectations.

We continued working down the Juan De Fuca canyon, exploring the bottom area between seven hundred and a thousand feet. The results were about the same, and my hopes of finding large concentrations of Pacific Ocean perch off the northwest Washington coast began to dwindle. One tow did have a good showing of large side-striped shrimp, a species known to be abundant in certain areas of Alaska. Considering that we were using 3.5-inch cod-end mesh, catching side-stripe shrimp was a promising development, but one that we would have to explore in the future using a smaller mesh shrimp net. Although the catches were modest, we had recorded new distribution records for several species of rockfish. I was beginning to feel more comfortable identifying the diverse rockfish family.

After several days of work along the continental slope between Cape Flattery and Destruction Island, we discussed attempting some tows off the Spit, which ran southwest from southern Vancouver Island. The fishermen on the crew who were familiar with the area felt the Spit was useless because of its many large boulders and rock ledges. We decided, however, to give it a try. We turned and ran almost directly offshore across the submarine canyon to the Spit's westward edge and to a depth slightly over 600 feet. From there, the continental slope fell steeply down to the Cascade Plain, a portion of the sea floor that ran from the top of the continental shelf to a depth of about 1,200 fathoms (7,400 feet) westward to a seamount region several hundred miles offshore. On the way to our destination, the captain ordered the crew to remove the trawl net and replace it with a large section of heavy chain. Otherwise, he noted, we could lose a lot of nets quickly if we got hung up on large rocks or other seabed obstructions. Our plan was to explore the region with the chain link and, if we found regions that we could pass through, to plot and subsequently deploy the net in those areas.

We began our explorations of the outer Spit (about forty miles west of Cape Flattery) just after noon. First, we swept over the areas of interest, checking the bathymetry and bottom topography using a paper-recording echo sounder. The echo sounder traces suggested there was an area with a relatively smooth bottom sloping seaward, just to the north of the tip of the Spit. The area was not very large, perhaps several miles long, at depths ranging from about 750 to 900 feet. The skipper and I discussed whether we should chance putting a trawl down. As I recall, he said, "Well, we're here to explore, so let's get on with the job." The crew went to work removing the chain link and putting the trawl back on the ends of the dandy-line gear leading from the back of the trawl doors to the net. When the job was completed, we selected a starting point at a depth of about 800 feet at the north end of what we perceived to be the clear area, set the doors and net out, spooled off about 2,000 feet of cable, and began our first Spit tow.

We were all a little anxious because there was still a fair chance that the cable could tow over an area that would trap the net, and that the position of the net on the seabed would not duplicate the area we had swept with the chain link. I kept a close watch on the clock—ten minutes, twenty minutes, and no sign of trouble. As we approached thirty minutes, I began to feel that we would make it. The last few minutes of the tow seemed like an eternity, but we lasted an hour, covering slightly less than three miles of seabed.

Late in the afternoon, the skipper gave the word to haul the net. We all waited patiently as the winches spooled in the several thousand feet of 5/8-inch wire cable. The entire crew was watching as the net approached the surface, either from the galley windows or from their assigned deck stations. Suddenly, we saw air (gas) bubbles rushing to the sea surface and within seconds we could see a large island of red rockfish forty or fifty yards off the stern. The crew judged the catch to weigh 12,000 to 15,000 pounds. The crew raised the net along the starboard side and lifted the catch lifted in "splits" that weighed 2,500 to 3,500 pounds each. As there was a fair swell running in on us, the skipper and crew had to handle the splits with care, making sure they didn't swing out of control across the stern. When the catch was onboard, we filled almost every checkerboard bin on deck. We had rockfish up to our hind ends! A variety of brilliant red and orange rockfish, with diverse colors, patterns, and shades, swamped the deck. The job of documenting the catch by species, sizes, and weights would take the remainder of the day and well into the night.

The catch validated my embryonic hypothesis that the Pacific Ocean perch was one of the dominant (if not the dominant) rockfish inhabiting the greater arc of the north Pacific Ocean. Its counterpart on the East Coast, the Atlantic Ocean perch, rapidly was becoming the dominant fish species eaten in the United States. Each year, fishermen landed several hundred million pounds a year of Atlantic perch, comprising several species. If my hunch was right, this expedition and subsequent explorations would find the ocean perch distributed along the upper continental shelf, from Oregon north to the Bering Sea, perhaps even westward along the Aleutians and south toward Japan.

The following morning, we carefully expanded our survey tows, first up the slope to depths from 600 feet and then down to depths of about 1,500 feet. Almost every tow, with the exception of two in which the net was severely torn, yielded four to eight tons of rockfish, ranging from juveniles to adults. Although the perch seldom exceeded a couple of pounds, some of the larger red rockfish species were well over twenty pounds. The diversity of the genus was amazing, with fifteen to twenty species in a tow, mostly colored in brilliant shades of red. With each tow, excellent catches of rockfish surfaced, mostly Pacific Ocean perch but also a smattering of black cod, Dover sole, rex sole, and a variety of sculpins, blennies, and smelts. My conviction that a great variety and abundance of rockfish inhabited the continental slope continued to grow.

Oregon fishermen's discovery of the perch on the slope had revealed the species potential, and our explorations simply documented what they had already believed for some time. We continued working the Spit for the better part of a week and located close to a dozen productive tow areas. One tow on top of the Spit yielded a fair quantity of petrale sole, the prime target species of the local trawl fishery. But the success did not come without cost. We shredded a number of nets during the survey and spent considerable time mending those that could be repaired.

When we were surveying the Spit, I heard the trawl boat skippers working off Astoria talking on the radio and called several of them to chat about our experiences. I had just signed off with the *Harold A.*, when Einar Pederson called to congratulate us on our "Spit work." He requested bearings and Loran data lines, which identified the position of our successful tows. The word quickly spread about the catches and tow areas that were free of obstruction. Over the years, the offshore tow region would become known as the "*Cobb* Spot."

In the following weeks, we made our way south, fishing the continental slope between 600 and 1,500 feet and finding quantities of rockfish and black cod concentrated in the deeper waters (900 to 1,200 feet). By the time the trip was over, I was thoroughly convinced that the Pacific Ocean perch and similar red rockfish were a major fishery resource in the slope waters of the northeast Pacific. But to prove it would require further surveys of the waters off northern Oregon, British Columbia, and Alaska. The *Cobb* sailed back to Seattle after several months of work, loaded with specimens of various rockfish and other species needing identification. As we moved into the strait, I took another long and thoughtful look at the island that I had lived on as a young boy. Then I returned to the galley and joined a game of cribbage to help pass the time.

The trawl cruise had burned up much of the summer, which isn't very long in Seattle, but the family managed several nice trips to the campgrounds just east of Mount Rainier. We had fallen in love with the large canyons and beautiful streams that drained the east slope of the mountains and joined to form the Cowlitz River, which flowed south into the Columbia River. We rolled out a few cheap sleeping bags that Ruby had purchased, built a nice campfire, and prayed that it wouldn't rain. We made a bed in the back of the Chevy, hung old cloth diapers over the windows to shut out the morning sun, and put Bob down for the night. In the mornings, I grabbed a pole and ran up to Panther Creek, hiked down the stream to where it joined the Ohanapecosh River, and worked the river a half-mile or so down to a large waterfall. I seldom returned without half a dozen nice cutthroats and, during the late 1940s and early '50s, I don't remember ever meeting another fisherman on that stretch of the river. While I was gone, Ruby would take Bob to the river next to the campgrounds and let him fish in several of the ponds. He didn't catch anything, but he splashed around the riverbank, got wet, threw rocks into the river, and kept happy. When I returned, we would fry bacon and eggs and brew coffee. Oh, how sweet it was!

Back to Alaska

The schedule for shrimp explorations included further work on the distribution and abundance of shrimp species inhabiting the inside waters of southeast Alaska. We planned to explore the many inlets and bays of the region, north to Juneau and then out of the Icy Straits to Glacier Bay. The surveys of the shrimp populations conducted through 1951, although interesting, had not turned up anything spectacular. Evidence of pandalid, a northern family of shrimp, was found in almost all of the surveyed areas, but the catches were small and not considered commercial. During the winter cruise in 1952, however, our catch rates of pink shrimp species began to improve as we moved north to the waters west of Juneau. After finding good signs of various pink shrimp in Port Armstrong, we headed west via the Icy Straits.

Our plan had been to survey Glacier Bay, but as we were running west in the strait, we spotted a small gillnet boat that was adrift. There was a stiff wind blowing, a moderate to heavy swell running in from the sea and a heavy chop. In the words of the crew, it was "messy weather." As we drew nearer, it was apparent that the vessel was taking on water and that only one man was aboard. Our captain hailed the skipper of the small boat and asked if we could be of help. He pleaded for us to tow him into Idaho Inlet, which was about a twenty-mile run from our position and took us off course, but we did so. His home was a small, isolated Norwegian community on the east side of the bay that had been established early in the century. We dropped him off in front of the village, where he managed to stabilize his boat and go ashore. He left us with short thanks, which I felt at the time was somewhat reserved, since we had rescued him from what could have been a disaster.

Later in the day, however, he invited us to take supper with his family. It was mid-winter, and we suspected that they were probably low on supplies. Furthermore, taking a small boat ashore after dark didn't seem very tempting, so we declined the invitation with thanks to our would-be host. The village was largely hidden by evergreens. There was about five feet of snow on the ground and banks of snow against the sides of the cabins, which were connected by a network of paths

265

dug through the snow. There were large woodpiles, and smoke rose from the chimneys. There also were numerous deer in and around the village; if the villagers hunted deer, it wasn't near home. The skipper and I talked over our schedule and decided that we would alter our plan and investigate the inlet, since circumstance had brought us there.

The next morning, just after sunup, we set out our beam trawl at the upper end of the inlet and towed out toward Icy Strait. We had shortened our tow time to thirty minutes so that we could more effectively sample, identify species, record catches, and preserve the species that needed further systematic studies. It was a cold morning, but the wind had died down and the waters of the inlet were calm. While we were towing, I completed my breakfast, put on my foul weather gear, and got ready to collect specimens and record the catch, which was taken at depths of thirty to fifty fathoms. The results of the cruise had not shown significant commercial potential and, as I watched the trawl warps being spooled in, I didn't really expect anything spectacular. I was wrong. As we lifted the first tow aboard, all I could see was the deep red color of a net teaming with shrimp.

The skipper and crew estimated that we had made a catch of about 5,000 pounds during the half-hour tow. It was made up almost entirely of pink shrimp that ran about forty to the pound. It would have been a good catch for any commercial fisherman, and we were elated. For several years, a commercial shrimp fishery had operated out the several bays near Petersburg and Wrangell, but this was the first good survey result we had achieved in southeast Alaska. Andy Granier, a commercial shrimper from Petersburg whom we had hired as a consultant, agreed that the catch was very encouraging. For the remainder of the day, we continued to make sets at different depths and areas of the inlets, with good catches of shrimp in almost all the tows.

The interesting thing was that there was almost no bycatch, that is, there was little evidence that there were significant quantities of finfish in the region. It seemed counterintuitive; one would expect that the concentrations of shrimp would attract a number of predators. But Andy explained that when shrimp fishing was good, bycatch was not a problem. When the day was over, we were pleased with the results and our documentation of the species, sizes, and abundance in the Idaho Inlet. During the remainder of the trip, we found several other bays and inlets where shrimp seemed plentiful, but not as much as in the Idaho Inlet. The trip, however, was just the tip of the iceberg. Subsequent surveys to the west would yield tremendous catches of pink shrimp and demonstrate that there was a large latent resource in the inside waters of southeast Alaska, the offshore waters of Kodiak, and west to the Shumagin Islands.

Shortly after our work in the Idaho Inlet, I received a radio call stating that Ruby had almost lost our second child and was confined to bed. I was taken to Juneau and caught the first flight to Seattle, which turned out to be a Pan Am Strato Cruiser. At the time, I thought it was some plane! The seats were spacious, and there was a lower section that housed a bar. I must say a passenger in tourist class was treated as well or better than first-class passengers thirty years later. Dinner was served on a tablecloth with silver or stainless-steel utensils, and included a salad, bread rolls, choice of entree, and dessert. I felt like a king. The one drawback was that the flight, which took more than five hours, seemed to last forever.

Number Two

When I returned home to Normandy, I found that Ruby had been on bed rest for almost a month. Her doctor said she would have to remain there until she gave birth. During my absence, her mother had flown up to help her out. It was a tough period, especially for Ruby, and the time seemed to drag as we waited for the baby's arrival. We sweated out April and on May 9 at 8:00 a.m. Ruby gave birth to our second child, Susan Lee. She was, of course, a beautiful child, perfectly formed, over six pounds with fuzzy hair, and it seemed like she had a smile on her face from the moment she was born. For the first two days after we brought her home from the hospital she never cried, and we began to think something was wrong. She found her vocal cords on the third day, however, and made up for lost time during the next week.

Do the Tags Stay On?

During May and early June, Don and I put the final touches on the plans for the second albacore survey. We scheduled the trip to retrace the pattern we had taken the previous year. From the observations we had made on the swimming speeds during our earlier excursion, I had concluded that the Peterson disk (or button) tags that were used on a variety of fish species were not appropriate for tuna. Tuna just moved too rapidly, and the disk tags, which were secured aft of the dorsal fin, were likely to be ripped out of the their backs by the force of the water. To test this hypothesis, I talked Don and the boss into spending the money to build a small water tunnel at the University of Washington Hydraulics Laboratory. With assistance from lab personnel, we designed a tunnel in which we could insert a twelve- to eighteen-pound albacore, fix it in place, and then gradually increase water speed while observing the behavior of the fixed Peterson tag.

During the test period, which extended over several weeks, I noticed a young man working around the several water troughs next to the Hydraulics Laboratory. I thought he might be a student, but he appeared somewhat older that the average undergraduate. One morning he watched as I slowly increased the water speed from eight to fifteen miles per hour (mph). As in earlier tests, the Peterson tag ripped out of the albacore's back at a velocity of just over sixteen mph. The young man introduced himself as Bill Miller, a graduate student in hydraulic engineering, and asked about the experiment and test that I was conducting. He observed that the flesh of the dead albacore was unlikely to have the integrity of a live fish. I agreed and noted that the relatively low speeds at which the tags were torn off suggested that some other tag design might be needed for fish with high swimming speeds. In addition, at speeds greater than ten mph, the disks began to shear outwards, increasing their resistance to the water flow. "I can't believe that they could last very long even on a live fish," I said.

Bill nodded, and we began to chat about our backgrounds. He was a World War II vet, had served in the navy for several years, was working toward a master's degree, and had spent several years in China. With that last comment, my mental circuits jumped to full attention. I told him that I too had served in China from 1944 to 1946. We began to exchange more details and realized that we had both come over the Hump at about the same time and had been stationed in Kunming, China. I mentioned Verne Benedict and told him he was living on Mercer Island. We had a good talk, and it felt good to know there were a few SACO vets who lived in and around Seattle.

I continued the work at the Hydraulics lab for several more days and then closed down the experiment. I had hoped to see Bill and extend our acquaintance, but he was not around during the final test stages. We would meet again in the future. The collected data was not conclusive evidence that the thousands of tags placed on tuna would be quickly lost to the sea, but it certainly raised a lot of doubts. The experiment was considered of such interest to the university that the results were published in the U.W.'s *Trends in Engineering* magazine and in the Department of Interior's *Commercial Fisheries Review*. Floy Tags, a new company that produced fish tags, shortly thereafter developed "spaghetti tags," which had much less resistance to the water. Tuna tagged with the new markers were recovered for several years, leading to greater knowledge of the migrations of these great ocean nomads.

Eventually, we learned that the larger mature adults, spawned in the western Pacific off Japan, and the juveniles, which moved with the currents sweeping across the northern Pacific Ocean, eventually make their way to the northeastern Pacific and enter the fishery as two- and three-year-old youngsters. Few, if any, mature fish are handled by the commercial fisheries located off the coasts of Oregon, Washington, and British Columbia. As the albacore grow and age, they move westward across the expansive Pacific and ultimately join the spawning populations inhabiting the central and western regions of the ocean. The trans-Pacific migrations encompass various life stages of the fish and constitute a complex process designed to access various food resources and propagate the species. Americans, Chinese, Japanese, Koreans, and a host of islanders in the central and western Pacific fish the immature tuna population in the north Pacific and in the coastal areas of the United States.

By the end of 1953, the *Cobb* surveys had completed significant explorations of the continental slope off Washington and Oregon and had begun to explore the waters of southeast Alaska. By then, I had enough information to go public with my hypothesis regarding the Pacific Ocean perch. I wrote several articles regarding their potential for local fishery magazines, noting that they were perhaps the most abundant species inhabiting the upper continental slope regions from northern Oregon to western Alaska.

While surveying an area off the Oregon coast, we had encountered a seemingly high concentration of Pacific Ocean perch just south of

Newport at a depth of about 750 feet. While the trawl traveled over the bottom, checks of the echo sounder showed myriad marks just above the seabed. When the net was hauled, gas bubbles began to cover a broad area, and then the net burst to the surface, creating a giant island of fish off the stern. As the air bladders of the perch continued to expand, the net stretched to its limit and then exploded, throwing fish into the air and across the surface of the sea. The skipper and crew estimated that more than 40,000 pounds of perch had been caught. The fish bag on the net just couldn't handle the pressure of thousands of air bladders expanding the volume of the net. After the Oregon trip, I was convinced that Pacific Ocean perch constituted one of the great slope resources in the north Pacific. It was just a matter of time before they would be a major target for one of the fishing nations in the region.

Domestic Matters

At home, our daughter started walking at the age of ten months, and by the fall of 1953, she was scampering around the yard with her brother. Ruby and I decided that our kids would enjoy a sandbox, so we built an eighteen-foot square box in the backyard. We then located a sand source, rented a trailer, and hauled in several loads of sand. The sandbox became a favorite gathering place for neighborhood small fry. Bob and our next-door neighbor's young daughter, Adrian, built forts and roads, or made pretend cakes and decorated them with flowers or rocks for hours on end. Sue, however, found the sandbox entertaining for only about thirty minutes, and then she was off to play with something else.

Note on World Fisheries

To keep abreast of the rapidly developing world fisheries, I began reading various fishery magazines and literature from Europe and Japan when translations were available. Although World War II had devastated the fishing fleets of Western Europe and Japan, it also had left a legacy of new navigational devices (radar, Loran, and advanced echo sounders), synthetic fibers (nylon), and advanced propulsion and refrigeration technology. This would pave the way for the globalization and expansion of fishing activities across all seas and oceans. Within months after the termination of hostilities, Japanese and European entrepreneurs began to direct their attention to redeveloping the coastal fisheries. By the early1950s, world catches of marine fish were expanding, and fishing fleets were busy harvesting the bounty of the North Sea, Norwegian Sea, and the other traditional fishing grounds of prewar Europe. The Japanese were busy rebuilding and converting vessels to harvest the abundance of fish inhabiting their local waters and the more northern waters of the western Pacific.

In the years prior to the war, Japan's aggressive fishing fleets had moved east across the Pacific and were catching ground fish and salmon along the outer Aleutian Islands. They also had ventured into the Bering Sea. These activities were terminated, of course, at the onset of World War II, and afterward, the expansion of Japanese fisheries was hemmed in by agreements that limited them to the western Pacific. But by the early 1950s, the agreements had been altered, and the Japanese started to return to the eastern Pacific to catch salmon and bottom fish. Their expansion of salmon gillnetting into the eastern Pacific was seen as a direct threat to American and Canadian fishermen, as many of the fish that they harvested were perceived as originating in U.S. or Canadian rivers.

Where did the salmon that emerged from the myriad rivers along the west coasts of the United States and Canada spend their time at sea? The prevailing belief had been that much of their ocean life was spent feeding in the coastal waters. Evidence from earlier Japanese fisheries and the postwar developments in the eastern Pacific, however, made it clear that we didn't know much about salmon migrations, feeding patterns, and general behavior in the ocean environment. There were stacks of master's and doctor's theses at the U.W. and other coastal universities on the subject of salmon life, but most of them dealt with their early life history in the streams, rivers, lakes, and estuaries.

The expansion of Japanese fisheries into the high seas rang a bell all the way to the White House and initiated one of the largest investigations ever conducted in the northeastern Pacific. The *Cobb* was redirected to examine the ocean distribution of the five species of salmon native to North America. The Bureau of Commercial Fisheries mobilized its scientific talent in the Pacific Northwest and Alaska to study the origins of the salmon that was harvested across the northern Pacific Ocean, the seasonal distribution of feeding salmon on the high seas, and the oceanographic features that influenced their distribution and feeding habits.

Switching Tracks

I watched these developments with great interest, both because they introduced a new dimension into investigations of ocean resources in the North Pacific and because they were bound to affect the nature of our group's explorations. I had become uncomfortable with the general leadership in our area of investigation. There was a growing tendency to minimize the biological and scientific component of our work and to focus instead on issues of commercial and economic significance. From my perspective, both aspects were intertwined and whatever findings we released needed to be based on a fundamental understanding of the biological factors that determined resource productivity and harvest sustainability. This philosophy, which did not seem to be shared by my supervisor, led me to investigate other job alternatives in the region.

I soon contacted the Washington Department of Fisheries (WDF) regarding a position to lead the trawl-fish studies for the state. A number of bottom fish trawlers operated out of Puget Sound and several of the coastal ports. Their landings amounted to close to twenty million pounds of mixed species. Very little was known regarding the species' life history, distribution, and migration or, more important, the consequences that fishing had on the various stocks. The job would give me added experience with a group dedicated to biological and ecological investigation of harvested marine resources. Ruby was less enthusiastic because the change also would result in a small decrease in my salary, and we had been struggling to keep ahead with our new house payments and the increasing cost of living. However, given the problems within the bureau, I resigned and accepted the job with the WDF. It was not an easy decision as the work with the exploratory group was challenging, but it gave me a set of experiences that was both exciting and invaluable.

A Northwest Interlude

I switched jobs in midsummer 1954. Our son Bob was about five years old and would start kindergarten in the fall. Sue had just turned three. The long winter rains faded away by early July, giving us the opportunity to make the best of the summer by investigating the wonderful parks that were spread across the state. We usually headed for the several camping areas on the east side of Mount Rainier, where there was good trout fishing, hiking, opportunities to feed birds and wildlife, and a wood supply for campers. We decided to take several weeks vacation before I started my new job. Bob had heard that one of his friends was going to be camping at Lake Wenatchee and convinced us to camp there as well. So we gathered up our gear and headed for the lake.

The drive from Seattle to the campground took several hours, but it was a warm and comfortable day so we moseyed along and arrived a few hours before sundown. We had started off on a weekend, which wasn't too smart since the campgrounds filled up early. Fortunately we got a campsite, but it was a long way from the lake and almost all the way across the campgrounds from Bob's friend, Jimmy. We put up the tent while Bob ran off to find Jimmy. As the sun began to set, the wind increased off the lake; it was cold and Ruby and I were having difficulty getting a fire started for dinner. I had to make a windbreak out of rocks just to keep the fire burning and to get enough heat to cook anything. Bob returned about an hour later. We were not sure how he found our tent, as it was pitch dark and the only light came from the occasional campfire. It was unpleasant outside, so we finished dinner and retreated to the tent, played cards for a while, and turned in for the night, expecting a better morning.

When we awoke in the morning, the wind had increased and a number of people were pulling out of the grounds. It was too windy and cold out to cook, so Ruby made breakfast on the camp stove inside the tent. By the time breakfast was over, it was obvious that no one was very happy about having to spend the day inside the tent, but Bob still hoped to have time with his friend. He left and returned in about an hour and said that Jimmy's family was headed back home because of the wind. Within an hour, we had taken the tent down and stored our gear in the car. We weren't sure where we were going, but there had to be something better. I turned the car east and headed over Stevens Pass. We had not previously visited eastern Washington, so this was fresh territory for the Alversons. We wound up the mountain road to the summit and then down the eastern slope of the Cascades to Leavenworth, a quaint tourist town with a Swiss motif. We stopped for a snack, looked around the village, then studied a map to find another camping area.

The map showed a road turning to the north across Bluett Pass that ultimately crossed the main east/west highway (Interstate 90) linking Seattle to Spokane. After considering our options, we headed down the highway to the junction that swung off to the south toward the pass. The narrow two-lane highway followed a river up toward the mountain pass, then climbed steeply as it rose to the summit. We kept our eyes peeled for a camping site, but didn't spot one. As we wound down the south side of the pass, however, we broke out of the firs into a small valley and saw a sign for Misty Creek Campground. The name sounded inviting, so we decided to investigate.

The grounds weren't well developed; there were no more than a dozen camp sites, and all but one were empty. We selected one site in a cluster of tall firs, near the stream. The next site upstream was occupied by a medium-sized camper and a tent, but we could not see the occupants. We put up our tent, unfolded the Coleman stove, stored our food, and gathered a supply of wood from a large pile available to campers. Bob collected firewood, and Sue filled the water bucket. They were delighted with the area even though there weren't any other kids at the camp.

Bob got out his fishing pole and was eager to go to the stream and start fishing. Sue, although not immediately, asked to join the fishing party. We didn't have another pole, so I fashioned one out of a tree limb, rigged it with a line, and Ruby found a small safety pin for a hook. We used salmon eggs as bait. Ruby and I watched as the kids made their way down stream, fishing each of the small pools in succession. It wasn't long before they both claimed to have bites, and then Bob pulled in a small trout—stretched, it may have been six inches long. Meanwhile Sue seemed quite content to lash away at the water with screams of delight when she thought she had a bite. We fished for an hour or more before they lost interest and decided to return to camp. Bob had caught two small trout and demanded that we cook them for his breakfast. Sue seemed satisfied with talking about the bites she got and the fish she almost caught.

Our next-door neighbors had returned and there were about ten adults, a mixed group of men and women. The men were mostly bearded and all sat around a campfire drinking beer. They looked over at us, but said nothing. The sun had set, and it was getting cold. We had gotten wet while fishing, and our feet were freezing. I started our fire and then built it up until it was quite a blaze. We filled a large dishpan with water and heated it, then took turns soaking our feet until we were all warm.

During the "hot tub" session, Ruby prepared a meal of pork and beans, salad, and cookies for dessert.

Next door the party had become increasingly loud, and the language had deteriorated into a four-letter word contest between the participants. To Ruby and me, their occasional glances seemed threatening and we discussed pulling up stakes, but the kids were quite happy and at ease, it was getting late, and we didn't have many options. We had no idea where to find another campground and not enough money to pay for a motel, so we stuck it out. After dinner, Bob and I added a lot of wood to the fire, and we roasted marshmallows and sang Boy and Girl Scout songs. We did our best to ignore the screams and arguments at the next camp and finally went into the tent and climbed into our sleeping bags.

The ruckus next door went on well into the night and threatened to erupt into a fight several times. Sometime after 2:00 a.m. things quieted down, and Ruby and I finally fell asleep. In the morning, the camp was calm. After I got a fire started, Ruby made bacon and eggs with fried toast. The deep smell of the evergreens and the noise of the brook made it almost serene. The kids were anxious to explore the area. We were chatting about where we might go when one of the women at the campsite next door called over and said, "I hope we didn't disturb you too much last night. We saw your children and the girls tried to calm the guys down, but they were having a great time." She told us that there was an old mine site on the dirt road that ran up the hill. Our concerns diminished, and we set off for the mine. When we returned, our neighboring campers invited us over for coffee, and we joined them for a friendly talk. The Misty Creek Campground experience turned out to be great for all of us, and we learned a little about fellow campers and ourselves. As we headed back to Seattle, Ruby and the kids sang all the way to Normandy Park while I dreamt about starting my new job.

CHAPTER VIII
THE WATERFRONT BEAT

An Informational Retrieval and Database

The job at the WDF involved keeping track of the catches and the impact on bottom-fish stocks caught by about a hundred trawlers based in Seattle, Everett, Anacortes, Bellingham, Blaine, and several other small coastal ports. The fleet fished from Oregon to Hecate Strait, Canada, and almost 500-mile stretch along the coastline. It had expanded a great deal during World War II, partly in response to military demands and a growing public need for vitamin A. The latter was abundant in the livers of most sharks, and hence gave birth to an extensive fishery for dogfish shark and other species along the Pacific coast of the United States and Canada. After the war, fish processors attempted to expand the markets for various bottom-fish species, promoting both fresh and frozen fish fillets. Due to the prices that buyers were willing to pay, however, the markets were weak. The fish houses bought cod and many types of rockfish for three to four cents a pound, and some skippers, or perhaps most of them, were discounting their catch prices just to maintain a market.

Historical information on the fishery included data collected via "pink slips," the landing documents that were required for tax purposes. A scientist named Fred Clever had done some excellent work involving the tagging and migrations of flounders. The WDF also had a smattering of life-history data on most of the dominant flounders, cod, lingcod, and several rockfish species. However, there was almost no information on the effort the fishermen expended to make their catches, the seasonal distribution and behavior patterns of the various target species, the age structure of the species in the catch, or the origin of catches according to specific fishing grounds and depth strata. All were vital components of any database that would give the WDF a minimum understanding of the consequences of fishing on supply abundance. The trawl fisheries landed a great variety of fishes, perhaps forty or more different species in any given landing. Occasionally, there were weird-looking specimens that were never reported on the "pink slip" catch records. It soon dawned on me that the first order of business was to develop a system to inventory what was being landed, in what quantities, from which fishing grounds, and, if feasible, to associate the quantities of the catches with the fishing effort expended to make them.

Several possibilities surfaced from the fisheries literature. I could (a) ask the state to require the fishermen to maintain logs containing the needed information, (b) request that the pink-slip sales record include a broader set of critical information, or (c) establish a system of sampling the vessel operators as they landed the catch. The last option called for one-on-one interviews with the skippers of the vessels, asking pertinent questions regarding the origin and composition of the catches made on different fishing grounds.

Various fisheries investigators had used logbooks with different degrees of success. Extracting information from the logs took considerable effort, and their reliability depended on the commitment of the skippers to the program. In addition, some skippers were concerned about matching the log data with sales slips. If a skipper was discounting his catch to maintain a market, the log records might be inconsistent with the sales (tax) slips. On the other hand, the "higher ups" were not interest in expanding the amount of information on the sales slips. Pink slips served one vital government goal: to determine landing tax levels. The government had little interest in complicating their records to aid a biologist's research. After discussing the issue with several of my peers, I decided to combine an interview system with a voluntary fishing log, which many of the skippers maintained for their own use.

The interview system and sampling the catches would require the collecting of tremendous amounts of information. The nagging question that remained was, how could we possibly store and process the data? During any year we collected landing records from several thousand fishing trips. Each interview form contained information on up to thirty species harvested from about sixteen different fishing grounds along the coast throughout the year, taken from depths ranging from about twenty-five to 200 fathoms. These records were then correlated to dates of fishing. The compact desktop computer had not yet arrived on the scene, but the Department of Fisheries used a new IBM card-sorting computer to tally vessel landings based on the pink slips, among other things. I asked the department's computer experts if they could put a scientific database on the IBM. Much to my delight, they were excited about the possibilities. I gave them a preliminary interview form and, within a week, they had designed a program that would sort the data by time, area, species, and depth, and also the effort expended by the fishermen. I was ecstatic and launched the interview program.

The offices for the marine research group were in the Bell Street Terminal, within easy walking distance of the various fish houses on Elliot Bay. There were perhaps five fish houses along the bay front that handled bottom fish. Several others were located up the Duwamish River toward the south end of town, and one was on the ship canal that connected Lake Union and Lake Washington to Puget Sound. Most of the boats offloaded their catches in Elliot Bay, then ran to the Lake Washington ship canal, where most of the boats moored in fresh water and took on fuel, fishing supplies, and food stores, etc. Within a month of starting my new job, I completed a final interview form with the help of Al Pruter and Heater Heyamoto. Both Al and Heater were veterans of World War II and had gone through the U.W. School of Fisheries. I had known Al well because we were in many of the same classes. He was a "brain" (nerd was not in our lexicon) and seemed to coast through his classes getting all A's. Heater was a year behind me, but was well known as one of the stars on the Husky baseball team.

I hoped to contact between 60 and 70 percent of the vessels landing in the state. Most of the bottom fish were offloaded in the Seattle/Everett area or at the north end of the Sound in the Bellingham, Anacortes, and Blaine region. Nick Passquale was hired to assist in the bottom-fish investigation and conduct interviews and sample landings in the northern Sound. I would handle the southern region. The vast majority of the trawl skippers who delivered their catches to the Seattle and Everett fish houses were of Norwegian or Yugoslavian descent, and I had met several of them when I was with the Bureau of Commercial Fisheries. Some had served on an informal group that helped plan and review the U.S. Bureau of Commercial Fisheries resource explorations work. It seemed logical to start the interview process with a few boat owners and operators whom I already knew.

My first interview was with Einar Peterson who owned the *Susan,* a boat of about sixty feet. He delivered his catch to the Main Fish Company located at the south end of the waterfront. As I walked south along Elliott Bay, I saw a number of trawlers offloading at several fish houses. It was my intent to go to the south end of the waterfront and then check each vessel as I returned back to the office. When I arrived at the Main Fish Company around 7:40 a.m., the unloading of the *Susan* had already begun. The vessel was sitting deep in the water, with what was obviously a good catch on board. The hatch had been removed, and a crane operator was lowering a large fish bucket into the hold. When the bucket was in place, the crew used fish pues to remove the fish from the ice, sorting out particular species. When the bucket was filled, the crew raised it to the dock level, emptied it into a fish wagon, took it to a set of scales, and then dumped it on an ice layer on the floor of the fish house. Depending on the quantity on board, it could take several hours to most of the day to offload the fish.

I watched the unloading operations for a few minutes, then made my way down a rusty old iron ladder to the deck. The boat was sitting about ten feet below the dock surface. Einar was inside the galley eating a roll, and he invited me in for coffee. After I told him about the interview program, he got up, went to the wheelhouse, and returned with his fish log. He had been fishing off Vancouver Island targeting true cod, snappers (rockfishes), petrale sole, and lingcod. In total, he had somewhat over 40,000 pounds on board—a good load. During the trip, which had lasted about six days, he had fished the Forty-Mile Bank, Swiftshore, Ucluelet, and the Esteban Grounds, as well as Cape Scott at the north end of the island. He and his crew of three would share about $1,600 after deducting the boat costs, including fuel, grub, and other miscellaneous expenses. The overall price per pound received by the fishermen was less than four cents—if the scale weights were honest.

Einar was an easy interview. He was a large man, with a reputation for driving himself and the crew hard. He was known as a "highliner," a skipper who continued to be one of the leading producers and an excellent seaman, and considered a top fisherman by his peers. As I asked questions, he looked through his log and spoke slowly with a moderate Norwegian accent. When he was not sure what was being asked, he would interrupt me and clear up the issue. He seemed genuinely interested in the program and was a willing participant. We finished up with coffee and talked a bit about the fisheries in general. I learned that he had fished cod, sardine, tuna, salmon, and shark. He was also a halibut line fisherman. Like most of the seasoned trawl skippers, he had experienced the gamut of fisheries off the west coast of the continental United States and Alaska. Einar had begun his days as a fisherman in Norway, hand-lining from a dory when he was ten years old.

Upon completion of the interview, Einar said that he would spread the word around the fleet and encourage other fishermen to participate. As I climbed the ladder back up to the dock, I saw that another trawler, the *Heather,* just astern of the *Susan.* Two men were on deck clearing the fish hatch. I yelled down, "Can I come onboard and talk to the skipper?" One of the men yelled back, "He will be up in a minute or two." Within a few seconds, the skipper climbed the ladder, and said, "I'm John Courage. What can I do for you?" I introduced myself and told him about the interview system we were introducing to the fleet. He didn't seem overly enthusiastic, but said, "Well, I'm going up for a beer and, if you want to join me, we can come back and look at my fish log in thirty minutes or so." I think he expected me to decline, but without hesitation I agreed, and we walked up to the tavern at the head of the dock.

We sat at the bar. Several rather rough-looking customers sat next to us, but a number of the men who worked the waterfront always looked a little skuzzy. John ordered a couple of beers, and I explained a little more about the interview program. He noted that he had been landing fish at the Main Fish Company for years and that no one had ever given a damn about the trawl industry, which was basically true. It was an industry with little or no regulations, other than the ban on landing halibut, salmon, or Dungeness crab. He then asked who would see the information and if the data would be shared with the fisheries enforcement group. I explained that this was part of an expanding state-run scientific program to collect information on the fishery and try to understand the impact of fishing on the various stocks. He probably wasn't very impressed, but he did seem more at ease as we returned to the boat. The *Heather* was several feet longer than the *Susan*, but had a slightly smaller catch on board. We went through the fish log and completed the interview in about fifteen minutes. My first two interviews had gone quite well.

Within a month I had interviewed the skippers of more than thirty-five different vessels, including some that had made a second trip. Gradually, I learned the names of the skippers and the vessels and could recognize the boats when they approached the dock. Most of the skippers were more than willing to participate, but some made it abundantly clear that they were not interested and wanted me to stay off their boats. Fishermen had an underlying suspicion of state and federal scientists, which continues to this day. I enjoyed being on the docks in the morning, although at times the seedy side of life showed itself. A number of winos worked the docks, attempting to bum fish from the boat owners. If they were successful, they often took their fish to one of the small waterfront restaurants and exchanged it for a drink of wine. Many of the trawl skippers, however, seemed to be sensitive to the less fortunate and gave fish to the winos even when they knew it wasn't to stave off hunger.

An Alien Had Landed

About once a month I drove to Bellingham and met with Nick Passquale to check on his progress with the interview program and review his research on the age and growth of English sole. He had become a good friend of the owner of Bornstein's Fish Company and had been given an office and lab space in its waterfront plant. A number of larger vessels, seventy feet or more in length, delivered their catches to the fish houses in Bellingham. Bornstein's handled more bottom fish than any other fish buyer in the state, and it had begun to buy and process a considerable amount of deepwater rockfishes, which were filleted, frozen, and sold as Pacific Ocean perch. The owner, Meyer Bornstein, was a short, thin man who loved his booze. When I was introduced to him, he pulled a bottle out from behind the desk and asked me to join him in a drink. I said, "Yes, but only one." I'd heard that every skipper who returned from a trip joined Meyer in his office for a drink. He was always helpful to our efforts, however, and went overboard to make our job easier.

After downing a short shot of Seagram's with Meyer, Nick and I returned to his office where he showed me an odd-looking flounder one of the skippers had asked him to identify. The fish was about eighteen inches long and had a series of raised star-like areas across its colored back. It had been caught about halfway up the west coast of Vancouver Island at a depth greater than 200 fathoms. Flounders and other flatfish start their lives as normal larva fishes, but before they settle to the seabed, one of their eyes migrates across the head to the other side, which becomes the "top" side of the fish. Depending on the species, they may be either right or left-handed. On most flounders, the side that lies against the seabed is white or gray while the upper side is variously colored. Nick had scanned the taxonomic texts in his library, but couldn't find anything closely resembling the unidentified flounder.

We checked again through Clemens and Wilby's *Fishes of the West Coast of Canada*, but found nothing that even resembled this critter. It was either a new species or a new distribution of a species that had not been reported in the region before. That afternoon, I took the fish to Dr. Welander at the U.W. School of Fisheries, and together we looked through several specimens in the museum. Two days later, he called to say that he had identified the fish. It was a species of flounder that was abundant off Japan and harvested in Japanese fisheries. It had not been reported, however, on the west coast of North America. Welander and I discussed the potential genesis of the strange flounder. The best theory we came up with was that the larva of a female flounder that had spawned off Japan had drifted westward across the Pacific. It had remained in the pelagic state for several months, and then settled on the seabed, among a collection of unfamiliar fishes, on a strange continent. Within the next few years, several more of the Asian intruders were off the coast of Vancouver.

A Summer Picnic

I had been with the WDF for about a year when a flyer arrived on my desk announcing a department picnic at Clear Lake, just to the south and east of Seattle. Ruby and I thought that it would be an opportune time to get better acquainted with other members of the department, such as the hatchery, enforcement, and administrative personnel. It would also allow our children to meet the offspring of some of my fellow researchers.

When we arrived at the picnic, more than a hundred people were sitting at rows of picnic tables, perhaps fifty to seventy-five yards from the lakefront. We sat at a table with several of my fellow biologists while Bob and Sue ran off to play with the other children. We met a number of the staff and chatted with those around us for more than an hour before the food was served.

By that time, the kids had returned and were eager to join the food line, which offered hot dogs, fried chicken, hamburgers, watermelon, various salads, and the normal picnic fare. The tables had olives, pickles, rolls, and butter. After we had filled our plates, I heard Ruby ask Bob, "What do you think you are doing?" Bob had found the olives and placed one on each of his ten fingers. He called out, "Look, look," as though he had accomplished something wonderful. The other children found the olive caper very amusing, but I suspect their parents were not so thrilled. We hadn't scored any brownie points for the Alverson family, I thought to myself. But then I saw that a number of the kids were placing olives on their own fingers, I dismissed the matter and enjoyed the meal.

When we finished eating, we decided to take a walk along the lake. It was a warm day, and the lake was filled with swimmers and, further offshore, water skiers. We had walked about fifty yards when Ruby said, "Someone out there is calling for help." A young boy was thrashing in the water about a hundred feet from shore. An older man was trying to get to him, but the would-be rescuer didn't seem to swim any better than the child he was attempting to help. I ran to the sandy edge of the beach, kicking off my shoes, and pulling off my socks and sweatshirt. I ran out onto a floating dock near the swimmers, dove into the water, and swam out to the young boy. I helped him over to a paddleboat, which had come to render help. He hung to the float of the boat, which took him ashore. Meanwhile, I'd reached the older man, who turned out to be the child's father, and assisted him ashore.

A young woman was standing on the dock, dripping wet and screaming that my dive off the dock had dumped her into the lake. She hurled a number of choice four-letter words at me because I got her wet. I walked by her and returned to Ruby and the children who had run along after me and picked up my clothing. As I was soaking wet, we decided to drive home for some dry clothes, but a shellfish scientist at our table said that he had an extra set of pants and a shirt in his car. He soon returned with the clothing, and I went to a nearby men's dressing room to change. Ruby gave me her heavy Indian woven coat to use as a towel. I dried off, put on the pants, which were four sizes too large, and returned to the table, where there was an ongoing discussion about the events at the lake. The distraught lady turned out to be the mother of the rescued youngster. She apparently was so scared and upset that she was "out of it."

Nevertheless, no one ever bothered to come and thank us. We left the picnic prematurely, as I didn't feel very comfortable in the oversized clothes and also had the shivers. When we returned home, which was about forty minutes away, Ruby noticed that her jacket was missing. I had left it in the men's outhouse. Since it was a rather expensive coat, Ruby jumped into the car and returned to Clear Lake while I took a hot bath. She went immediately to the outhouse, but it was not there. As she started to walk to the picnic area, a boy told her that he had seen a man take the coat back to his car. Ruby found the car's owner, who had the jacket but wanted to know why a woman's coat was in the men's outhouse. Ruby tried to explain, but he didn't seem to believe her story. It was a day that we and our friends would always remember.

By Design or Accident

In the late summer of 1955, someone from the *Seattle Times* called me and said that he was planning an article on the trawl fishery. The caller asked if I would take a *Times* photographer to the waterfront to shoot photos of the trawlers and the offloading operations. The photographer came to our offices at the Bell Street Terminal early the next day. To keep track of the vessels that were offloading on the Seattle waterfront, I called the Trawl Fisherman's Union for a list of the expected vessels and the locations they would deliver their fish. Unfortunately, only one boat was due, the *Trust*, which was to deliver her catch to the Main Fish Company. That constituted a definite problem. The skipper, Sig Olson, was an ornery Norwegian who had refused to participate in the interview program. Occasionally, he would ask me down for a cup of coffee, but never offered any information. I told the *Times* photographer about Sig, but we decided to see if he would let us take some pictures of his vessel.

The *Trust* was an old halibut schooner with the house aft and the work area mid-ship. When we arrived at the boat, there were two deck hands unloading the catch. I yelled down and asked if we could come aboard. One of the men looked up and said, "I don't think Sig wants any company, and he's not in a very good mood." The mood of a skipper frequently was related to the success, or lack of success, of the fishing trip, so I misunderstood the message the crewman was attempting to convey. "Well, Sig is never really in a good mood," I hollered back, and made my way down the ladder to the deck. The photographer followed, moving slowly down the ladder as he held his camera off to one side. When we were both on deck, one of the fishermen looked at us and shook his head. I still didn't get the message.

Sig was standing in front of the stove pouring himself a cup of coffee. A voluptuous young woman wearing only a slip, with one strap off her shoulder, was at the back of the galley. She said, "Hi," just like we were old friends. Sig said nothing for a few seconds, then looked at me and said, "Sit down and have a cup of coffee." I asked him if the photographer could take some shots of the boat and the offloading activity. He nodded his head in the affirmative. The photographer, as surprised as I was, went out on deck and began taking a few shots while I stayed and chatted with Sig. Before I'd asked a question, he went to the wheelhouse, got his log, and said, "OK, what do you want to know?" While I filled out the interview form, the young woman sat on the other side of the table drinking coffee and smoking a cigarette. She never bothered to get dressed, and Sig never mentioned her during our conversation. We suspected that he was entertaining one of Seattle's ladies of the night, but we couldn't be sure. After all, he was a married man. As we left, she posed in a sexy stance and said, "Have a good day, fellows. Sig has my card if you want to get in touch."

From that day on, Sig became an ardent supporter of the interview program. We never mentioned the girl, and we seemed to get along fabulously well. I often wondered what he thought when I showed up with a photographer. The *Times* article on trawlers described the fishery but nothing more. The reporter was very discreet.

Surprise in the Deep

Since the discovery of large concentrations of Pacific Ocean perch and other rockfishes, as well as Dover sole and sable fish, in the deeper waters off the continental slope, West Coast trawlers had begun to explore deeper and deeper. Fishing at depths between 200 and 300 fathoms became common, particularly during the winter months when the rate of catches in deep water improved. In the early winter of 1953, trawlers exploring the deeper water off Vancouver Island, on what the fishermen referred to as the Esteban Grounds (the name referred to an adjacent land feature), discovered large concentrations of petrale sole. The most desired of all the flounder species, petrale grew to a relatively large size, was broad, had a nice white texture, and could be filleted easily. Fishermen received a whopping seven cents per pound for petrale. The Esteban Grounds catch was surprising in that petrale traditionally was fished in the spring, summer, and early fall months and usually at depths between forty and ninety fathoms.

The *Susan* was one of the first trawlers to land in Seattle with a large catch of petrale from the Esteban Grounds, although several good hauls already had been delivered to Bornstein's in Bellingham. According to Nick, the entire catch of petrale in Bellingham was made up of mature fish, predominantly females. "They are big fish," he said, "full of eggs and sperm and most so ripe that the reproductive products are spilling out of the ovaries as they are landed." Apparently, the fishermen had found spawning concentrations of petrale sole in the deep water off Vancouver Island. Other reports indicated that similar concentrations were discovered on the Spit grounds west of the Strait of Juan De Fuca, which had been surveyed by the *Cobb* several years earlier. I was anxious to ask Einar what he knew about the catches from the Esteban Grounds.

The *Susan* was lying low in the water; it was evident that Einar had a good load on board. He told me that he had never seen such a large concentration of petrale sole "We caught about 50,000 pounds of petrale in seven tows," he said, "all at depths between 210 and 220 fathoms." He added, "A good number of the Washington trawls are now fishing the Esteban Grounds or heading for the Grounds. There are petrale sole all over the area, from 180 fathoms to 250 fathoms." As the vessel was offloaded, I measured, sexed, and took otoliths (ear bones) from several hundred fish. Nick's description was right on target. On average, the fish were larger than the ones we had sampled during the summer months; all were mature and the females' egg sacks were either broken or very extended.

Later examination of the ear bones showed that the fish ranged from about six to fourteen years of age. Existing life-history descriptions of petrale sole suggested that spawning occurred in the spring on the outer continental shelf. The literature did not mention that the species migrated to deeper waters in the winter months or that large commercial concentrations existed along the continental slope. Now, there was evidence that during the winter months the species spawned in water that was at least 150 fathoms (900 feet) deeper than their locations in the spring, summer, and fall. News of the petrale discovery swept through the trawl fleets in a matter of hours, and petrale landings surged over the next several weeks. It was an economic windfall for the fishing fleets, but I was concerned that the petrale sole in the spawning grounds would become vulnerable to exploitation and that the fleet might overfish the stock and have a negative impact on the abundance of the spawning population.

When I returned to the office, I called Nick in Bellingham to get the latest catch information and then called Max Chatwin, a Canadian ground-fish (or bottom-fish) biologist. He was aware of the situation off the Vancouver coast and expressed the same concerns that I had. He noted, correctly, that we couldn't be sure whether the deep-water populations were a component of the inshore population or an independent stock that inhabited those waters. That question could be answered only if we could tag a number of the spawning fish and recover them during subsequent feeding periods. Washington State, however, did not have the funds in its budget to charter a vessel for such a project, and to the Canadians, it was not of sufficient priority to divert money from important salmon studies. But Max and I believed that the tagging project was important to the understanding of the life history of the species and would help develop a conservation management plan, if one was needed.

If we couldn't find funds to charter a vessel, there were always skippers who would take us out to the grounds, gratis. But there wasn't a chance that we could get the work done until early 1954 and, by that time, charter money would be available. We began to seek out vessel owners who would be interested in a short charter and then put out a bid. The charter was awarded to Johnny Courage, owner of the *Heather*. We were unable to get out on the grounds until May, but we did place over a thousand tags. We had hoped to tag for two years in a row and, by happenstance, I mentioned our need for a vessel in 1955 to some of my friends at the Exploratory Fishing Group in the Department of Interior. Fortunately, they felt the project could reveal important scientific information regarding one of the prized flounder species harvested by West Coast fishermen. The BCF group offered to allocate ten days of the *Cobb's* time to the tagging project. Max and I were elated. We scheduled the cruise for early March or April 1955, and both of us would carry out the tagging work.

We departed Seattle in early April and headed up to Neah Bay. It was about a ten-hour run out to the ocean, and we arrived late in the evening. The skipper decided to anchor overnight and head for the Esteban Deep Grounds early the next morning. We figured to be on the grounds around noon as it was about a six-hour trip to Vancouver Island. We left Neah Bay around 6:00 a.m., made our way out past Tatoosh Island, and headed northwest, several miles off Vancouver Island. The weather was not great, and a relatively large swell was running in from the northwest. The *Cobb*, which tended to be heavy forward, was rising sharply into the oncoming swells and then plunging into the next wave, jarring the whole vessel. We were taking green water over the bow and heavy spray on the bridge windows. It was an uncomfortable ride, but tolerable.

We had been underway for several hours when the captain noted that a small trawler about a half-mile ahead of us apparently was headed for the Esteban Deep. I watched the boat with binoculars for some time as it plowed into the oncoming sea and heaved sharply from side to side. The trawler, which appeared to be under seventy feet, was having a much tougher ride than the *Cobb*, and the crew had to be miserable. Over the next hour, we gained ground on the small trawler and, when I could clearly make out her lines, I said to the *Cobb's* skipper, "That's the *Susan*, I'll give them a call." Over the radio, I asked, "Is that you, Einar, a few hundred yards ahead of us?" In a few seconds, Einar's voice came through, and he said, "You guys are really taking a beating. What are you doing out in this weather?" If we were taking a beating, what was it like on the *Susan*? But Einar was a tough Norwegian, completely at home in "jackass" weather.

I told him we were headed for the Esteban Deep to tag as many petrale as possible over the next week or so. "Yeah, everyone knows your research plan," Einar said. "I talked to a couple of boats that were fishing the Esteban Grounds yesterday, and they were doing well along the 220-fathom contour." He concluded by saying he would see us out on the grounds, but he didn't think anyone else was fishing the area because the swell had gotten so bad. We arrived on the grounds about noon, took Loran readings, and started to tow at 220 fathoms. The swell was fifteen to twenty feet high, and the wind was blowing twenty-five to thirty-five miles an hour. However, the 400 fathoms of cable we had let out and the heavy trawl doors seemed to steady the *Cobb*. The commercial boats would tow their nets for two or more hours, but we shortened our fishing time to about thirty minutes so the fish would be in good shape.

When the gear was hauled and the net was on board, we had made a catch of several thousand pounds, mostly petrale sole, although there were a few rockfish and Dover sole. I was surprised to find that the petrale sole were in very good condition. Although they had been hauled rapidly to the surface from some 1,300 or more feet, they were not scaled and swam actively in the holding boxes. Flounders, unlike rockfish or cod, are blessed with a type of air bladder that allows the fish to make rapid adjustments to changes in depth. Max and I worked from noon to sundown, measuring, sexing, and tagging fish because we had been told earlier that all of the fish were mature and in a spawning state. Once released, the fish seemed very lively and immediately swam toward the seabed, but we couldn't be sure that they had not been physiologically damaged. During the first day we tagged several hundred fish and, over the next week or so, we tagged close to 2,000 fish. All told, in 1954 and 1955 we placed button tags on about 3,500 fish.

Trawl fishermen from both British Columbia and Washington knew about the research effort and were anxious to see the results, so we did not have to offer a reward for tag returns. Trawl vessel operators began returning tags within several weeks after the tagging effort had ended. Our first question was quickly answered. It was obvious that a good number of fish had survived the tagging ordeal. As the year wore on, the number of recovered tagged fish continued to grow. Max and I realized that the recovery pattern would change the perception of the petrale sole migratory pattern and add a new dimension to our understanding of their seasonal behavior. After several years the migratory story became clear, and we published our results in the *Canadian Journal of Fisheries*.

After spawning, the Esteban Deep stock migrated quickly up the continental slope to the shallower waters of the continental shelf, where the fish commenced feeding. The tag-recovery pattern showed that they had entered feeding grounds along the Vancouver Island coast within a month after tagging. Almost all the tag recoveries occurred predominately in traditional inshore fishing grounds north of the Esteban Grounds. Several months after the tagging experiment, trawlers were making recoveries as far north as Cape Scott, and later in the year, north of Vancouver Island in Hecate Strait. The following winter the migration pattern became even more interesting when we recovered tags from the same spawning grounds where we'd tagged the fish a year earlier. The population of petrale sole that spawned off Esteban traveled inshore and several hundred miles to the north during their feeding migration. Then, in the late fall, they migrated back to the south and returned to the same spawning grounds.

The spawning adults moved quickly into the deep water during the winter and returned to shallower areas after spawning. The males seemed to arrive on the grounds early and remain there for several months, while the females spawned and immediately returned to the inshore feeding grounds. The petrale's migration to specific spawning areas and the integrity of its spawning population were not unlike those of salmon, which return to their streams or rivers of origin. We suspected that the migration to deep water allowed the fish to spawn their eggs and sperm in a current that slowly moved the surviving larva to the north and inshore. After settling on the seabed, the young petrale grew to maturity and subsequently joined a spawning population. They then migrated back to the south, offsetting the northward drift of the larva and ensuring the integrity of the spawning population. The results of the study ultimately led the U.S. and Canadian governments to close the Esteban Grounds to winter fishing.

The Sinking of the *Santa Maria*

The *Santa Maria* was a typical seine-trawler, house forward, with the winches and working area in the rear. The captain, Tom Jurkovitch, was just over six feet tall with sharp facial features that had been weathered by the sea and time. Unlike many of his Yugoslavian countrymen, he was soft-spoken but firm. He delivered his catch to Eardlies Fish Company near the center of the Seattle waterfront and was very cooperative during interviews. He was also very proud of his vessel, which he kept clean, and even prouder of the wine he made each year. Before an interview began, we would always have a toast with a small glass of his wine, which he poured from a gallon jug. Tom and I became good professional friends and, after several months of interviews, he gave me a gallon of his homemade wine. I didn't want to take it, but had the definite feeling that he would feel insulted if I refused. I took it home, and Ruby and I tried a small glass before dinner. It was excellent, a dry deep red wine that left a pleasant taste in the mouth.

About a month later, the department decided to charter the *Santa Maria* for some test drags in various areas of southern Puget Sound. Various sport fishing groups had accused the trawlers of catching and killing large numbers of salmon in the Sound. Don Johnson, the director of research, asked me if their concerns were legitimate. I had seen hundreds of trawl hauls emptied onto the decks of fishing vessels, but had only seen several dozen salmon in all the tows. "They are very uncommon elements in the catches," I said. "The best way to settle the question is to charter a trawler, invite some of the leaders of the sports groups to accompany us, and let them choose the fishing areas where they feel salmon are being caught during trawl operations." We made arrangements with Tom and, on a clear summer day, we sailed out of Seattle with a small number of sportsmen and Milo Moore, the new director of the Department of Fisheries.

The run to the south Sound took a couple of hours. We started our fishing in Carr Inlet, just south of the Tacoma Narrows where we made about four thirty-minute tows. The catches were composed of several species of flounders, a few rockfish, dogfish, and a variety of invertebrates, including several species of shrimp, squid, and sea cucumbers. No salmon were encountered. We then headed south to Case Inlet. On the way, the cook made an excellent lunch, which began with a glass of the skipper's red wine. The results in Case Inlet were much like those in Carr Inlet, except that many of the flounders were ulcerated. We didn't know why at the time, but we found out later that one source of the ulcers was the various toxins dumped into the Sound by paper mills and other industrial activity. By the time we completed our work in Case Inlet, we still had not caught a single salmon. But the sport fishermen still wanted to check out Budd and Totten Inlets at the south end of the Sound near Olympia.

When we arrived at the entrance to Budd Inlet, we made a short tow, perhaps twenty minutes long, and then began to recover the net. It didn't take long to reel the net in as we made the tow in relatively shallow water, not more than forty fathoms. The net was quite heavy, and the cod end hung straight down from the stern. As the crew began to haul in the gear, we could see that the net was loaded with dogfish shark. We had about 14,000 pounds of dogfish shark ranging from a little over one to three feet in length. I don't recall a single other species in the catch. It took several hours to offload the dogfish and clear those caught in the webbing. "If that is what the trawlers catch," the sportsmen noted, "they are more than welcome to fish the area." We then started our run north to Seattle and arrived at the Ballard Locks late that evening.

I didn't see the *Santa Maria* for several months and thought Tom took off the trawl gear and rigged the boat to fish salmon during the late summer and early fall periods. At that time of the year, a number of the trawlers seined for salmon in the northern Puget Sound and southeast Alaska and then returned to trawling later. Late in the fall of 1956 the *Santa Maria* was scheduled to land her catch at Eardlies. Several other trawlers also were due that morning. There was a light drizzle, which was characteristic of the area during at least half the year. I checked in at Main Fish Company and interviewed the skippers of the *Panther* and *Heather*, then headed for Eardlies. The *Mitkof* was lying at the dock, but there was no *Santa Maria*. I asked the union representative who attended the fish scales when the vessel was expected. "It should have been here by now," he answered. "It's a couple of hours late." Ivar Angle one of the local vessel skippers, told me during an interview that the *Santa Maria* might have had a problem because there was a lot of radio chatter about a navy LST (landing ship tank) running down a boat in upper Puget Sound.

With several other vessels to check on, I continued working my way north and interviewed three other skippers. Returning to Eardlies, I found that a crowd had gathered in the office upstairs, and a number of Coast Guard officers and crewmen were busy talking to the fish house owners. It wasn't long before the story of the *Santa Maria's* fate began to take shape. It had been hit broadside by a navy LST and rolled under the keel. What was not clear, however, was why, or if, the trawler had crossed in front of the navy ship, only that all but one of its crew had died in the collision. The crewman was questioned at a meeting later that day, but he looked completely out of touch with reality. He seemed to be still struggling to survive and groping to reach the surface of the sea. He was unresponsive to the investigation.

There was an official Coast Guard hearing. I don't recall how the matter was sorted out, but I believe that the conclusion was that the crewman at the wheel of the *Santa Maria* had fallen asleep. The vessel then veered off course, crossing in front of the LST. The trawler heeled over and broke up as it ran under the keel of the LST. Tom and most of his crew died within minutes. Their deaths reminded me that fishing is a tough business and has one of the highest fatality rates of any profession in the country. A week or so later, I went down to Tacoma and talked for some time with Tom's wife. The gallon of wine that he had given us lasted for almost two years, and we thought of him each time we sipped a glass of "Tom's Best."

The Southerly Migration

Each fall after Thanksgiving, Ruby and I would plan our traditional trip south to Los Angeles and San Diego to visit our parents and other relatives in southern California. We had come to dread the trip because we almost always ran into snow in the mountains of southern Oregon. Putting chains on the car and taking them off wasn't fun, but it was necessary. The 1955 trip was special because Ruby's dad had been sick for several years, and he wanted us to make a visit. We loaded up the car with Bob and Sue (who were then about seven and four) and presents for our parents and started down old Highway 99 about a week before Christmas. It had been raining hard for the entire week before we left, and we were anxious to get started before the whole state sank into mud.

When we reached Portland, the news on the radio announced that south of Salem, Highway 99 was flooding but passable, and that there were difficulties in northern California. We thought about turning back, but the kids started to cry, so we continued south. Ruby thought it might be her last opportunity to see her father. As we approached Albany, Oregon, we could see a long line of stopped cars. We pulled over behind a large semi-truck, whose driver told me that there were twelve to eighteen inches of water on the roadway, and cars were being transported in a convoy across the flooded area. About fifteen minutes later, a convoy pickup truck came through from the south with a hundred or more cars.

The pickup then turned and proceeded to the front of our line, the driver telling each car he passed not to go over five miles an hour. Because of the occasional heavy rain showers, the trip had been slow going all the way, and we were falling behind schedule. About a quarter of a mile down the road, we ran into water that gradually got deeper until it was high up on the wheels. It took us almost thirty minutes to cross the half-mile or so of flooded roadway. When we reached higher ground, we pulled up to a state police car and asked the officer what he knew about the highway to the south. He told us that 99 would be closed north of Grants Pass and that the route east and down the other side of the mountains was closed because of heavy snow. We seemed to have no choice but to return to Seattle. The police officer said that, if we were going back, we had better get in line as he expected they would close the road within an hour or so. I did a one-eighty and got in line to be taken back across the flooded area.

The kids went into a panic because they wouldn't see their cousins and other relatives and would have to wait to get their gifts. Furthermore, it wasn't certain that we could get back to the north, which made our thoughts of Christmas somewhat bleak. We had been in line only a few minutes when a car filled with college students pulled up behind us. It had California license plates, so I asked the driver where he came from. They were UCLA students from Bellingham, Washington, and they had driven straight through from Los Angeles. They had passed through Grants Pass several hours earlier and had no trouble reaching Albany. After a quick family conference, we decided to take our chances, and we reversed our course again. It got dark an hour or so south of Eugene, and the rain continued to fall, making our trip slow and driving miserable. Nevertheless, we plodded on. Just before we got to Grants Pass, the traffic dwindled and the going seemed to improve. We were traveling about sixty miles an hour when we hit about two inches of water on the road. The car hydroplaned for about fifty feet and then veered slightly to the right. I thought for sure we would go off the highway, but I managed to straighten out the car and get back to the middle of our lane. After that, I drove the remaining ten miles to Grants Pass at a reduced speed.

By the time we reached the pass, we were about five hours behind schedule. We plodded on, heading east along the upper Rogue River, which was also swollen from the heavy rain. Everything went well from there down to Shasta. Somewhere between there and Redding, a small section of 99 had washed out, but a short detour took us around the blockage. We stopped in Dunsmuir sometime in the early morning and checked with the state patrol about the roads to the south. The rains had been heavy all the way to Sacramento, and we learned about another setback over the radio. Highway 99 West had been closed due to a bridge washout somewhere south of Red Bluff. Highway 99 East was partly washed out, but traffic was still making it through. The state patrol suggested that we turn back. "Well," I said, "whatever is ahead couldn't be as bad as what we have already seen," so we decided to push on.

About forty miles south of Red Bluff, the traffic came to a stop. A small river had flooded its banks and eroded half the highway. A crew was attempting to stabilize the situation, and trucks were dumping gravel and small rocks onto the roadbed. Slightly more than an hour later the traffic was allowed to edge past the blockage clinging to the right side of the highway and away from the washout.

From there on, the trip began to improve. The rains stopped, and traffic moved at or in excess of the speed limit. We arrived at Ruby's folks' house around midnight. We had been on the road for close to thirty-eight hours, about eight hours more than we had planned. Everyone seemed glad that we had held our course. Our plan was to spend a couple of nights at Ruby's parents' home, head for San Diego, spend a few days with my parents, and then start the long trip back home. Ruby's brother, Noel, Jr., his wife, Anne, and their four children joined us on Christmas morning. Ruby's father gave each of his six grandchildren a $1,000 savings bond, which thrilled them. It was a wonderful holiday, and we always felt that God had somehow helped us to stay the course because Ruby's father passed away the following February. It was the last Christmas meeting of the Lane family.

The kids had a great time visiting the relatives, but it was becoming increasingly apparent that we were slowly drifting away from the Alverson and Lane clans. Furthermore, the thirty or more hours we spent on the road in a nonstop marathon were getting too difficult. In subsequent trips, when we were somewhat more flush, we would spend a night at Red Bluff. As the children grew, the trips to the south became fewer and fewer.

The Vessel Seizure

In April 1956, the state fisheries headquarters requested that I go immediately to Bellingham to check the catch of the *Sunbeam*. The vessel had been seized in Neah Bay, reportedly for illegally fishing petrale sole on the Esteban Grounds. The area had been closed to trawl fishing as a result of the studies we had conducted earlier, I protested that the government was using a scientist for enforcement. Besides, I wasn't interested in abandoning my family on a Sunday. My arguments fell on deaf ears, and I was told that the scientists were only expected to identify the species of fish on the vessel. Early the next morning, I left for Bellingham, where Nick filled me in on the details. When two state fish wardens climbed onto the *Sunbeam*, the crew locked them in the radio room and started to run the vessel offshore. Apparently, the skipper had second thoughts, because he reversed his course, went back into the strait, and took the vessel to Bellingham. The WDF had been contacted to identify each species in the *Sunbeam*'s hold because the wardens were not taxonomic experts.

After Nick told me the story of the vessel seizure, he said, "We have to unload the catch of about 45,000 pounds of fish and make a log of everything on board." That task alone was bad enough, but the crew had refused to help, so Nick and I would have to take off the fish hatch, remove the top ice cover, and load the fish tubs that were lowered into the hold. We were nervous because if we damaged the vessel or the catch, we were likely to be sued. We carefully removed the fish hatch and began to unload the catch. As we proceeded without difficulty, the crew quit watching us. I like to believe they stayed around at the beginning just to make sure that we didn't injure ourselves; that's the best twist I can put on the scenario. We worked from about 8:30 a.m. until just after noon, offloading about 20,000 pounds of petrale into the fish tubs, which we then lifted up to the fish house deck. We were both tired and chilled so we took a short break, drank some coffee, and ate some sandwiches.

We worked until about 5:00 p.m. to complete the job. The vessel had about 38,000 pounds of petrale on board, plus a few thousand pounds of other species. We had examined each flounder as we dug it out of the ice to make sure that it was properly identified. When we were done, we gave a copy of our records to the state fish officer. He thanked us, and I departed for Seattle knowing that Nick and I would be state witnesses when the case went to court. I felt very uneasy about the whole matter; it wouldn't help our interview program since we would be labeled as associates of the enforcement agents. The *Sunbeam's* owner wasn't all that popular, however, and a lot of the fishermen supported the arrest. The next day the *Seattle Times* and *Post-Intelligencer* covered the vessel seizure, and the news was the talk of the waterfront. Every skipper I interviewed wanted to know all the details.

The trial was scheduled for late spring in Port Angeles, Washington. My concern was that the state might have a hard time proving its case since the arresting officers couldn't sort out a petrale sole from a flathead sole. Even worse, the local courts seemed to have difficulty convicting fishermen. Local sentiment was with the hard-working fishermen, and most people didn't think the state should be punishing its citizens for activities outside state waters. Of course, the state laws were based on illegal landings, but it was a fuzzy issue. At any rate, I remained on good terms with the vessel owners and at the same time was doing what was necessary to support the state's case.

When the case went to trial, at least half of the industry bet that the *Sunbeam's* skipper and owners would beat the rap. The state made a good case documenting the illegal catch and the kidnapping, noting that the skipper had locked the fish wardens in the radio room and taken them to sea, where they had become seasick. The case was the subject of an unbelievable rumor mill; there was talk that the skipper had planned to put the wardens in a small boat and then take the catch to California. The sentiment of the audience seemed to be moving with the prosecution. The defendants never challenged the state regarding the illegal catch, so I wasn't called to testify. Instead, their lawyers hammered home two major points: first, the state had no jurisdiction over a catch taken off the coast of Canada, and second, the captain had not tried to kidnap the fish wardens. He had merely run out west to Tatoosh Island to pick up a bearing that was necessary to head up the strait toward Port Angeles.

The defense scored on both issues. The state did not have an expert in maritime law or anyone who knew anything about vessel navigation.

The prosecutor did not even challenge the defense regarding the skipper's alleged run to Tatoosh Island for a bearing, which was total nonsense. I looked at my feet, knowing the defense was winning the day. Furthermore, the state's argument was very unconvincing. When the state rested its case, I had a sinking feeling that we would lose the case because of an inept prosecutor who had little knowledge of fishing and marine law. I can't remember whether the case was decided by judge or jury, only that we lost and that the defendants got their fish and probably some compensation as well. Conservation lost the day.

Light Weights

During the summer of 1957 a number of the trawl skippers complained that they were getting "short-weighted" by the fish buyers. I had heard these claims since I started the personal interview system. Some skippers always felt that they were cheated on the scales; they knew their boats and how much fish they carried. Cheating hardly seemed likely as all of the fish were moved in wagons across the state-monitored scales, and the weights were recorded by a trawl association member. Furthermore, the negotiated price for the bottom fish was so low that cheating didn't seem worth the effort. Regardless, the growing number of complaints gave some credence to captains' claims. By mid-summer almost every skipper who was offloading spent time with the association representative checking the scale weights. One day Einar said to me, "Well, I guess my boat is getting smaller, because it doesn't hold near as much as it did last year. Ya think all the boats have gotten smaller?"

The rumblings went on through much of the summer. It seemed to me that, if the scales were certified by the state and supervised by union representatives, it would be difficult to short weigh the vessel owners. I did not say anything about the matter, however, since access to the plants and vessels was essential to the conduct of my work. I didn't want to take sides and get caught in the middle of the debate. The issue almost always came up during my vessel interviews. I listened carefully to the skippers and said, "Well, if it is going on, I don't have the slightest clue as to how." They probably wanted to hear something more supportive, but I was like the three monkeys who saw nothing, heard nothing, and said nothing.

The end of this story came quickly. I woke up one morning, read the newspaper and behold! The head of the trawlers' union had been short weighting his own members and taking kickbacks from some of the buyers. Of course, we were all shocked, but the vessel owners who paid to have the scale weights checked were outraged. From that point on, the political and economic strength of the trawlers association waned.

Breakfast at the Wharf

For years, groups of fishermen had gathered each morning at the Wharf Coffee Shop to exchange news and sea stories, tell dirty and sometimes filthy jokes, and assist each other in their business. Several of the trawl skippers I interviewed urged me to join one of the groups. They set up a morning coffee club, which included John Wedin, editor of the *Fishermen's News*, and frequently Milo Moore, director of the Washington Department of Fisheries. Moore strongly believed that fish-farming techniques could bring the salmon runs to historical levels, and he was an avid critic of fisheries biologists. He was quite popular with the fishermen in the region, and I suspect that he earned support on both counts. I had not had any run-ins with Milo but was quite aware of his animosity toward scientists. Thus, I was reluctant to join the morning call to Mass. John was a close friend, however, and he urged me to come to the morning sessions, so I decided to give them a try.

The coffee klatch usually started around seven in the morning and included ten to twenty participants. When I arrived for my first session, perhaps a dozen fishermen were sitting at the table: a few bottom-fish trawlers, salmon gill-netters, trollers, and seiners. I had met a few members of the group, but most were strangers. John Wedin introduced me to the various fishermen. A few people who worked at fish-supply houses and shipyards were also present. Milo Moore was not there and that put me more at ease. John led the discussion, which focused on the recent short-weight scandal. The trawl fishermen dominated most of the conversation.

When they had milked the topic dry, John asked me to talk about the interview program and explain what we hoped to get out of the data. I sketched out the rudiments of the study and then responded to several questions. Everyone was generally polite, and the trawl fishermen made some complimentary comments on the state's interest in trawl fishers and the results of the studies on petrale sole migrations. Other topics bantered around included prices paid to fishermen, new government regulations, the continuing debate over Indian fisheries, trends in salmon-run sizes, and global fishery developments. Most of these topics sustained a good number of meetings. To my surprise, a number of the fishermen were very well informed and could articulate their concerns clearly, despite their strong Norwegian or Slavic accents.

The next time I joined the fishermen's coffee hour, Milo Moore was present. He greeted me in a friendly manner, probably because we had been together on the *Santa Maria* and I was putting together a report for him on the distribution, abundance, and behavior of dogfish shark in the Pacific Northwest. John Wedin, who had selected dogfish as a topic for the morning discussion, also appreciated my being there.

During the World War II years, a high demand for vitamin D had led to an extensive fishery for the species. The war in Europe had disrupted the historical supply of cod liver oil. As a result, fishing for shark livers that were rich in vitamin D suddenly flourished along the entire Pacific coast of the United States and Canada. Sharks were harvested with line gear, set nets, and trawls. They were taken on board and gutted, their livers were stripped, and the carcasses then thrown back into the sea. Dogfish shark were particularly vulnerable because they were abundant, tended to form large aggregates, and feed near the seabed. During the latter years of the war and in the late 1940s, catches soared to well over fifty million pounds, and it was evident that the fishery was having a significant impact on the dogfish population. The conservation of the exploited sharks became a mute issue, however, because the development of synthetically produced vitamin D greatly lowered the price for shark livers, and the fishery for dogfish and other sharks dwindled away.

In its stead, a sport fishery that had developed for salmon in the inshore and offshore waters of Puget Sound and the Washington and Canadian coasts became the topic of discussion. As the population of dogfish increased, sport fishermen were increasingly plagued by the dogfishes' tendency to grab their salmon bait. The sharks particularly liked herring, which was extensively used as bait when the fishermen "mooched" for salmon; this entailed lowering the bait, frequently near the bottom, and then slowly spinning the herring behind the lead weight. The sport-fishing groups wanted the state to initiate a dogfish-control program or promote a commercial fishery for the "pest." The commercial salmon fishermen also supported the idea because the dogfish constantly fouled their gill nets and occasionally ended up in seines and on the lures of salmon trollers.

The morning session went well, but Milo took the opportunity to inform the group that biologists had screwed up the state's salmon and other fisheries. It was useless to enter into a debate with the director of fisheries. Milo's remarks were quite popular with the attending fishermen, and I couldn't really win the day even if I won the argument on merit. Milo had already caused several biologists to resign and seek work elsewhere, and I didn't want to join the exodus before I'd worked up the data from the Observer Program. The program was more than three years old and needed analysis. As Milo made his derogatory comments about biologists, the fishermen glanced at me, expecting some sort of defense. I remained silent but would, in time, respond. Since Milo related his dislike of biologists at each coffee session, I decided to add a little humor to the following week's meeting.

I asked Ruby to buy me a set of earmuffs and, at the next session, I waited patiently for Milo to make his weekly attack. Unfortunately, he never raised the topic, but time was on my side. The following week, he couldn't get through the meeting without his usual spiel. As soon as he began to deride the scientists, I pulled out my earmuffs and covered my ears. That got a great howl out of the participants, including Milo. I was surprised to find that he had a good sense of humor. When I returned to the office, Milo made no comment about the incident, but it did nothing to quell his favorite topic. I knew that it was just a matter of time before the anti-scientist atmosphere at the department became unacceptable.

Rhythms in the Sea

For the rest of the spring and well into summer, I worked steadily to organize and analyze the data collected in the Observer Program. I sorted the data by species, fishing grounds, depth zones, fishing effort expended (in hours), and catch for each of the four seasons. During the several years we had collected data from the fishermen, I detected what appeared to be repeating patterns of behavior. Every spring, the trawlers targeting English sole concentrated their fishing effort south of Cape Flattery at depths ranging between fifty-eight and sixty-four fathoms. The fish seemed to aggregate in a narrow depth range along regions with sandy and muddy bottoms. Frequently, the skippers reported that their entire catch was taken along one depth contour, e.g., one fathom. Dover sole seemed to frequent shallower depths during the summer and fall months than they did in the winter and early spring. The fishermen were well aware of these seasonal movements and shifted their fishing patterns to respond to the changes in the bathymetric (depth) distribution. However, there was no detailed description of the complex of bottom fish inhabiting the shelf and slope regions off the west coasts of the United States and Canada.

The data sorting took several days, and IBM summary printouts were stacked in piles more than two feet high on my desk. In a matter of hours, the computer had evaluated a mass of data that would have taken months, if not years, to compile and summarize manually. The computer expedited and enhanced scientific analysis to levels that were previously unthinkable. The scientist's job involved identifying the information to be gathered, designing the sampling methodology, and collecting and storing data in a manner that could provide outputs, i.e., addressed fundamental scientific questions or client inquires. Sorting and collecting the thousands of data inputs into identified variables, time sectors, and geographic areas was left to the computer. The power of the new computers compared to the computers used to break codes during World War II was tremendous, but the developments over the rest of the century would be even more extraordinary.

My review of the computer outputs began with a sort of all of the data sheets dealing with flatfish. Four species were important elements of the trawl catch: English sole, Dover sole, petrale sole, and starry flounder. The adults inhabited the continental shelf and slope waters between about ten and well over 200 fathoms. Of course, not all four of these species of flounder were caught on the more than twenty identified fishing grounds.

Small numbers of English sole were caught on most of the fishing grounds, but most of the catch landed in Washington was taken from south of Cape Flattery to Destruction Island (about forty miles south of the cape) and in Hecate Strait, north of Vancouver Island. When the seasonal catch data was plotted, it became clear that the English sole on both grounds were moving to deeper water during winter months and migrating into the shallower areas of the shelf during the summer months. The dominant catches during the summer were taken in depths from twenty to fifty fathoms, while in the winter the catches were predominately taken at depths from sixty to seventy fathoms. The data also showed that in the summer, the fish fanned out over a broad area of the shelf, while in the winter they were concentrated within a narrower depth zone. The inshore/offshore movements of the English sole was repeated for each of the three years for which data was available, confirming a seasonal depth migration that the fishermen had noted in their discussions with me.

Next, I looked at the seasonal movements of petrale sole. The Esteban tagging experiment had given us a good idea about the petrale sole migrations, but I wanted to see how well the seasonal catch data would bear out the earlier findings. The graphs clearly showed the petrale sole's distribution in deep water during the winter, but also showed that a portion of the catch was taken along the shelf at depths from fifty to about ninety fathoms. The split in the distribution of petrale sole during the winter months could be explained by the fact that the fish moved into deep water for several months. As their eggs became ripe, the fish moved to the deeper water spawning grounds and, after spawning, returned to the shallower feeding areas along the continental shelf. The seasonal movement of the adult portion of the population supported the results of the Esteban Deep study.

Dover sole generally occupied the deeper waters of the shelf and the upper slope to depths of about 300 fathoms. Fishermen working off the northern California coast discovered their concentrations in deeper waters. As the trawl fisheries off Oregon and Washington entered deeper waters, fishermen found an abundance of Dover sole, ocean perch, and sablefish. The seasonal catch data showed that Dover sole, like English and petrale sole, were migrating from deeper waters of the continental slope to the outer shelf during the winter, summer, and fall months. In late fall, they would return to the deeper offshore waters. All three of the flounder species showed the distinct offshore/inshore movements, but each seemed to inhabit different sections of the shelf and slope waters. The extent of the inshore/offshore movement was greatest for those species generally inhabiting the deeper portions of the shelf and slope.

When I began to look at the starry flounder data, I expected to see the same pattern, but restricted to the shallower portions of the shelf. Most of the starry flounders taken off the Washington coast during the study period were caught during the summer and fall months at depths between twenty and sixty fathoms. During the winter months, they were absent from the catch even though there was a considerable fishing effort between forty and 200 fathoms. They were simply absent in the depth range sampled by the fishery. Was it possible that they migrated to shallower depths during the winter? It seemed unlikely that the species would move to an area where they would be exposed to heavy winter storms. It was time to go back to the literature.

After scanning my collection of reprints, I found several papers that indicated that starry flounder were found in shallow waters in bays and near river mouths during the winter period. The catch data of a few vessels that fished the inside waters of southeast Alaska during the winter, and were outside my defined study area revealed they occasionally took large quantities of starry flounder near the Stikine River. The fish were known to inhabit saline and fresh water environments; perhaps they sought the protection of the bays and estuaries. The suggested bathymetric movement of the starry flounder differed from the well-structured pattern of the other examined fish, but then each species has found its own solution to survival.

The vast majority of the cod taken by Washington fishermen was from Hecate Strait, north of Vancouver Island, and from Cape Flattery south to Destruction Island off the northern Washington coast. This allowed me to narrow my study of the cod to two major fishing grounds. A Canadian scientist named Keith Ketchen had shown that the cod stocks appeared to move to deeper waters on the shelf during the winter months and into shallower waters in the summer and fall. The interview data clearly showed the inshore/offshore movement among portions of the cod stocks. They would move into the shallower waters during the spring and summer months, returning to the deeper areas of the shelf during late fall and winter. The cod stocks off northern Vancouver Island and off the Washington coast behaved in a similar way. The seasonal depth migrations were repeated each year.

Lingcod, Pacific Ocean perch, and sablefish also showed inshore/offshore seasonal migrations. The lingcod seasonal pattern was confined to the intercontinental shelf, while portions of the sablefish and Pacific Ocean perch stocks moved from the deeper waters of the upper slope well onto the shelf. On closer examination of the catch of each species by depth, it was evident that the average depth of all the examined species, except the starry flounder, was shallower in the summer months. Some species moved over an extensive depth range, and others moved in a somewhat narrower depth zone. The summer depth distribution was more extensive than the winter period. It appeared that almost all of the bottom fish formed denser concentrations during the winter when they were confined to a rather narrow portion of the annual depth-distribution range. During the spring and summer months, portions of the stocks of each species would move into shallower water, but a segment would remain in the deeper portion of their depth range. It was like one side of an accordion moving away from the other during the warm weather and then closing up again in the winter. The entire population of bottom fish was in constant flux, shifting inshore and offshore and, with some species, from south to north, responding to seasonal oceanographic signals.

The interview data had revealed an extensive seasonal movement of biological material, cycling inshore and offshore, but in response to what? The offshore winter movement from the shallow waters of the shelf could be a migration designed to avoid the impact of large winter waves. Could the migration of several species to depths well over 100 fathoms (600 feet) be explained in the same manner? The petrale sole excursion to deep water was brief, while portions of the same population remained inshore on the continental shelf. It was apparently a spawning migration, after which the species quickly returned to feed in the shelf waters.

After studying the region's physical oceanographic features, several cogent hypotheses emerged. Species, such as Dover sole, Pacific Ocean perch, and sablefish, that were most abundant in the deeper waters beyond the shelf edge, migrated shoreward in the spring and summer as the deeper slope waters welled up onto the shelf. They retreated during the winter as the upwelling process diminished.

The species that usually inhabited the shelf waters aggregated at greater depths where the wave action had minimal bottom effect and spawning aggregations could carry out reproduction successfully.

The movement of the petrale sole down the continental slope during the winter placed it in a position where the deeper waters moved northward, carrying the eggs and larva onto the shelf. There, they metamorphosed and settled on the seabed. As they entered the adult population, the fish migrated back to the south and spawned in the same areas as their parents, thus retaining the integrity of the population, much like salmon.

Starry flounder found an alternative spawning strategy. They generally inhabited the shallow waters of the shelf and found refuge in bays, inlets, and estuaries during the winter, where they could spawn without battling the winter storms.

There was no simplistic unifying scenario to explain the seasonal depth migrations. Individual species responded to a complex set of physical, biological and evolutionary signals.

My fellow workers Heater Heyamoto and Al Pruter, as well as Don Kauffman, the director of marine research, reviewed the study results. They urged me to publish the findings in the *Pacific Marine Fisheries Commission Bulletin*. I agreed, but the growing anti-biologist sentiment in the Department of Fisheries had soured the working environment. It had been a good place to hone and sharpen my scientific skills, and I had published several articles in the WDF's research bulletin. But I knew that it was time to leave; it was simply a matter of when.

Petrale—A Convenient Pen Name

In addition to the scientific writings, I had published a feature article in *Pacific Fisherman* magazine regarding the potential of Pacific Ocean perch. In preparing the story, I conducted in-depth discussions with DeWitt Gilbert, editor of the magazine and a number of other fishery-oriented publications. DeWitt was very knowledgeable about conservation efforts in the Pacific Northwest and well read on fisheries developments around the world. He was interested in my view on the extensive Japanese fishing efforts in the Bering Sea, where they fished for bottom fish, salmon, crabs, and herring, and in the eastern Pacific, where they fished for salmon. Japan's fishing activities had raised "holy hell" with U.S. fishermen, particularly Alaska's salmon and halibut fishermen. The International North Pacific Fisheries Commission (INPFC) had been formed during the mid-1950s to gain some understanding of the nature of and control over foreign fishing efforts in the northeastern Pacific. The commissioners were prominent individuals from the United States, Canada, and Japan.

The United States and Canada had enunciated the "Abstention Principle," which stated that foreign and distant water fisheries should not target stocks of fish that were fully utilized by a coastal state. The principle gave some protection, even if just nominal, to domestic interests and helped to stay the anger of the Northwest and Alaska fishermen.

Gilbert and I also chatted about the rapid post-World War II development of fisheries in the north Atlantic. The traditional north Atlantic fishing nations had returned to their historical fishing grounds with new and better vessels. The old steam trawlers were rapidly being phased out and replaced with modern diesel-freezer trawlers that ranged in length from 160 to 200 feet. They were giants compared to the small trawlers that worked the Pacific coast. Of even greater interest was the unexpected development of a large Soviet fleet that not only extended its operations to the traditional western European fishing grounds, but also developed a distant water fleet that was fishing off the eastern seaboard of the United States. At the end of our talks, DeWitt asked me to write a monthly column for *Pacific Fisherman*. I thought about it for a few minutes and said that I would love to but was concerned that someone at work might take issue with my views. That problem could be eliminated, DeWitt noted, if I wrote under a pseudonym that he would never reveal. Thus "Petrale," a monthly column presenting new developments in fisheries and commentaries on current events in fisheries, was born.

"Petrale" offered a great opportunity to present global fishery developments to a West Coast audience and comment on developments and issues of concern to local fishermen. I frequently found myself chatting with fishermen, biologists, and fishery managers about what they read in "Petrale." I even participated in a morning coffee session whose participants debated the identity of "Petrale." I continued to write for DeWitt long after I had left the Department of Fisheries and somehow my authorship remained secret over the entire period. The column paid $25 a month, which seems like small change, but it allowed Ruby and me to keep our heads above water while I worked for the state, which paid relatively low wages compared to the federal government.

The Great Adventure

In May 1957, a flyer from the United Nations Food and Agriculture Organization (FAO) announced a conference on world fishing methods to be held in October in Hamburg, Germany. The conference outline listed special sections on fishing gear and methods, nets, twines and selectivity, and behavior of fish to trawls, etc. During one of his visits to Seattle, Hilmar Christianson, the head of FAO's Fishing Gear Section, had encouraged me to attend the conference and present a paper on trawl fishing in the Pacific Northwest. It sounded great, but the cost was out of my league, about $1,200, including air travel, hotels, and food. DeWitt Gilbert said *Pacific Fisherman* would pay a couple of hundred dollars for a report on the meeting. It was a small start, but it got me to check out the flights from Seattle to Hamburg. The best I could do at the time was a ticket for about $600, but it allowed for stops in about eight different European cities.

Several days later, I went over to Ballard, the Norwegian sector of Seattle, and talked to Bob Stewart, owner of the Ballard Oil Dock, where a number of the local trawlers moored their boats. Bob had read about the conference and when I brought it up he asked, "Do you want to go?" I responded, "Hell, yes, but the department is not about to cough up the money and probably won't let me go." "Well", Bob said, "if you will prepare a report covering the meeting for our fishermen, Ballard Oil will ante up $500." Suddenly the impossible became possible, and I made every effort to put the trip together.

The first thing I had to do was get permission from the department, namely my supervisor, Don Kauffman; Don Johnson, the director of research; and the WDF director, Milo Moore. The first question from the first two in the chain of command was whether I was willing to pay for the cost of travel. When I told them about the two possible sponsors, much of their concern seemed to wane. When I told them that I had been invited to present a technical paper, they approved my going to the director. Don Johnson said, "If he approves, then you can start planning the trip."

The trip was still a big "if" because Milo was no fan of biologists. I tried to think of all the reasons why the department should approve the trip, but most of them seemed weak. I decided just to tell him about the conference and how the trip would be financed. To my surprise, Milo was very positive and inquired about my itinerary. He then requested that I add Brno, a city about eighty miles from Prague in what is now the Czech Republic, and Rome, Italy, to my trip. Some advanced aquaculture work was being done in a government laboratory near Brno. Milo also wanted me to look into the fish culture and aquaculture work underway in Italy. He had turned a fantasy into a reality.

I started making arrangements for the trip and working on a paper describing the local trawl fisheries. Before I had finalized my plans, a friend in the BCF called about a meeting of the International Council for the Exploration of the Seas (ICES). It would be held in Bergen, Norway, just prior to the UN meeting in Hamburg. If I were interested, he would arrange for an invitation for me to participate as an observer. The council was well known internationally for its contributions to ocean science studies and its advice regarding management of the northeast Atlantic fishery resources. The meeting sounded like a great opportunity to get exposed to the scientific methodologies employed in Europe and meet some of the world's leading fisheries biologists. I asked Don Johnson about attending the ICES meeting, and he said, "Well, I would like to have the chance to go and you shouldn't miss the opportunity as long as you are going all that way." Bergen was added to the schedule. An eight-day trip had been stretched to six weeks.

Twelve European cities were included in my schedule: Copenhagen, Denmark; Oslo, Bergen, and Stavanger, Norway; Hamburg, Germany; Amsterdam, Netherlands; Prague and Brno, Czechoslovakia; Vienna, Austria; Rome, Italy; and London and Aberdeen, United Kingdom. It was hard to believe the department had approved the travel, and I was excited for the next four months thinking about the trip. I finalized my travel arrangements and contacted the scientists and laboratories in Europe that would be important to visit during the trip.

Throughout most of the summer, my time was spent organizing and defining the various outputs that would be desirable for the observer data analysis. Before starting the study, it was important to compare the interview data with data of each catch and origin of catch taken from the pink sales slips. I completed this analysis and published a paper in the department's *Research Briefs*, which clearly showed a major difference in the two data sources. It was clear that when responding to the pink slip questions, fishermen tended to clump together their fishing areas to ease their jobs and squeeze the information into the abbreviated space available.

By the time the fall rains returned to Seattle, the position of scientist at the WDF was in a downward spiral. Al Pruter, who was one of my closer friends, had taken a job with the Halibut Commission, and the rumors of other departures were rampant. Every scientist in the department seemed to be trying to find work elsewhere. Don Powell, who was still with the BCF, asked if I had any interest in returning to the federal government. We agreed that I would get back in touch after my European trip. By October, I had completed the paper describing the small-boat trawl fishery that operated along the West Coast. It was one of the few areas where the trawl was set and recovered from the stern of the vessel. Most of Atlantic coast and European trawlers set and retrieved the trawl from the waist (midship) just forward of the house, which was located at the vessel's stern and ran forward a considerable way. Thus, the working area was forward of the house on the Atlantic vessels and on the stern in the Pacific. It wasn't until I started preparing the paper and educating myself about trawling elsewhere that I realized how much larger the high-seas trawlers were in many other areas of the world.

On a drizzly morning in October 1957, with passport, visas, and ticket in hand, I boarded a United Airlines flight to Los Angeles, where I connected to an SAS flight to Copenhagen. The plane was a four-engine propeller DC-6, I believe. Although my ticket was coach, there was more than adequate space for my legs, and I was quite comfortable. I wasn't clear about the route but knew it would be a long flight. We seemed to be fed continuously. We ate off white linen, and the food was excellent. The first leg of the flight took us north across the United States into Canada, where we landed in Manitoba late in the afternoon to refuel. We had been in the air almost six hours.

We then headed on north and east across Baffin Bay and made a second stop for fuel in Stromfiord, Greenland. As I recall, there was a relatively high cloud cover butting up against the ice sheet that covers a great portion of the massive island. The pilot brought the plane down under the cloud cover and then ran almost due east into an inlet with mountains towering on both sides of the plane. If the field were socked in, we wouldn't have enough room to turn back. But there was no fog; the plane dropped down on the runway, and we taxied to the fueling area. The passengers left the plane and entered a small holding area where local souvenirs were sold. I bought my first piece of travel trivia, a silver spoon for a collection Ruby wanted to begin.

From Stromfiord, we flew east across the Greenland ice cap and then just south of Iceland. I didn't sleep at all on the flight, although I made an effort; my motor was running too fast. As we approached the European mainland, I thought of the thousands of U.S. bombers that had rained terror on Germany in an effort to destroy their factories, military installations, and the morale of the general population. It was more than a decade after the war, but my memories were still strong. We landed in Copenhagen early in the morning, where I was to clear customs and catch a train for Oslo.

As soon as I checked through immigration, I boarded a bus that took me downtown to the main train station. It took sometime to figure out which track the train would leave from and to get my luggage to the departure area. It all seemed very confusing, but somehow I ended up in the right car, along with a group of young Norwegians. After traveling a short time, we boarded a train barge, which took us across the Baltic Sea to Sweden. By then I was fighting two opposing battles—the desire to acquaint myself with the Nordic landscape and the need to sleep. I had been awake for more than thirty-four hours, and my eyelids kept getting heavier. Sleep soon won out over curiosity, and when I awoke we had already passed through customs and immigration. No one had bothered to wake me and request my passport, so either they thought I was a local or the process wasn't very formal. At any rate, the view out the window suggested we were heading up the east coast of Oslo fjord. In Oslo, I went straight to the hotel, had dinner, and went to sleep. I had an early departure to Bergen the next morning.

I boarded the train to Bergen and the ICES meeting. The train struggled all morning as it wound up the mountains and finally stopped at a pass at the summit. There was snow on the ground, and it was quite cold outside. From that point on, the train wound down the west slope of the rugged mountains. We arrived in Bergen late in the afternoon, and I took a taxi to the Viking Hotel. That evening I met several of the American delegates, who told me about the next day's meetings and the various scientific sessions. They were courteous to me, but there was a formality that suggested that rank had its place. Still, they asked me to join them for breakfast.

The next morning, I attended a special session on the behavior and migrations of flatfish. I had read some of the speaker's publications and the subject seemed interesting. The meetings were relatively formal, technical and, for the most part, not very exciting; in fact, some verged on boring. Over the next few days, however, I met a number of Europe's elite marine scientists, whose works I had read. It was always a challenge to match a face with the vision that was stimulated by the works of a particular author. Most of the time, I was wrong; they were older, younger, friendlier, more reserved and, at times, even the wrong gender. My memories of the meeting are little more than a blurred vision of moving from one room to the next, going to several receptions, and visiting some of the sights of Bergen. If I learned anything, it was that European fish scientists were modeling and formulating the mathematics of fish population, attempting to sort out the consequences caused by natural events from those induced by human intervention.

It seemed that I had no sooner arrived than I was on my way to the airport to catch a plane to Copenhagen and then on to Hamburg, Germany, the site of the United Nations FAO meeting on Fishing Gear and Methods. As I rode into town on a bus, I could still see the scars left by World War II. There were remains of some bombed-out buildings, piles of brick here and there, and empty spaces where buildings had been torn down. At the same time, there was evidence of renewal and new construction. On the way to the hotel, I ran into Peter Schmidt, grandson of the founder of the Olympic Brewery in Olympia, Washington. Peter had graduated from the University of Michigan with a degree in marine architecture, returned to Seattle, and established a small shipyard, MARCO, that specialized in fishing vessel construction and deck equipment. The company had gained world recognition for its development of a powered block that facilitated retrieval of seine nets from the sea. We chatted for a few minutes and then decided to share a hotel room.

When we arrived at the large pavilion housing the meeting and toured the fishing gear displays, we were both impressed by the tremendous advancements in net designs and fabrication, diesel engines, powered deck machinery, and navigation. The war that killed millions and left northern Europe in ruins also created a heritage of technology that would replace steam engines with diesel and cotton and other natural fiber nets with synthetic nets. There was an array of new synthetic twines and navigational and acoustical depth- and fish-finders. Within a decade the new technology and the global demand for animal protein would foster a modern high-seas fishing fleet that would invade every major ocean of the world, from pole to pole. I was probably more mesmerized than Peter, who kept more abreast of global fish developments. We had the opportunity to go aboard several of the German large side trawlers (approaching 200 feet) and a new Russian factory trawler (about 260 feet). They both caught and processed fish on board. The Russian vessel was so big it looked like a cruise boat or small whaler. The vessels dwarfed anything that existed in the Pacific Northwest; I was almost embarrassed to give my talk on small West Coast stern trawlers.

Over the next few days, as I moved about the exhibits and listened to the formal presentations of scientists and technicians, it rapidly became apparent that fishing was undergoing a technological revolution. The role of the world's oceans in supplying needed animal protein was escalating in western and eastern Europe, Japan, and Korea. The conference participants were so enthusiastic about the growth of ocean fisheries that they gave the impression of being caught up in a fever. It wasn't a surprise to see western Europe and parts of Asia involved in the race to the sea. What was a surprise, however, was the sudden emergence of the eastern bloc nations, which had no substantial history in world fisheries prior to World War II. The Soviets talked about their ten-year plan and their intent to greatly expand high-seas fishing using mother-ship operations and a fleet of modern, self-contained catcher/processor ships. They also talked about their efforts to learn more about the oceans' great capacity to produce food for mankind. Most Americans were unaware of, and/or took little interest in, the race to the sea.

The United States had been one of the major fishing nations of the world since it gained its independence. The nation's fisheries got a shot in the arm during World War II as the demand for fish and fish oil to feed to country's service personnel gave added impetus to coastal fisheries. When the war ended, however, the U.S. demand for fish products increasingly was filled by imports from Europe and Japan. Their fishing fleets delivered fish products at lower prices and their products were frequently better. Major sectors of the U.S. fishing fleet had become stagnant and were economically constrained. For the most part, the industry was unable to participate in the race that had energized national fisheries in many areas of the world. The potential of the oceans to feed an ever-expanding global population, however, was not lost on or ignored by all of the nation's leading ocean scientists. Some of them, including Web Chapman and Benny Schaefer, attempted to inform the U.S. Congress about the growing importance of the global ocean and the sudden emergence of the Soviet Union as a major player in the field of fisheries and oceanography.

I gave my talk on the next to last day of the meeting, humbled by the presence of many of the world's authorities on fisheries and ocean sciences, most of whom I knew only through their literary contributions to fishery science. The next morning, I headed for Amsterdam; one of the Dutch scientists had asked me to visit their national laboratory in Imouden. The head of the lab was one of the world's leading experts in oyster culture, and several of the scientists had made major contributions in methods of enumerating fish populations. The visit gave me a glimpse into western Europe's methods for monitoring fish stock population trends. Their fisheries dwarfed those on the West Coast of the United States, as did their commitment to understanding the consequences of fishing activities on the oceans' living resources. I met Director Korringa (I had read some of his works on oyster culture), and several scientists gave me details on some of the lab studies on flounder migrations. They had read the work that Chatwin and I had published in the Canadian fisheries journal. Unfortunately, I had only one day for the visit, and we had hardly broken the ice when it was time to return to Amsterdam and begin the next leg of the trip to Prague and Brno.

I was up early the next morning for my flight to Prague and checked to make sure I had my tickets, visa, and correspondence with the Czech Fisheries Laboratory in Brno. I felt a little uneasy about going behind the Iron Curtain, as U.S. relations with the Soviet bloc were shaky at best. Furthermore, the lab's communications with me had been rather curt, but they had said I was welcome to visit their facility. Milo Moore encouraged me to visit Brno because the scientists there had developed advanced methods of speeding up the reproductive cycle of various species of fresh-water, cultured fish. The immigration officers at the Prague airport looked through my documents and asked me several questions about the nature of my work. The process seemed to take an awfully long time, but it was probably exaggerated by my nervousness. Once my passport and visa had been stamped, I felt more at ease. I made my way to the section of the airport that handled in-country flights and took a seat. The Brno flight didn't take off for a couple of hours, so there was time for a cup of coffee and a look through a souvenir shop, where I purchased another silver spoon for Ruby's collection

The plane to Brno appeared to be a DC-3 or an old C-47. The interior was dirty, parts of the seat coverings were torn, and the plane seemed to be rundown. The passengers getting on the plane were carrying all sorts of packages; there simply wasn't enough room in the overhead area for all the items, so they were placed in the aisle and no one complained. In a roped-in area behind the door, the stewardess was helping to stow various wildlife, including a dog and some chickens. When the door was closed, the pilot taxied down the runway, turned into the wind, and gave the engines full throttle. There was no revving of the engines before the takeoff. Within seconds we were off the ground, chickens and all. An elderly woman sitting next to me tried to communicate with me. I believe she asked if I could speak German, but we made no progress, although perhaps she understood that I was American. Despite the appearance of the plane, it lumbered on and landed safely in Brno, a little more than a hundred miles to the east and slightly south of Prague.

Again, my nerves were on edge. Almost no one at the airport could speak English, but the taxi driver looked at the address of the hotel the government Intourist people had arranged for me and headed to town. Brno was a relatively large manufacturing city, not particularly attractive, and I didn't know what to expect in the way of accommodations. Tourists going to Czechoslovakia were required to pay the cost of hotels, travel, etc., before entering and were given a set of certificates to use instead of cash. Upon arrival, travelers took whatever had been booked. It turned out to be a pleasant surprise when the driver pulled up to the hotel. It was not a large building, but it was clean and attractive. At the front desk, a blue-eyed blonde, who looked like she had just stepped out of *Vogue*, asked in excellent English if she could help me. I relaxed, knowing that there was someone at the hotel with whom I could communicate. She arranged for me to visit the Fisheries Laboratory the next morning.

My room was pleasant and nicely furnished, and I lay down to rest for a few minutes, reflecting on the different world I was in. Besides the language problem, there was the question of my acceptance at the lab and my concerns about wandering around alone in a then-hardcore communist country. My thoughts returned to the attractive woman at the desk. Why should a beautiful woman, who spoke such fluent English, work in a small hotel? Was it possible that she worked for the government authorities? At 7:00 p.m. I went down to the dining room. The menu was in Czech and French, and I could understand just enough to make out chateaubriand. It sounded good, so I pointed it out to the waiter, who smiled and left for the kitchen. The dinner was one of the finest I have ever consumed. Along with the steak there were braised potatoes, fresh string beans, excellent bread, and a glass of local red wine. The waiter knew a smattering of English and attempted to chat as he served the meal. After dinner, I went to my room and decided the hospitality of the Czechs had survived even under communism.

After breakfast the next morning, I sat in the hotel lobby, waiting for the driver from the lab. My concerns about my reception there were still with me. On arrival, I was taken to the director's office and told that the scientist who had corresponded with me was in Prague on business. I became a little tense, but they showed me around the lab, talked a little about their work with accelerating maturity in cultured fish and gave me a book on the lab's work. I was introduced to several of the top scientists and to two Soviet investigators who were there on a training mission. At least, that is what I was told.

Nevertheless, it was clear that the lab personnel were uneasy about my visit. Was it because the scientist who had invited me was not there? Or was it not the right time to host an American scientist while two Russians were on staff? Or was the interpreter having a difficult time and unhappy with his task? Perhaps it was all a figment of my imagination; I would never know. At any rate, my visit was over before noon, although I had anticipated spending the day at the lab.

I was not sure how to spend the rest of my time in Brno and felt a little depressed. My flight back to Prague wasn't until late the next afternoon, but I decided to get back there as soon as possible. I explained my dilemma to the blonde receptionist at the hotel. After finding out that my next destination was Vienna, she said, "There is no reason to return to Prague, you can catch a train that will take you to Vienna. It is less than a hundred miles to the south, and it will only take you several hours. You can take a train tomorrow morning, and you will enjoy the scenery." She helped me book my travel and get the train ticket. In the morning, I would be off on a new adventure and the thought of getting back to the west was comforting.

Several people sitting next to me on the train tried to make conversation, but the few English words they knew and my absolute unfamiliarity with Czech rendered the attempts useless. After more than an hour, the train pulled into a small town, where officers boarded and began checking passengers' travel documents. Two officers made their way up the rows of seats in the car, reviewing each individual's papers. When they arrived at my seat, one of the officers said something in Czech, which I interpreted as a request for my passport and visa. He looked at them quickly and then stuck out his hand. At first I had no idea what he wanted, but then he said, "billet," which I recognized as a request for my ticket. After taking it, he began to examine my documents. He seemed to have a problem, because he kept staring at the visa and looking back at me. When he called the other officer over, I knew they were not satisfied with my papers. The second officer studied the passport and visa for several minutes and then told me in broken English to get my luggage and follow him. He took me into an office in the station and asked me to sit and wait. He then returned to the train, which left without me about twenty minutes later. My mind searched for a rational explanation.

For some reason, I was not being allowed to proceed, and I remembered everything I had ever read about westerners being arrested for no reason in communist countries. My documents had not been returned to me, and from time to time, several of the officials in the room would look at me and then return to their work. I reviewed the past twenty-four hours: the rather shabby treatment at the Fisheries Laboratory, the attractive blonde coaching me to take the train, and my sudden removal from the train. Had I been set up? I assumed they were waiting for someone who spoke good English to interrogate me. No one talked to me for the better part of an hour, but eventually a young lady gave me a cup of coffee. After almost two hours, which seemed an eternity, a train headed in the opposite direction stopped at the station, and an officer motioned for me to follow. We walked several cars back and stored my luggage, and I was shown to a seat. The officer sat next to me and managed to convey that we were en route to Prague. I assumed that I would be arrested on some trumped-up charge and began to imagine my fate.

The officer chatted with a number of the passengers sitting near us, and a number of them looked my way. I could tell nothing from their expressions; they just seemed curious. Later one of the passengers offered me an orange, which I took and ate. It seemed a friendly gesture considering that I must have been considered the "unclean." During the long trip to Prague, I thought of my wife and family and that it might be some time before I would see them again. When we finally arrived, the officer led me to a military car that took me, not to prison, but to the Czech Intourist Office. There a very pleasant man told me that, according to the law, I could exit the country only by the means and route identified on my visa, which he noted was by plane from Prague to Vienna. They had been notified by phone that I was being brought back to Prague and had made reservations for me at a first-class hotel. They also said that the officer had been sent with me to ensure that I arrived safely at Intourist and that they could alter my visa so that I could take the train to Vienna.

I was so shaken up by the events of the day that another train ride was out of the question. The officers had merely followed the rules and had gone out of their way to take care of me. It was, however, too late. My imagination had run wild, and in my mind, I had already spent ten years in a Czech prison during the trip back to Prague. I just wanted to get back to the West. I left for Vienna the next morning on a Pan American flight. As we cleared the field, I felt a sense of relief even though I recognized that my torture had been self-inflicted.

The remainder of the trip went well. I arrived in Rome two days later and was taken to several fish farms in the south. One farm had been operated since the time of the Roman Empire. The ponds, formed with old Roman brick, covered ten or more acres and enclosed a lagoon that was perhaps several hundred acres. The crop was mostly mullet, which had also been a favorite in the ancient empire. The culture techniques were quite simple and didn't seem to have changed much in eighteen to twenty centuries. I spent a day at the FAO, where I met some of the leading marine scientists of the time. Later in the week, I took in Rome's tourist sites, including the Coliseum, old Roman baths, Circus

Maximus, Forum, Piazza del Popolo, Spanish Steps, and numerous other attractions before traveling on to the United Kingdom.

From London, I went east to the great trawl city of Grimsby and to the famous British Fisheries Laboratory at Lowestoft. After a day of visiting with the lab's scientists, I headed north to visit a Scottish laboratory in Aberdeen. The port had one of the largest fish auctions in the region, and the catches of the vessels were sold off every morning between approximately 5:00 and 7:00 a.m. When I got there, about ten steam trawlers were tied along the extensive dock area. Steam from the trawlers' stacks was drifting up into the early morning sky. It made a fine photo; from my angle I could capture the steam from the row of vessels sitting against the front of the dock. It was an opportune time to capture the scene because within half a decade, these trawlers would disappear from the fleet as diesel engines phased out steam as a propulsion source.

I returned to Seattle and wrote several reports about my trip, one for the state and one for Ballard Oil, which had helped sponsor my trip, and several articles for *Pacific Fisherman*. It had been a great personal experience and introduced me to a number of Europe's leading marine biologists. The reports were well received, thus adding to my recognition within the scientific and Seattle community.

Time for a Haircut

Things had not improved for the biologists working for the Washington Department of Fisheries, and several more key scientists had moved on to other jobs. I had made up my mind before leaving for Europe to start looking for a position with another agency. My first thought was to check back with Don Powell, who had been running the Exploratory Fishing unit in Seattle. Don stressed the necessity of collecting and evaluating the data in a scientific and professional manner. When I contacted Don, he opened up the conversation by asking if I would like to take over his job at the Bureau of Commercial Fisheries, Fish, and Wildlife Service. Don had received a promotion and was going to Washington, D.C., to head up the national program. Ocean exploration had been my great "love affair," but I had abandoned it because the local leadership was not committed to the application of scientific methods to the survey work. That had all changed, and I couldn't believe my luck. I told Don what was going on over at the WDF and that I could tidy up my work for the state in about six weeks.

Milo Moore's disdain for scientists had become increasingly vocal, and he seemed determined to pare down the department's research component and build up its hatchery and fish-culture sectors. On a personal level, I had gotten along quite well with the controversial director, and he had supported my trip to Europe while others under him had not. Nevertheless, the working environment for biologists was rapidly deteriorating. One morning in the winter of 1958, I was asked to go to Moore's office. There, I found Don Johnson, head of the research department, Dick Presey, who headed up the salmon studies, and Don Kauffman, head of marine research, all seated in front of the director's desk. I didn't have the slightest idea what Milo had in store for the group, and I'm not sure he did either. He started off with a rambling commentary about how the scientists had ruined the department and why they were responsible for the deteriorating salmon runs. No one interrupted his tongue-lashing, and since I wasn't involved in salmon research, I remained silent along with my peers.

After about thirty minutes, he suddenly changed his tone and his comments became quite personal. He accused the scientists of being liars, dishonest, and a bunch of "long hairs." My colleagues' faces were frozen, but I'd had enough and started toward the door. "Where are you going?" Milo asked. "To get a haircut," I responded. I went down to my office and wrote my resignation. Within two weeks, I had moved my belongings and computer outputs from nearly five years of study to the U.S. Fish and Wildlife Montlake Center, where I had started my career some eight years earlier. I never really understood Milo because he was always friendly to me even after I stomped out of his office. Perhaps he had achieved what he had set out to do—trim down the number of scientists at the WDF. After I left, the flight of biologists—including Dick Presey and Don Johnson—from the department continued.

The small trawler Kristine, 1950's

Chapter IX
The Exploding Fish Scene

During my absence from the Fish and Wildlife Service, fisheries throughout the world moved into high gear. The Japanese returned to the Northeast Pacific and were fishing for salmon, king crab, bottom fish, herring, snails, and shrimp. Their operation in the region began in the early 1950s, and their fleet gradually increased each year. On the one hand, U.S. and Canadian fishermen were fit to be tied about the Japanese salmon operations and the Alaskans were upset about the king crab fishery. The Japanese bottom-fish operations were, however, less offensive to our ground fishermen, other than the halibut line fishermen. The trawl fisheries off the contiguous Pacific states were a long way from the Bering Sea and did not compete for the same markets to any great degree. The high-seas, salmon, drift-net fisheries, however, had kindled a significant political fire along the U.S. and Canadian Pacific coasts. The American fishermen and processors were convinced that the Japanese were harvesting thousands of fish that were of U.S. and Canadian origin.

The salmon industry, one of the more politically potent elements in U.S. fisheries, had put pressure on Congress to exclude foreign high-seas salmon fisheries in the eastern Pacific. Many hoped that the scientific investigations would demonstrate that the U.S. and Canadian fishermen were fully utilizing the surplus salmon runs in the northeastern Pacific and that the Abstention Principle would force the Japanese high-seas gillnet fisheries to the west. But it was not so simple. The Japanese fielded excellent scientists who buried the Americans and Canadians in requests for answers to difficult questions that postponed their retreat from salmon fishing off Canada and the United States. The studies on the origins of salmon feeding on the high seas were only a few years old and, although we had begun to get a handle on the high-seas distribution of salmon, we had not yet proved that the eastern Pacific stocks were fully utilized. Thus, for the time being, there was a stalemate; the political chess game for the valuable fish resources of the region raged on as the Japanese continued to expand their non-salmon fisheries in the Bering Sea.

The U.S. fishery problems, however, were not confined to the Pacific region. Distant-water, trawl-fishing fleets from a number of western European countries, as well as the USSR, Poland, and East Germany, had begun to expand their operations on a global scale. U.S. and Canadian trawlers and small local boats watched as large factory trawlers from Europe swarmed over their traditional fishing grounds off New England and eastern Canada. For more than a hundred years, the internationally recognized control over adjacent seas had been three nautical miles. This meant that most of the productive fishing grounds next to coastal states were accessible to foreign fisheries. The northwest Atlantic region also had an international fisheries commission, but its decision-making process and operational modes were much different than those governing foreign fisheries off the Pacific coast of the country.

Other segments of the U.S. fishing industry were not overly enthusiastic about the Abstention Principle. The shrimp and tuna sectors were opposed to the principle's concepts, particularly the fishermen who operated off Mexico and Central and South America. Similarly, there was no big push for the principle within the New England community, which fished the productive Grand Banks off Canada. U.S. fishery policy became increasingly strained as different sectors sought to respond to the rapid changes in the industry. The presence of catcher mother-ships operating in our backyard, exploiting fish sought by U.S. fishermen, was causing increasing stress between factions of the domestic fleet. The situation was further aggravated because U.S. fishermen were having a difficult time competing with large government-supported ships from behind the Iron Curtain and with western European vessels that the Americans claimed were built with generous government subsidies.

Ground-fish imports into the United States were increasing, while the U.S. trawl fleet was finding it difficult to fish for anything besides the fresh market.

The Domestic Scene

My return to the government fishery agency came at a time when Ruby and I were finding it more and more difficult to deal with our financial burden. Bob and Sue were about nine and six, and soon outgrew their clothes as they sprouted up like weeds. We needed a new car and to finish furnishing our new home. We had managed to grade and put in our front lawn but could not afford to buy plantings for the yard.

Before we moved into our new home, our circle of friends included Verne Benedict, with whom I had served in the war, a few of my U.W. classmates, and neighbors in the Cedar River Park housing project. We also corresponded with Jim Murphy and Chris Vesper who lived in Los Angeles and visited both during trips to California. Normandy Park had grown rapidly since our arrival in 1951 and was filling up with young families. A new home had been built next door, and the Smiths moved in with three children aged three and six. The Donovans built directly behind us and added three children between four and ten years old to the local population. Across the street to the north, the Berrymans joined the neighborhood with four more children. Within a few years, we had developed a community network of friends. In addition to the families noted above, we formed bonds with the Braicks, who lived on the street behind us, and later the Osbuns and the Slatvigs joined the "clan." Like the Alversons, most of them had migrated to Seattle from other states and had no local relatives. Thus, we tended to join together on holidays, at dances at the local community club and other social events, and at the local beach. These new acquaintances, including the young ones, would remain our friends for the rest of our lives.

Most of the Normandy Park men were either pilots or engineers who worked at Boeing. All of them seemed to be in a better financial situation than we were. The move back to the "feds" involved a substantial pay increase that would offset a four-year drain on our savings. We had been supplementing our monthly salary with occasional draws from our savings, and we were close to rock bottom. The return to the exploratory group constituted a significant turn in our fortunes and the beginning of a professional career with the federal government. We soon would have the means to finish furnishing our home.

Back to the Sea

By the time I returned to the BCF, several new scientists had joined the exploratory unit, and the budget had been increased to accommodate additional work. During my absence, the bureau continued its survey work on bottom fish off Oregon and Washington and pink shrimp populations off Alaska. The *Cobb* had conducted a deep-water survey work off Oregon and Washington and shrimp studies in the Prince William Sound, Kodiak regions, Cook Inlet, and Alaska. Promising populations of several species of northern shrimp had been located around Prince William Sound, and large concentrations of pink shrimp were reported off Kodiak Island and to the west, near the Shumagin Islands. In fact, the quantities of pink and other northern shrimps in the Gulf of Alaska, west of Kodiak, suggested that it was possible to develop an extensive U.S. shrimp fishery in the region. The work off Oregon and Washington continued to demonstrate that the continental slope region hosted a rich population of rockfishes and black cod. There was also an increasing interest in the potential shrimp population in the Oregon region, and BCF had planned a shrimp survey there even before my return to the Exploratory unit.

I decided to get back into the swing of things immediately and joined the cruise to the south. We made our early shrimp surveys off Alaska using beam trawls and traps. We decided to switch, however, to the Gulf of Mexico shrimp trawl in future surveys. Shrimping was a major industry in the gulf, and nets for their capture had evolved over the years. The nets were constructed to fish near the bottom with the net opening set at a relatively low height. They worked well as the shrimp in the gulf lived close to the seabed and buried themselves in the mud-substrate during portions of the day. Unfortunately, it took us several years to realize that the northern shrimps had a more pelagic behavior and, at times, inhabited waters well off the bottom. Despite the inefficiency of our sampling tool, we discovered several relatively large concentrations of pink shrimp off the Oregon and later the Washington coasts.

Most of the concentrations of pink shrimp were found at depths between fifty and eighty fathoms, but at times we ran into good catches of side-striped shrimp in deeper water. A small pink shrimp fishery had operated out of Petersburg, Alaska, since the late 1800s, but had not expanded into the inside waters of southeast Alaska or off the contiguous West Coast states, where the same species was known to live. Following publication of the *Cobb's* findings, there was an increased interest in establishing commercial operations in the area. Several problems, however, confronted the development of a viable commercial operation. Pink shrimp were relatively small, ranging between fifty and more than a hundred shrimp per pound, depending on the age and sex composition of the catch. In Alaska, the carapaces of the shrimp were removed by hand, leaving the small white meat often used in shrimp salads. The process was labor intensive and high-volume production depended on the availability of a mechanical shucking process. Thus, the industry imported automatic shucking machines from the Gulf States and designed new or modified equipment to handle the small Pacific pink shrimp. Eventually, fisheries developed in a number of coastal ports from Washington to Northern California and later in offshore waters of Alaska.

During the development stages of the fishery off Washington, Chet Peterson, owner of a small coastal trawler called the *Mylark*, visited my Montlake office. Chet lived in Westport, Washington, and wanted to get into shrimp fishing since the bottom-fish market wasn't great. We discussed the recent survey findings and the fishing, rigging, and potential catch rates. I told him that the potential catch rates for shrimp seemed good, but that there was risk involved. Chet said, "Fishing is always a gamble, and I suspect that's why a lot of us are in the fishing business." Within a month, he had rigged the *Mylark* for shrimping and was delivering his catch to a plant in Westport. Chet later went north to Alaska and became one of the leading shrimpers operating out of Kodiak.

By the end of 1958, our scientific group had completed studies on pink shrimp spanning Oregon to the Aleutians and published a review of our findings. The results of the work done in the fifties led to a database that allowed the West Coast shrimp fisheries to flourish from northern California to Alaska. In their heyday, they produced more than 150 million pounds annually—much of which came from grounds originally surveyed by scientists on the *John N. Cobb*.

After we completed the shrimp work in early winter, we returned to our staging-area dock at the foot of Stone Way on the Lake Washington ship canal. During the winter months, we usually had the vessel hauled, cleaned, and checked; the work was quite routine, and the cost well within our planned budget. We took the vessel to the Pacific Fisherman's Shipyard, which was smack in the middle of "snoozeville" (Ballard, the Norwegian "fiefdom" in Seattle). The vessel had not been up on the "ways" for more than forty-eight hours before the shipyard foreman called. "Hey," he said, "you have a hole on the port side of the vessel. Some of the planks outside of the freezer have dry rot, and we have to explore and see how bad a problem you have." His words sent shivers down my spine. They meant that the wood rot had been found in the exterior shell, which could require major rebuilding of the vessel's hull.

I crossed my fingers and hoped for the best. The discovery could not have come at a worse time for our program. The Atomic Energy Commission (AEC) asked if we would participate in a survey by the marine fish and invertebrate program in the Chukchi Sea, just north of Bering Strait. As a part of Project Plowshare, a program that harnessed nuclear energy to build structures, the AEC had considered the possibility of using an atomic weapon to create a harbor south of Cape Lisbon, the "Chariot Site." Before carrying out this ambitious and somewhat unpopular project, the commission scheduled a number of biological studies to identify the flora and fauna in the region that might be affected if an atomic bomb was detonated. The marine resources of the Chukchi Sea, which was bounded by northern Alaska on one side, Russia on the other, and the polar icecap to the north, had almost no history of exploration.

It was a dream come true, and the entire staff was excited by the possibilities. The work was scheduled for late July and August after the ice had retreated far enough to the north. The immediate questions facing us were the extent of the dry-rot problem on the *Cobb* and the time it would take to repair it. Three days after hearing about the problem, I made a trip to the shipyard. The seven-foot hole had grown about fifteen feet forward and eight feet up and down between the guardrail and the keel. It was obvious that it had also gone into the ship's major ribs just aft of midship. The foreman said it might be another week before the shipyard determined the full extent of the problem.

The dry rot turned out to be worse than we could imagine. Rotten wood had been ripped away from about twenty-five feet of the portside, and it extended from the keel to the deck guardrail. Eight of the major ribs had to be replaced and new planking added to replace the wood that been removed. The problem started around the large freezer that had been built against the skin of the vessel. There was simply no air circulation to allow adjacent wood to dry. To remedy the situation, the area had to be redesigned with space between the freezer and the hold. Confident that the work would be completed in time for the Chukchi Sea trip in July, we added the task of building a wet lab into the hold area. The total cost of this work almost equaled the cost of the vessel nine years earlier. I had to contact headquarters in Washington for help. It was eventually given, but not before most of the work had been completed.

In preparation for the AEC work, I hired a young Ph.D. candidate from the University of Washington, Walter (Wally) Pereyra, who had majored in invertebrate biology. He was considered one of the top graduate students at the U.W. School of Fisheries. Wally was assigned the job of developing a sampling system, collecting, identifying, and evaluating the distribution and abundance patterns of everything from shrimp to tube worms. The fish species of the Chukchi were not well known, so I contacted a friend who worked in Alaska, Norman Willimosky, a fish taxonomist who had published a number of papers on the fish species inhabiting the waters off Alaska. To fill out the scientific staff, I arranged for Ford Wilkey, a marine mammal and seabird specialist with the Fish and Wildlife Service, to join the cruise. We would go north with four scientists; Norm would join us in the Eskimo village of Kotzebue, situated to the north and east of the Bering Strait. While the problems of preparing for the Chukchi survey kept us busy, another issue captured my attention.

Invasion

In March 1959, Stan Patty, the marine writer for the *Seattle Times*, told me that the U.S. Coast Guard was reporting that about one hundred Soviet vessels were fishing the large continental shelf off of western Alaska. The fleet appeared to be comprised of trawlers, mother-ship processors, support ships, and perhaps several whaling ships. We should have anticipated the arrival of the Soviet fleet in the region, as we were aware that they had been conducting fishery/biological surveys in the area. The success of the Japanese fishing fleet was well known in Moscow. This development had the whole West Coast fishing community on edge, and there was all sorts of speculation regarding which species of fish and/or shellfish the Russians were targeting.

Stan asked me to examine some pictures he had obtained from the coast guard. When I arrived at his office, he was reviewing the photographs with one of the paper's editors. I joined them and began looking over the pictures of vessels and fishing operations. I could identify most of the craft in the photos from vessels or pictures I had seen at the United Nations conference in Hamburg or elsewhere in Europe. The Soviet entourage seemed to include a number of catcher vessels about 125 feet in length. Several large mother-ships, which received and processed the catch, were over 300 feet long. There also were larger catcher-freezer trawlers about 180 feet long and a ship that appeared to be some sort of patrol vessel. Pulling a picture from the bottom of the stack of photos, Stan asked, "Is this a whaler?" It was one of the Soviet's new catcher-processor trawlers (Pushkin class), about 270 feet in length, a carbon copy of the vessel I had visited at Bremerhaven in 1957. The construction of this new class of catcher-trawlers took place in the western Soviet Union and in Polish shipyards. The vessels apparently had been transferred along with others to Nakhodka, just north of Vladivostok. It was not clear which species of bottom fish were being fished, but from the photos, I guessed that they were mainly flounders.

The arrival of the Soviets in the Bering Sea, with a fleet considerably larger than the Japanese fleet that had been in the area since the early 1950s, raised the hair on the necks of fishermen from California to Alaska. Several scientists didn't believe that the fishery resources could sustain such a large fishing fleet in the region. Others were concerned that both the Japanese and Soviets were catching large quantities of young halibut that had been fished by both American and Canadian fishermen in the Gulf of Alaska. Early studies by our own government just after World War II, however, had shown that large quantities of flounders and pollack inhabited the region.

Some even speculated that the Soviet fishing fleet was there to cover up spying activities. Several weeks after the fleet arrived, a caller from the bureau's office in Washington, D.C., asked if we could estimate the potential catch of the Japanese and Soviet fleets in the region. I met with Tack Mirahara, a Japanese-American scientist studying the foreign crab fisheries in the Bering Sea, who was very familiar with the Japanese bottom fleet. We put our heads together and came up with a minimum estimate of 250 million pounds of assorted flounders, cod, pollack, and rockfish a year. The figure was almost twice that landed by the entire U.S. trawl fleet fishing off the West Coast. Many of our colleagues and peers called our estimate ridiculous and much too high, but the scientist in charge of the bureau's research effort in Washington, D.C., felt that it was a reasonable figure. When catch data was finally obtained, it turned out that our estimate was lower than the actual catch made that year, 1959.

To better understand the fleet and its targeted species, our regional director asked me to join a coast guard flight over the Soviet fishing operations. We flew a seaplane across the Shelikof Strait to the Alaska mainland and then west to the Bering Sea. The plane was somewhat smaller than the old PBYs, and when it shifted from side to side so we could look out the windows for fishing vessels, the pilot would adjust for the change in weight distribution. After several hours in the air we spotted a Soviet mother-ship with several catcher trawlers tied alongside. The pilot made a run over the vessels, and we saw flounders on the decks. Further on, we ran into several larger freezer trawlers but did not spot any of the large mother-ship factory trawlers that I had hoped to see.

The Soviet movement into the Bering Sea was part of a worldwide expansion of ocean fisheries caused by the depletion of stocks in the home waters of many countries and the search for richer fishing grounds in distant oceans. The Bering Sea was an obvious target for the Soviets since it was relatively close to many of their Siberian ports, and their rapidly expanding fleet required fishing new and more productive grounds. What was happening off the Alaska coast also was taking place off eastern Canada and the United States, North Africa, Argentina, and parts of Asia. The distant-water fleets operated under the long-standing legal concept of "freedom of the seas," and everything outside of the three-mile limit was up for grabs. They were "colonizing" the world's oceans from pole to pole in every major ocean. Coastal fishermen in the United States and many other countries were furious as the stocks that they had traditionally fished declined rapidly due to what was termed "pulse fishing"—intensely fishing in an area and then moving on to more productive regions.

The Chariot Mission

By June, the work on the vessel was largely completed, and we had gathered a range of sampling devices, including a bottom trawl, mid-water trawl, small mesh shrimp trawl, bottom dredge, gill nets, and assorted oceanographic devices to sample the salinity and temperature of the bottom sediments. The vessel had a new captain, Peter (Pete) Larsen, who had considerable experience in Alaskan waters, although he had not worked north of Bering Strait. He noted that the Loran lines didn't cover the area and that we would have to do the best we could with radar, radio-direction finders, and dead reckoning. We planned a series of track lines that would run perpendicular from the coastline that moved northwesterly from Kotzebue Sound toward Cape Lisburne. The station pattern would take us from about 66 to 69 degrees north latitude, well into the Arctic Circle.

Pacific Fisherman Shipyard completed the work in early July, and we began loading supplies on the *Cobb* before she was out of the shipyard. The vessel sailed from its moorage at Stone Way after mid-July. I had a number of loose ends to tie up and decided to board the vessel in the Pribilof Islands. I had developed a rather bad allergy, and the doctor didn't think the trip was a good idea, but it had been giving me a bad time for almost six months and nothing helped. Several days later, Ruby took me to the airport along with all my pills and I was off. I arrived a day before the *Cobb* and stayed in one of the government's guest facilities.

The next morning the superintendent showed me around the island, including the haul-out breeding areas for the fur seals that made their home on the small islands situated in the middle of the Bering Sea. The Pribilof Islands were also home to a number of Arctic birds and a large number of fox. Other than some short shrubs and stunted trees, there was little in the way of flora. Early in the afternoon the *Cobb* sailed up the east shore of the island and anchored just off the small harbor. A small launch took me out to the ship, and within the hour we were heading for Nome, south of the Bering Strait. We planned to investigate any area where we found Soviet vessels to determine the composition of bottom fish on the grounds where they were fishing. We were particularly interested in finding out whether there was halibut in the area.

After several hours, we saw a vessel on the horizon. As we approached, we could make out what appeared to be a Russian mother-ship with several catcher vessel boats tied to its starboard side. I discussed the issue with Captain Larsen, and we decided to fly a signal flag requesting permission to board. The flag was run up the yardarm when we were about a quarter of a mile from the mother-ship, but there was no response. Nevertheless, we continued to close in on the vessel and noted that it had a large floating bumper set out on the port side. We inched up against the bumper, and crew members came to the ship's guardrail and peered down at us. We made motions indicating that we would like to come on board, but it was clear that they did not want company. We threw some magazines up to the ship, and they tossed down some strong Russian cigarettes. One of our crewmen threw up a carton of Lucky Strikes and, after we waited five or ten minutes, it was obvious that either no one on board spoke English or they weren't in the mood for any official dialogue.

We were in international waters, and there was no reason to push the issue any further. We moved away from the vessel and headed north. As we did so, we could see a catcher boat heading for the mother-ship with a good quantity of flounder on deck. We set out our trawl net to sample the fish population in the region. The water depth was relatively shallow, about 230 feet, the sea surface temperature was about 7 degrees Fahrenheit and there was a modest chop running to the northwest. We towed for thirty minutes and then hauled in the trawl. The cod end of the net was hoisted on board and dumped into the sampling box we had constructed and secured on the aft deck. The one-ton catch was largely comprised of yellow-fin sole and several other species of flounder (flathead and rock sole). There were also three young halibut weighing perhaps four pounds each. We measured the halibut and then made four other test hauls. The results were much the same as the first tow. In all, we caught about ten halibut. The information on the halibut catches was radioed to our office in Montlake and then sent on to the International Pacific Halibut Commission. Confirmation that the Russian fleet was fishing in areas where they were likely to catch young halibut would heighten the concerns of U.S. and Canadian fishermen.

As we ran north through light seas, the crew and scientists passed the time playing cribbage and poker and reading magazines. At the end of the first day at sea I realized that my nose was clear, there was no sign of phlegm in my lungs, and I felt great. The allergies that had plagued me for six months vanished in a matter of hours. A dozen rounds of antibiotics had done nothing, but a new environment and an exciting mission had stimulated mystical curing powers.

After leaving the Soviet vessel, we did not encounter another ship as we made our way up to Nome to top off the fuel tanks. We did not expect to have to replenish our water and fuel tanks again until we had finished our survey in the Chukchi Sea. Shortly after noon we anchored about a quarter of a mile off the small harbor. There were no port facilities that would accommodate a vessel of the *Cobb's* size. The servicing of the vessel would take several hours, and the cook wanted to see if he could get any fresh fruit or vegetables in town. We headed to the docking area in our larger motor launch. All of the scientists wanted to see Nome and have the chance to buy some carved walrus ivory. The King Island Eskimos who sold their wares in Nome were well known for their ivory work. Everyone bought a few pieces, and I managed to get a beautiful carved kayak with a hunter in the seat area and a seal lying on the bow in front of him. It was the first week of August, the sun was out, and the temperature was in the mid-seventies. We were surprised to see a number of young natives swimming in the harbor, even though the surface water was only in the low fifties. We enjoyed Nome and its people, but time was limited, and we were back on board by late afternoon, heading for the Bering Strait.

As we entered the strait, there was a palpable sense of excitement among the crew and scientists. We wanted to see Diomede Island, which was just west of Little Diomede and held by the Soviets. The natives who occupied both islands were from the same Eskimo group. We spotted the larger island, but it quickly faded from view, and we ran northeast along the north side of Seward Peninsula toward Kotzebue Sound. We intended to initiate our track line sampling stations there, but first we were scheduled to pick up Norm Willimosky in the town of Kotzebue. The pickup was not easy as a broad, shallow, mud bank extended well out from Kotzebue. We made radio contact with Norm and asked him to meet us on the beach off Cape Blossom. The skipper ran the *Cobb* into about four fathoms and set anchor. The University of Washington's ship, the *Brown Bear*, was lying just east of us, in somewhat shallower water, waiting to take on fuel. We scanned the beach and thought we saw some movement along the shore, but it was not at all clear that it was Norm. Regardless, the mate and I set off in the launch. A number of sand and mud banks along the route made travel tedious and, when we reached the beach, we found that the moving objects had been caribou. We waited an hour or so then headed back to the *Cobb*.

During our absence, the skipper had decided to top off the water tanks. A small shallow draft vessel came out of Kotzebue Harbor, tied up next to us, and proceeded to offload water into our tanks. Norm had taken advantage of the water boat's trip and hitched a ride out to the *Cobb*. As we pulled anchor and moved offshore, we got a call from the *Brown Bear* reporting that she was aground on the mud flat. We returned but couldn't get near enough to free the vessel. Later, a small tug helped move it off the mud flat. Both of our vessels were soon underway and, within several hours, we arrived at our first sampling station.

We had planned to set the bottom-trawl dredge at each station and the gill nets at selected stations. As we set out the first trawl, we were anxious to see what the tow would yield. We fished the net for thirty minutes and then pulled it up from the seabed. The Chukchi Sea is relatively shallow; maximum depths within our sampling area never exceeded thirty-five fathoms. The first station was located in about eighteen fathoms on a sandy-silt bottom. When the net arrived on deck, the crew emptied the fish and invertebrates they'd caught into the sampling box, and all of the scientists began sorting through the catch. It was comprised of several species of flounder, various species of smelt and tomcod, and a large assortment of invertebrates.

Since the entire Chukchi Sea was covered with surface ice for a good part of the year, we did not expect to find great quantities of bottom fish. We were surprised, however, at the small size of the flounders we captured. We thought that they might be juveniles, but on examination of their ear bones, it was clear that they were adults. The short period of ice-free conditions apparently was not conducive to rapid growth. In the southern Bering Sea, the same species was three to four times larger. Although the quantity and size of bottom fish were disappointing, the catch of invertebrates was spectacular, both in variety and quantity. We were amazed at the size of some of the zooplankton. The amphipods were both numerous and gigantic. With Norm's help, we managed to identify and preserve samples of the fish within forty to fifty minutes. Wally spent hours collecting and identifying whatever he could among the species of invertebrates and putting samples of what he could not identify into preservatives.

As we moved offshore, the diversity of both fish and invertebrate species increased, but the abundance of fish species remained low. We had run offshore perhaps eighty miles and the depth had increased slowly to about thirty-two fathoms. After about three days, we were close to the international date line that split the Chukchi Sea. We didn't want to run on the Soviet side of the line, even though it was technically in international waters. We were about to swing north to our next transit line when Pete called me to the bridge and said, "Take the glasses, and look out toward the horizon just off the starboard side." I picked up the binoculars and saw a number of whales surfacing and diving, so we stayed on course. In about an hour we were in the middle of a giant pod of gray whales. Ford, our marine-mammal scientist, attempted a count and said, "There have to be close to a hundred animals in the pod." There were whales ranging from about thirty-five to seventy feet all around the vessel. They came so close that when they spouted, we could smell their foul breath. They seemed to be feeding right on the bottom because as they came up to blow and take in new air, the deep-blue water turned a muddy color and formed a cloudy, brown circle on the ocean surface. Ford said that whales were known to feed in the Bering Sea, but that such a large pod north of the Bering Strait was quite a sight.

Chatting with the skipper, I said, "Let's see if we can find out what they are feeding on." Pete answered, "Good idea," so we set out the small, meshed trawl and towed it along the seabed for half an hour. When the trawl was retrieved, we emptied the net into the sampling box and nearly filled it with giant amphipods. They looked like oversized potato bugs, so fat, they were almost round. The whales were diving to the seabed and then apparently scooping the large zooplankton right off the bottom. When they had made their haul, they came to the surface, extruded the mud, refilled their lungs, and headed back for another harvest. We stayed with the pod for the better part of an hour, taking pictures and watching the feeding behavior.

We continued on the track line for more than a week, and the scientists worked with little rest, using the variety of sampling gear on board to identify the catches. It was almost mid-August, and daylight lasted twenty-four hours. Late in the evening the sun would approach the horizon, move east, and rise again later in the eastern sky. We set and pulled in our gear for eighteen hours a day and then drifted or anchored for a few hours. Somehow we didn't seem to need as much sleep during the long daylight periods. Norm, who was prone to seasickness, lay down on a cot in the hold, but never let us down. He would lift his head to examine any strange fish and give it a taxonomic identity.

After somewhat over a week, we ran shoreward on the track line that supposedly would bring us into Cape Thompson, which was adjacent to the Chariot Site. A relatively large shore camp had been built where a harbor was to be blasted if the Atomic Energy Commission decided to proceed with the project. The sighting of Cape Thompson on the inshore leg of the track line would confirm that our positioning in the remote area, where normal navigational aids were lacking, had been reasonably accurate. Furthermore, the ship's mail had been sent to the shore camp. Pete remained on the bridge as we sighted land, scanning the rather featureless beach line with binoculars. In a matter of minutes, he pointed out a large bluff, and, as we moved farther shoreward, we could see hundreds of seabirds working off the cliff. We were right on track.

We arrived in the early afternoon, and everyone was anxious to pick up the mail. From the bridge a little over a quarter of a mile offshore, the surf rolling onto the beach didn't look very high. I noted to Pete that the surf was smaller the closer we got to land, which was behind Cape Thompson. We pondered the situation for a while and then put a small skiff with a six-horsepower motor over the side. Ford Wilkey and I volunteered to make the run into the camp. As we approached the shore, it became apparent that a small swell was running onto the beach but, given my experience making such landings when I was a youth, I had no real concern.

As we ran into the outer surf line, I positioned the boat between swells and began to approach the beach, which was relatively steep. The waves would break, rush up the beach, and then quickly retreat seaward. A wave was about to break behind us, so I opened up the throttle and sped toward the beach. In a matter of seconds, the skiff came to a sudden halt as we ran up on an offshore sand bar. I only had time to yell at Ford, "Lookout!" before the trailing wave crashed over our skiff, flipped it over, and shoved it up the beach. We both struggled to our feet and waded inshore while camp personnel rescued the skiff from the water.

The water wasn't freezing, but both of us were chilled to the bone. At the camp we immediately took hot showers and waited as our clothes dried. We also were given the mail, but camp personnel suggested that we not try to get back to the ship until morning, when the surf was expected to subside. They gave us a good dinner and sleeping quarters for the night. They also got in touch with Pete via radio and let him know about our situation. During the evening, camp personnel cleaned and ran the motor in fresh water so that it would be ready for our departure. After breakfast the next morning, we went back to the beach and, with help, dragged the skiff down to the water line. The camp had a novel arrangement for launching shore craft, which included ropes and a pulley secured to an anchor and buoy located offshore. We tied a rope to the front of the vessel and watched the surf; when all seemed right, the camp personnel towed us rapidly across the surf line. The motor started with the first pull, and we were off. When we came alongside the *Cobb*, we took quite a ribbing. Still, they were glad to have the mail.

With the skiff stowed, the mail on board, and one blast of the ship's horn, we pulled anchor and moved out to sea. The skipper set a course heading for a track line that would take us more than fifty miles offshore and then return to the northwest corner of Alaska near Cape Lisburne. Just south of the cape, there was a small sand jetty, referred to on the map as Point Hope, that ran a mile or more seaward. We were told that a small Eskimo band lived on the point along with a man and wife missionary team. Our sampling schedule was ahead of time, so we planned a visit the Point Hope Eskimos and sample the stream and estuary located at the southeast end of the peninsula. As we ran in on the shoreward track line, we caught fair quantities of several shrimp species and, just offshore from Cape Lisburne, considerable quantities of scallops.

Bird activity increased sharply the closer we got to the cape, and we suspected there might be populations of pelagic fish species in the vicinity. We set out the gill nets and made a good catch of herring in their first year of maturity. After completing the last station just west of the cape, we ran in behind the long sand spit and anchored just offshore of the Point Hope Eskimo village. It was late afternoon, so we decided to go ashore the next morning. Within an hour, however, a small boat with two natives and a Caucasian man arrived for a visit. We helped the three over the rail and invited them in for a cup of coffee. They were two of the village elders and the Methodist missionary. After a short talk, we invited them to come back later and join us for dinner.

They returned at a quarter to seven, along with the missionary's wife. She was a very attractive brunette with long hair. Everyone on board was glad to see a woman, and one could sense a change in the tenor and quality of the evening discussion. The missionary and his wife had taken on a two-year assignment at the Point Hope village and had been there about eleven months. The two elders asked us about the position of the ice to the north. The Point Hope Eskimos were a hunting community; after the ice retreated to the north, they were forced to do their hunting on land. We were surprised that they still used their sleds and dogs even though the snow had gone. They just pulled the sleds over the tundra.

The cook prepared an excellent dinner of steak, mashed potatoes, green beans, and salad, topped off with pie and ice cream. When the meal was over, the elders and the missionaries asked us to be their guests at a songfest and dance the following evening. We told them about our work and said that, if they had no objection, we would like to sample along the spit using a beach seine the next morning. They were quite familiar with the Chariot project, but were very concerned about potential repercussions on their village. Nevertheless, they had no objection to our sampling. They told us that we would likely catch some whitefish, but little else. One of the Eskimos asked me to visit him for lunch the next day, and I agreed.

Early the next morning we launched one of the small boats and headed east up along the sand spit with the beach seine. At the spit's inner origin, a small river drained the adjacent tundra. We decided to set the beach seine just below the area where the river entered into the ocean. We had seen bird activity and what appeared to be small fish breaking the surface as we approached the site. Before we set the net, we watched numerous large ground squirrels running about, seemingly with no fear despite our presence. We set the beach seine and then began to haul the net ashore. We had retrieved about a third of the net when we saw a great commotion in its midst. It wasn't long before we realized that we had caught a young walrus. He was perhaps a little over a hundred pounds and enraged about being tangled in the net. He thrashed about, tearing a small hole in the seine. He made a sudden lurch, freed himself, and snorted as he headed out to sea. What remained of the net caught a dozen or so small whitefish and a small sculpin. With our shredded net in hand, we returned to the *Cobb* with the catch of the day.

Just before noon, the engineer and I went ashore to visit the village. We had both been invited to lunch, but with different hosts. A number of small homes with sod over the tops had been dug into the earth, and a large green building served as the village social center. The cold winds were so strong that it was difficult to build above ground and keep a house warm. My host led me down a path to the wooden door below ground level that was the entrance to his home. The home consisted of a combined living room and kitchen and a smaller room where he slept. He had several jars of fish that seemed to be preserved in some type of liquid. He proudly pulled out two fresh whitefish, which he told me had been caught the day before. They were headed and gutted and I thought he would cook them for our lunch. No such luck. He dipped the fish in seal oil and handed me one of the delicacies. I had eaten lots of raw fish as a boy, but not dipped in seal oil! I did my best to down my lunch and smile. I think I could have handled the whitefish raw, but the seal oil left me reeling. After lunch, my host showed me around the rest of the village, but it was hard to focus because I was on the verge of vomiting. Somehow I held down the food until I got back to the ship.

By the end of the afternoon, I was still feeling rough, but I wanted to see the native dances. Most of the crew and all of the scientists went ashore to watch the entertainment. We brought with us three cases of oranges from the cooler. "The best gift you could offer the local Eskimos is fresh fruit," the missionary had said. "They seldom have the opportunity to have such a treat." When we got to the green building, the villagers were sitting in a large semicircle around the outer edge of the hall. We occupied the remaining portion of the circle. There were a large number of children, none of whom were racing around the room or causing any sort of commotion. They sat with their parents waiting for the ceremonies, most with smiles on their faces. Before the drum dances started, we passed out the oranges to everyone in the room. The villagers marched single file to the orange boxes, received their gifts, and then returned to their seats. The children were quiet and orderly as they came forward. The villagers entertained us well into the night and even performed the traditional blanket toss of people into the air.

We left early the next morning to finish off the two northern legs of the track line. Pete wanted to complete the work as soon as it was feasible and head south because the boat was floating high in the water. We had run about thirty miles offshore when Pete called me to the bridge and said, "Look just off the port bow about three hundred yards ahead." I couldn't believe it: there was a polar bear with a cub swimming almost due north. The ice was thirty miles in that direction. "Well," I said to Pete, "I don't know how she got this far south, but she seems to know exactly where the ice is." As we ran up on the two bears, they were about to cross our bow. Pete got ready to slow the vessel, but they turned and swam around the stern. As soon as they were clear of the *Cobb*, they turned and resumed their original course. "I don't know what sort of navigation they are using," Pete said, "but it's sure a lot better than ours." We finished the last station late in August, and the vessel looked like a toy duck floating high in the water. We had lost considerable ballast as we used up the fuel and water.

Before we headed south for the Bering Strait, Pete lowered the boom to deck level and tied it down in an effort to improve the ship's stability. He then removed all equipment and other gear from the deck and placed it into the hold. We were scheduled to take fuel at Unimak Pass, and everyone hoped we'd get across the Bering Sea before a fall storm set in. We ran south to Nome, where Ford and I left the vessel and took a plane to Seattle. The *Cobb* arrived safely in Seattle in early September.

Since my return to the Fish and Wildlife service, the budget for resource exploration had increased substantially, so I set out to add quality scientists to the staff. To begin with, I wanted to acquire Al Pruter and Heater Heyamoto. Al had graduated Phi Beta Kappa, had a good background in mathematics, and interacted well with the industry. Heater had worked on the salmon troll fishery, had good field experience, and also worked well with his peers. This was important because the entire group had to live for aboard a ship for extended periods. Within a year, I had lured both to our group and later would add Dr. Mike Tillman, a bright young student from the University of Washington, and Dick McNeely, a leading gear specialist from the exploratory unit in Mississippi. The scientific team was coming together well. As a result of the Chariot mission, Norm and I published a paper describing the slow growth rate of the flounders that inhabit the Chukchi Sea,

Wally published one on the abundance and diversity of invertebrates in the region, and Ford published one on birds and marine mammal encounters. The morale of the group was good, and pride permeated the entire staff.

Out of the Past

Life on the home front also was going well. The children joined the local swim club after Ruby took out a loan on our insurance to get the needed $300 application fee. Camping and fishing were a part of our spring, summer, and fall. Our social life centered around the local neighborhood and visits to my World War II friend Vern Benedict, who also had two children. I also stayed in touch with several other war buddies, including Chris Vesper and Jim Murphy, who both lived in Los Angeles, and Duncan, who had married and divorced a girl he had dated during the war. After the divorce, he stopped calling and I had not heard from him in close to five years.

In 1960, two unexpected events left me somewhat depressed and melancholy. Chris Vesper's wife called to tell us that he had died from a massive heart attack. Chris was the first friend I had in the service, and he was only in his early thirties. A lot of memories flashed through my mind, and it was difficult for me accept his death. Later in the year, I received an unexpected call from Duncan, who was in town slinging hash at a restaurant in the Pioneer Square area. I headed downtown to meet him during his break around 3:00 p.m., which gave us time to sit, have coffee, and re-live our China days. When I arrived, he was just coming out of the kitchen. The restaurant was the classic "greasy spoon," and its patrons all looked like they could use a good meal and a bath. Dunc and I went to a corner booth, and he filled me in on his post-war days.

"I have had a hell of a time converting back into civilian life," he said. "I got married to Betty, and she was a great woman, but I couldn't get off the booze." He had tried to dry out several times, but it never took. "Maybe I never wanted it to work," he said. Dunc looked drawn, thin, and much older than his age. I asked if there was anything I could do to help and he said, "Well, you could have me out to dinner, and I will behave." He was alluding to his excessive use of swear words. I invited him for the next evening, warning Ruby that the language might be a little raw. Dunc was one of those people you always wanted beside you

342

when you were in combat, but he was a little rough around the edges and could get mean when he was drinking. He showed up promptly at seven, clean-shaven and looking better than he had the previous day. Ruby made a roast, mashed potatoes, gravy, peas, a salad, and a cherry pie. The evening went well, but it was obvious that Dunc was no longer interested in getting off the booze. He made it clear that he was happy moving around the country, saying, "You can always find someone who will hire you to cook."

Dunc stayed until about 11:00 p.m. and said that it might be sometime until I heard from him again. I drove him to the Pioneer Square area. He held my hand for some time and said, "You're doing okay, Lee, hang in there." Then he headed down First Avenue. I returned home rather solemn, knowing that Dunc had lost his way and had no urge to seek out a better life. The following night, our phone rang about 2:30 a.m. It was Duncan. He was so drunk he could hardly talk and asked if I would pick him up. I could hear a number of loud voices in the background and wasn't anxious to bring him home in that state. I told him to hit the sack and sober up. He responded, "Yep, but just thought I would check and see if you would give an old friend a hand." He hung up, and I never heard from him again. I should have gone to his aid, and I have always felt a sense of guilt about my response. A close war friend needed help, and my failure to act has nagged me throughout my life.

The Continuing Love Affair

I managed to squeeze in one more survey before getting bogged down in administrative matters and various national and international meetings. In the early 1960s I joined a trip to survey the bottom-fish complex off the northern end of Vancouver Island. We worked several days just south of Cape Scott, making good catches of various rockfish. The area was relatively rocky, however, and we lost several nets before we moved northwest, south of the Scott Islands. We made four or five tows the first day, landing some excellent catches of various rockfish, mostly ocean perch. We anchored about a hundred yards south of one of the islands that extends east- ward from the cape. The wind decreased, the sea smoothed out into a gentle swell, and the late afternoon sun felt warm. It was one of those evenings that you dream about, but which seldom occur. The island was covered with beautiful old-growth firs, which sat quietly in the cool of the evening, beckoning us to come ashore.

The mate, Hans Janguard, and I rowed to shore to see what we could find along the beach. Between us and the beach was a large kelp bed, which was piled high with driftwood, parts of fish webbing, and flotsam and jetsam. As we made our way through the kelp, a large male Steller sea lion surfaced a hundred feet or so away and snorted into the wind, I guess to let us know we were in his territory. We acknowledged his rights, gave him a wide berth, and landed several hundred feet farther up the beach than we had planned. The beach was covered with driftwood, ranging in size from six inches around to large fir trees. There was an ample supply of fir bark, so we built a beach fire in a clear area.

The beach contained a treasure trove of fishing floats of all sizes, but the real surprise was the abundance of large Japanese glass floats that were close to eighteen inches in diameter. The Japanese tuna fleet used the floats to support their long lines as they fished in surface waters across the Pacific. At times, the floats broke free and drifted easterly across the north Pacific, landing on beaches from Alaska to California. We collected a half dozen or so large glass floats, stacked them next to the fire, and then sat on the beach watching the sun set. Then, taking advantage of what was left of the daylight, we loaded our skiff and headed back to the *Cobb*. We spotted a drifting glass float and added it to our collection. It was still surrounded by the heavy webbing that once secured it to the main line and, attached to its bottom, was a large gooseneck barnacle. Given the barnacle's size, we speculated that the float must have been drifting in the ocean currents for well over a year. When we climbed over the rail and onto the deck, we felt like pirates returning with a bounty of gold doubloons. The evening has never left my memory, and Ruby keeps the floats in our yard as a reminder of the good life.

At the onset of the new decade, the BCF was looking ahead to two main investigation areas. First, the AEC asked about our ability to sample the continental slope fauna, from an inshore station on the shelf to a depth of more than 1,000 fathoms (over a mile down). The commission was conducting a detailed study of the flora and fauna of the Columbia River, examining the amounts of radionuclides that were carried down the river in sediment and taken up by the biota of the region. The staff was particularly interested in the amounts and levels of radioactive substances that were transported into the ocean environment from the Columbia River estuary and the adjacent regions.

In our earlier albacore studies, we found that the Columbia River plume extended many miles seaward from the river mouth. I called key staff members together to help evaluate our capacity to carry out the AEC mission with success. Dick McNeely and Pete Larsen said they could not use the standard otter-trawl towing techniques, which required the use of two trawl warps (wire lines). They thought, however, that they could jury rig a method that would release all the cable from one drum and then transfer the load to the cable and davit on the other side of the stern. It would require using a bridle arrangement with the bottom trawl, but they were convinced that it would work.

We were concerned about the exploration of the water column above the bottom and below the shallow surface layer. Our exploration suffered because we didn't have effective sampling gear. The Europeans had experimented with mid-water trawls for a decade or so, but they lacked instruments that could provide precise information on the operational depth of the trawl. We recorded an amazing number of traces on the echo sounder and though we could only speculate, they seemed to be created by fish schools off the seabed. I brought the fishing-gear specialists together, and we spent an afternoon kicking around different possibilities. For the most part, the Europeans were using relatively small trawls and using speed to overtake their target prey. The alternative was to build a large net that the fish would not detect or respond to until they were far back in the trawl. I opted for the giant net. McNeely would design and build the net, making either an acoustical or solid wire link to the trawl. The AEC had commissioned us to explore the mid-waters, and it was now up to McNeely and his gear group to find the method.

In addition to these two new areas of investigations, we continued our shrimp investigation work off Alaska, working from Kodiak Island west toward Unimak Pass. Since the onset of the shrimp work in the early 1950s, we had moved throughout southeast Alaska, from the gulf to Prince William Sound and then west toward the Aleutians, covering more than 1,500 miles of Alaskan coast. The government had been pressured to document the juvenile halibut population in the region, so the BCF budgeted for a major effort to survey the entire Gulf of Alaska shelf and upper slope from 1961 to 1963. We would survey the southern portion of the Gulf using the *Cobb*, and the commission would survey the remaining areas of the gulf using chartered vessels.

The early 1960s were a volatile period in global fisheries development. Distant-water foreign fleets were moving into all of the world's oceans, searching out underused resources and often competing with local fisheries for the available fish stocks. A range of European vessels, including those from communist bloc nations, as well as Japanese and Korean vessels were fishing off Alaska and New England. U.S. scientists had limited knowledge of the ground-fish resources off Alaska, with the exception of halibut. As foreign fisheries moved south into the Gulf of Alaska and the waters off Canada, there was an increasing political outcry for governments to take action. But the United States, to a certain extent, was hampered by its own policies.

Later Congress established a commission to evaluate U.S. ocean policy, and leading scientists on the commission extolled the great resources of the oceans, suggesting that they could yield 500 to 600 million tons of fish annually. Furthermore, they believed that the exploitation of the oceans could support a rapidly increasing world population. U.S. policy supported making ocean resources available to other nations if they were not being utilized by the adjacent coastal nations. In addition, the United States had almost no way to monitor the activities of the distant-water fleets off its coasts, so the BCF could only speculate on the impact of foreign operations on domestic fisheries. The 1961–63 Gulf of Alaska halibut/bottom fish survey would provide a greatly increased understanding of the fishery resources of the region, although it would come a little late.

Our exploratory unit was asked to help answer a number of questions important to U.S. policy. In an effort to better understand the geographic and depth distribution and abundance of ground fish in the northeast Pacific, we pulled together all the information we possessed on the region. This included the BCF's historical survey work, which dated back to the late 1940s, and the surveys conducted by the Halibut Commission, established by the United States and Canada, to study and manage Pacific halibut stocks, during the early 1960s. This data was integrated with coast guard surveillance information on the numbers, types, and sizes of foreign fishing vessels operating in the waters off Alaska, allowing us to estimate how much fish had been removed from the region. We received relatively fair catch information from Japan and South Korea, but almost nothing from the eastern bloc nations.

The development of the large mid-water trawl had gone well. McNeely and his crew had created a depth-monitoring system using a cable that incorporated a conductor within its core. A pressure-sensing device located on the trawl door sent information through the conductor to the bridge of the ship, where pressure readings were converted to depth.

We conducted the initial trials of the net surveys off the Washington, Oregon, and California coasts. When unidentified echo traces were observed on the depth finder, the trawl was set in an attempt to capture and identify the mid-water targets. Although a number of species were captured, we quickly realized that the numerous echo traces we had previously noted on the echo sounder were mostly Pacific hake. The mid-water trawl exceeded our expectations, and we were making catches of 10,000 to 15,000 pounds during thirty-minute hauls. Within a couple of years, we had enough information to estimate the size of the adult hake population off Oregon and Washington, which was close to a million metric tons. The local bottom-trawl fisheries frequently took large hauls of hake when the aggregations moved down onto the seabed. Unfortunately, fish buyers had little interest in hake, and the fish generally was considered a nuisance that had to be sorted out, cleared from the decks, and thrown back into the sea.

The large mid-water trawl also turned out to be a very good sampling tool for larger pelagic species. In subsequent years, we would catch a number of ocean species, including mackerel, jack mackerel, bonito, black cod, varieties of rockfish and ocean sunfish, as well as a number of rare pelagic species. When operating off San Diego, we managed the most spectacular catch ever taken by the *Cobb* trawl. It was early in the afternoon, and we set the net on an echo trace that appeared to be a large anchovy school. After a few minutes, the towing wires began to straighten out. We were fishing well off the seabed, so we knew we had not snagged the net on rough bottom. As the force pulling the trawl on the tow cables increased, the vessel came to a halt, and we began to be towed backward. Within a matter of minutes the cable suddenly went limp, and we realized that we had lost at least a part of the net. We started to recover the net, but as the towing cables were pulled in, it became apparent that we had lost the trawl. The monster in the net had even parted the cables forward of the spreading doors. We had nothing left except the wire. Sometime later we found out that we had captured a U.S. submarine and had been operating in a military war-game area! In reality, the submarine caught us—a slightly embarrassing situation.

The *Cobb* pelagic trawl became well known in the industry. Although there was no real U.S. market for hake at that time, there was considerable interest in the technical advances in the mid-water trawling technique. Still, I was somewhat caught off guard when we received a call from a Mr. Janguard in San Diego, who said that he was going to use a similar trawl to catch albacore tuna. At first I was highly skeptical, but we had taken some rapid-swimming pelagic fish in the net, such as mackerel, bonito, and jack mackerel. I felt that the best bet would be to capture the albacore early in the morning or evening when the net would be less visible. Janguard, who operated a shipyard in San Diego, was the owner of *Tuna Clipper* and, as it turned out, was also the brother of the *Cobb's* mate. He had a reputation of being a forward-thinking and inventive fisherman. We discussed the idea for some time, and he decided to try the net out during the summer. He asked me to join him for the trial. It sounded exciting, so I agreed.

I joined the fishing vessel *Cape Flattery* in San Francisco in mid-July, and we headed north along the coast. The vessel was a converted navy ship that had been converted to a tuna seiner sometime after World War II. It was about 170 feet in length and, with its bait tanks filled with water, sat rather low in the ocean. The vessel rode easily in the moderate swells that were running in on the coast. Janguard, who owned and skippered the vessel, asked me to take a turn at the wheel between 10:00 p.m. and 2:00 a.m. the next morning. The moon was almost full and vision was excellent, the sea was calm, and it was a perfect time to stand wheel watch. Perhaps twenty or thirty miles north of San Francisco Bay, I noticed the port and starboard lights of a ship heading seaward and allowed it to pass on the coastal side of our vessel. In the moonlight, it appeared to be a small coastal freighter about 260 feet in length.

The freighter was the last vessel that I spotted on my watch; sharply at 2:00 a.m. a crewmember came up to the bridge to take over the wheel watch. He said that we would take bait in Moro Bay the next morning and then start scouting for albacore schools in the offshore waters between the bay and San Francisco. I had fished albacore using troll lines on the

Cobb, but never using a pole and line. I hoped Janguard would give me an opportunity to fish in the racks at the stern of the ship, but that would probably depend on whether we could attract a large school to the vessel. In Moro Bay, we set one of the large seine skiffs and worked our way through the bay. The skipper allowed me to join the baiting venture; we had been on the water for only ten or fifteen minutes when one of the crew spotted bait breaking the surface. We ran up ahead of the school and began to set out our lampara seine, a net with two long wings and a small mesh bag at the center. We quickly encircled the baitfish and began hauling in the two wings of the net. As we closed the net, I could see bait that seemed to be anchovy jumping on the surface. We "dried the net up" (drained the water off the fish) and then started to brail the anchovy into the fish well on the seine boat. I don't know how much fish we caught in the one set, but the fish well looked full, and the skiff operator headed back to the *Cape Flattery*.

Shortly after lunch we headed offshore, hoping to encounter a school of albacore, but we spent the rest of the afternoon scanning the horizon to no avail. The next day the crew was up before dawn, but there had been some trouble with the pumps and for a few minutes the skipper thought we might have to abandon ship. Whatever the problem, it was in hand by the time I arrived on deck around 5:45 a.m. Janguard wanted to find a good school before trying out the large two-boat trawl. Luck was with us and within an hour we had found one.

I watched as the "chummer" standing atop the bait tank took handfuls of live anchovy and tossed them off and on either side of the stern. Soon the men in the racks at the stern began hauling in fish. The heavy poles were about eight feet long, each with a relatively short line and a bare hook on the end. When the fish got into a feeding frenzy, they hit anything that made contact with the water. Janguard motioned for me to take a pole, and I quickly climbed into the racks off the side of the vessel. Seconds after flipping the line into the water, I had a fish. If you kept the fish's head up, then it practically swam out of the water. As that first fish flipped over my shoulder and onto the deck, the adrenaline surged through my body.

The tuna were boiling in the water, and every time the line and hook hit the sea, another fish was caught. The fish were coming on board so fast that the deck behind the racks was filled with fish; some of the crew cleared the area so that more albacore could be taken aboard. I don't know how long we stayed in contact with the school, but the thrill stayed with me until we climbed from the racks. While we fished, sharks—ranging about four to twelve feet in length—occasionally pushed their snouts against the bottom of the racks. The fishermen took the ends of their poles and push them through the holes in the grate, forcing the sharks to move away.

We moved on and began looking for other schools. I was hoping that we would get to set the large net. However, the companion vessel that was to work with us did not show up, and the mission was scraped. As far as I know Janguard never tested the large two-boat trawls, but his idea was not lost to other fishermen who, several decades later, began to use a similar vessel to fish tuna in the north Atlantic.

The survey work off the Columbia River, using a single towing wire and bridle trawl, had also gone well. We successfully sampled the fish and invertebrate populations along the canyon walls at depths from 300 to 6,000 feet. After several years of work assessing invertebrate species, Walter Pereyra put together a fascinating story regarding the migration of one of the Tanner crab species. For mating purposes, the adult crabs would concentrate on the continental slope at depths from 1,800 to 2,400 feet during the spring. After the mating ritual, the female adults moved several hundred feet down the slope, while the male population remained in the original area. When the fertilized eggs were shed, the larva forms were carried off the shelf. Following metamorphosis, the young crabs settled on the seabed of the Cascade Plain (a bottom topographic feature) at depths of about a mile. As the young crabs grew and approached maturity, they spent their lives migrating up the continental slope. Eventually they joined the adult population more than 4,000 feet above the abyssal plane where they had started their shoreward migration. Knowledge of the Tanner crab behavior gave birth to a dance called the "crab walk," which was performed by the exploratory group at the annual Christmas party.

A Fragile Treaty

The International North Pacific Fisheries Commission (INPFC), which included Japan, the United States, and Canada, had been established with the provision that it would stay in effect for a period of ten years, after which any party to the convention could ask for changes. The decade had moved by quickly, and it was clear that the fishing communities in the Pacific Northwest and Alaska were unhappy with the commission's actions or, rather, savagely angry over its lack of action. The fisherman expected the scientific efforts to demonstrate that the stocks of salmon and halibut off the United States and Canada met the criteria for abstention. The American and Canadian scientists felt that a good scientific case had been made for salmon in the Northeastern Pacific, but the Japanese were not convinced. To make matters worse, neither the American nor the Canadian scientists representing the INPFC believed the data supported the abstention for halibut in the Bering Sea.

Following the formation of the commission, all three parties to the treaty had initiated significant research efforts regarding the distribution of salmon in the ocean environment. A number of questions plagued the researchers. How far did the different salmon species range offshore? What were their oceanic migration routes? What did they eat, and what were their growth patterns at sea? What species preyed on the maturing salmon, and what were their levels of mortality at sea? More detailed information was needed on the distribution and migration patterns of juvenile halibut on the continental shelf; and were the halibut in the Bering Sea related to the stock that was exploited in the Gulf of Alaska and further to the south?

One of the commission's greatest achievements was its contribution to the understanding of pelagic fish resources and the oceanography of the region. Its scientists had added a new understanding of the ocean current patterns and demonstrated that salmon from both continents migrated far out into the north Pacific. There, species that had originated in the rivers of North America and Asia intermingled and grazed on zooplankton, squid, and small pelagic fishes. The commission also helped to foster a greater appreciation and understanding of the region's extensive ground-fish populations.

Despite the substantial increase in knowledge, in the early 1960s most fishermen and fish processors viewed the commission as a failure. They expected the dissemination of information to result in the Japanese salmon fisheries being moved to the west and keep them from "intercepting" salmon of North American origin. Neither of these developments had taken place, and when the scientists of all three countries concluded that they could not show that the Bering Sea halibut was overfished, the pot began to boil over. At one INPFC meeting in Tokyo, the industry had hung all of the U.S. commissioners in effigy. The scientists were spared, but we had lost a great deal of credibility.

In my opinion, many of the American and Canadian industry members, the press, and the lay public did not fully understand that the commission had no authority to impose regulations on fishermen from any of the three parties to the convention. Each member nation could only evaluate the science and the politics of the situation and make recommendations to the other members. The INPFC suffered the same procedural limitations that had led to the failure of many of other international commissions established to govern ocean fisheries. Changes occurred only as the result of a consensus among the members or unilateral action. It is not my intent to go into a lengthy discussion of the problem, except to say that the discord led to an attempt to renegotiate the convention in the early 1960s.

In preparation for the INPFC negotiations, I collected all the information on the ground-fish stocks inhabiting the continental shelf and slope from California to the Bering Sea. I had already summarized the resource survey data for the region but needed to review the historical literature. This included a number of scientific reports by scientists working for the old Bureau of Fisheries, coastal state agencies, and the Canadian fish agencies. I also managed to find several Japanese reports on their pre-World War II fisheries in the Bering Sea. Finally, I made an appointment with Dr. Don Bevan at the University of Washington.

Don had been a classmate of mine back in the late 1940s and spent several years in postdoctoral studies at the University of Moscow. He spoke Russian quite well and read it even better. Don identified a number of translated reports describing early Soviet surveys in the Bering Sea. He also had a translated copy of the Ph.D. thesis of a Soviet scientist named Moiseev that described the ground-fish stocks off the Siberian coast. Using these data sources, I put together a comprehensive review of the ground-fish species of the northeastern Pacific Ocean, including their distribution, abundance, and behavior. We would use the 200-plus-page report as our ground-fish information base during the negotiations.

Scheduled for June 1963, the first negotiation meetings were held in Washington, D.C., and all sides positioned themselves for a substantive session. Ambassador Benjamin Smith led our delegation, which also included two senators and one member of the House of Representatives. In all, there were about eighteen members, including the contingent of U.S. scientists. The scientists, however, contributed little in the first session, during which each side laid out the difficulties it had endured under the convention. After the major disagreements had been aired, the negotiations were put on hold until the fall. One thing did emerge from the meeting: the interpretation of the available science by Japanese scientists was often different than the picture painted by U.S. and Canadian scientists.

It was easy to see that each nation viewed the data in light of its own self-interest. The obvious level of advocacy science seemed pervasive in all delegations. To achieve any progress, all sides needed to reach a common understanding and then attempt to narrow the differences in interpretation of the known facts. I discussed the matter with Don McKernan, head of the Bureau of Commercial Fisheries. He agreed that a meeting between the INPFC scientists from the three countries should take place prior to the fall negotiations, which were scheduled to be held in Tokyo. It seemed only fair to select a site that would minimize travel costs, at least that is what I told myself, so we agreed on a ten-day session in Honolulu just prior to the Tokyo meeting.

The scientists gathered in Honolulu in early September. It was my first trip to Hawaii since the mid-1930s. I had not seen my childhood haunts for almost thirty years, and I was anxious to rekindle some old memories. I expected to be able to do a little exploring during the evenings or on weekends. We were housed in the Hilton Hotel on the beach just north of the Fort De Russy grounds, which separated the hotel from the main stretch of Waikiki Beach. The rooms were excellent and the gardens and pools made living easy. On the first day, I met Dr. Fukuda, the head of the Japanese ground-fish scientists, and Dr. Ketchen from Canada, whom I considered a mentor and the best marine biologist among the North Americans. Dr. Fukuda had a rather round face and a large growth on one ear. He looked rather stoic and perhaps unfriendly at first glance, but when he smiled or grinned, he melted away all my preconceived ideas about our Japanese counterparts.

The first day's meeting went smoothly, largely because we were establishing a work schedule, identifying which scientific documents would be submitted, and naming the scientists who would be involved with the topical issues. The planning session concluded sometime in the early afternoon, and we were free. I put on some shorts, walked through the military grounds and then down Kalakaua, the main street in front of the beach. A number of new hotels had been built, but the three old-timers—the Moana, Royal Hawaiian, and Halikalani—were still standing proud. On the strip, I rented a car and drove around Diamond Head and south toward my former neighborhood. It wasn't long before I saw a sign for Wailupe Circle. It still was a government facility, but it had been taken over by the U.S. Coast Guard. The large building on the water had been torn down, and the base now housed coast guard personnel. The grounds, swimming pool, and the homes all looked like they had never been touched; it seemed like I had stepped thirty years back in time. A flood of memories took over for a few moments, and I searched the area behind the building for the entrance to the cave we had explored. New homes had been built behind Wailupe Circle, however, so I couldn't see the cave. I could make out the top of the large hill where the army set up a machine-gun nest during war exercises in the 1930s.

As I left the area, I checked out the large fish farm that used to be south of the base. It, too, had disappeared, or rather had been filled in and was now supporting a number of houses. I was even more surprised to find the Hinds Clark Dairy gone and the whole green valley filled with new homes. I stopped at a restaurant and had a cup of coffee. The room seemed to be filled with several locals, mostly Japanese and Chinese. When they used some "pidgin" phrases, I knew that I was back home. Returning to Waikiki, I followed the road along the golf course, which had been there when I was a boy, then cut across the back of Diamond Head and into Kimuke, finding the grammar school I had attended. My trip down memory lane was complete, so I returned to the hotel. My final goal would be to see if any of my childhood friends still lived in the area.

The next day of the meeting was rather tedious. We inched through the technical documents, each side waiting for its presentation to be translated from Japanese to English or from English to Japanese. The process was time consuming, but it became particularly difficult when a disagreement surfaced. To make things worse, different technical and scientific nuances further slowed the pace of discussions. Sometime after lunch, when Dr. Ketchen and I presented our interpretation of the impact of fishing on flounder stock, Dr. Fukuda described it as unacceptable. It was difficult for us to figure out the nature of his concern. When he presented his understanding of the data, it didn't sound any different than the U.S./Canadian scenario. We would then detail our understanding more carefully, only to be challenged again. The debate lasted the better part of two hours, at which point we found out that we had agreed since the beginning. The interpreters were having a hard time with the technical explanations. Little by little, however, we began to make progress and eventually found agreement on a number of issues.

During the noon break, I called the local Bureau Biological Laboratory and talked to several of girls who had grown up in the islands. Fortunately, one of them knew the Puehow family and told me that Dede and Buster were living on the island, Dede in Honolulu and Buster in Kailua on the windward side. She wasn't sure exactly where Dede lived and didn't know her married name, so we could not locate her through the phone book. She did know one of her cousins, Lailani, and gave me her phone number.

That evening I went over to see Lailani, who lived just on the north side of Diamond Head. She was perhaps forty and remembered the Alverson family who had lived in Hilo. We chatted about Hilo in the 1930s, and she had a pretty good idea about the whereabouts of most of the Breakwater Gang. Chicken Little still lived in Hilo, as did Buddy. She thought Buddy was the fire chief in Hilo but wasn't sure what Chicken Little was doing. Alexander had gone off to war and became the "top dog" in one of the army's band units. He had returned to Hilo, but had since gone "stateside." She confirmed that Dede and Buster were living on Oahu,

The following day there was a message from Dede, and I called her when the work day was over. At first, we were both a little tense. After all, time had transformed us from children to adults; thirty years of diverse experiences had altered our physical and mental perspectives. She remembered a freckled face and a frequently sunburned kid, and I recalled a short girl with black hair, with whom we frequently played and swam off the Hilo breakwater. Once we got through the initial small talk, Dede suggested that she and her husband meet me for dinner at the Bare

Foot Bar on Waikiki. At that time it was located in what we called the Kiamana Hotel, which roughly translated to "the local people's hotel."

That evening, we sat down and began to exchange some history. Dede asked what I had done since we left Hilo a quarter of a century ago. I filled her in on my days in San Diego, the war years, and my move to Seattle. She had remained in Hilo until the war and then moved to Oahu, where jobs were more abundant. She briefed me about Buddy, who was still on the Big Island. We had an enjoyable visit over dinner and a few mai tai's, but I sensed a level of uneasiness and suspected that our past was of little importance to her. After all, there was no reason that she should have felt anything more than curiosity about a childhood playmate. I wondered about Buddy, who had been a closer friend, but decided not to get in touch with him until I could make a trip over to Hilo.

The INPFC technical talks had gone well, and we clarified many of the technical issues. I left the meeting with a certain fondness and respect for Dr. Fukuda. While the large growth on his ear was distracting, he had an excellent sense of humor, a great smile, and a good understanding of the fish populations being exploited off Alaska. He did not believe, however, that the data showed that the species harvested by the Japanese were being overfished. The Canadians and our delegation were convinced otherwise, although the database was not very good. Most of the available catch data had been collected and summarized by the Japanese. Information on the catches by the Soviets, Poles, and Koreans was sketchy at best. Still, although there was a great deal of uncertainty in the data, some of the catch-trend and fishing-effort information convinced us that the flounder and halibut populations of the area were taking a heavy beating. The meeting allowed us to agree on the information we could jointly rely on and to sort out some semantic differences. Progress, though, was slow.

Shortly after the Honolulu meeting, we flew to Tokyo to renew the formal negotiations. We were housed in beautiful rooms at the new Tokyo Hilton Hotel. Prior to starting the sessions with the Japanese, we were invited to a dinner party at the U.S. Embassy to meet with Ambassador Reichauer.

The meeting in Japan was my first opportunity to watch Kunio Yanisawa in action. Kunio worked for the international section of the Japanese Fisheries Agency and spoke impeccable English, certainly better than I did. Thin and always dressed in a dark suit and tie, he had gained a reputation as the number one nemesis of the U.S. delegation. He uncannily found the shortcomings in technical and philosophical arguments, and his manner of speaking sometimes seemed abrasive. Some of the members of the U.S. delegation gave him unbecoming nicknames, including the little arrogant SOB, Tojo Piss Ant, etc. But after listening to him a number of times, I sensed that our discomfort stemmed largely from the fact that he was simply very good at his job. He frequently presented his arguments in a highly rational manner that forced us to reevaluate and/or rethink our position. The global view on ocean jurisdiction and the rights of fishing nations rapidly were changing in favor of coastal states, and the Japanese were playing a delaying game to protect their distant-water fishing interests. Kunio was simply a master at doing what he could do to slow the process down. The longer I knew him, the more I respected how well he fought what he knew would be a losing battle.

The meetings moved forward at a snail's pace; it seemed that every topic had to be viewed from each national vantage point and then the shortcomings in each position were repeatedly shredded. A great deal of redundancy caused by protocol and international diplomacy extended the meeting, probably to the delight of those who enjoyed the social scene, which was filled with one cocktail party and dinner after another. Each national section had its own official party, and then there were the lavish parties thrown by different sectors of the Japanese industry. These were really spectacular and included one course after another of artistic foods, almost always presented by beautiful young women dressed in kimonos. The dinners frequently included Kabuki dancers, singers, and/or magicians and, before the end of the evening, a small gift from the hosts. It would be dishonest to suggest that I was offended by this opulence, but I was somewhat embarrassed, realizing that we could not reciprocate in kind when they visited in our country.

The parties provided an opportunity to see the personalities of our Japanese counterparts up close. Kunio had a good sense of humor and enjoyed his sake. He was always somewhat guarded, however, and never allowed himself to be influenced by anything he drank. The social affairs helped to bridge the cultural differences, but when the meetings were called back to order, the divergent character of the national positions was reinstated immediately. At the end of almost four weeks, we had made almost no progress, and the session was called to a halt with an agreement to meet again in Canada early the following year. One evening at dinner, Kunio asked me about my World War II experiences, and I gave him a cursory summary of my tour in China. He noted that he was only seventeen when the war ended and had been conscripted to work in an ammunition dump outside of Tokyo. By the time the war was over, he said, Tokyo and many other Japanese cities had taken a real beating from the air raids. We would continue to work with Kunio at the negotiations in Ottawa and at all future INPFC meetings. The longer I knew him, the more I respected his professional abilities.

The Ottawa negotiations turned out poorly for me. We had been in session for several days when I began to get the shakes, and Don McKernan suggested that I return to the hotel to get some rest. The shakes grew more severe, and I had pain above both eyes. Someone in the hotel gave me a thermometer; my temperature was just over 103 degrees. Within about five hours, however, it was back down to normal, and I thought I had beaten whatever was causing the problem. No such luck! Several hours later, I had the shakes again, and my temperature was 104 degrees, so a doctor was summoned. He took my temperature, asked no questions, and sent me to the hospital, where my fever cycled between 103 degrees and normal over a ten-hour period. I remained in the hospital for four days; they kept taking samples of my blood to the virus lab, but no diagnosis was forthcoming. On the fifth day, the fever disappeared, so after talking with the doctor, I made arrangements to fly home the next morning. During the flight back, I thought my fever had returned and that I was getting the chills. In Seattle, Ruby met me at the airport and helped me to the car and then home. She immediately took my temperature and it was normal; my plane symptoms most likely were psychosomatic. During my unknown sickness, I lost eighteen pounds in ten days, and it took me several weeks to recover. The cause of the cyclic temperature was never diagnosed.

The Ottawa negotiations also ended in a stalemate, but events outside the meetings led to an easing of the disagreement with Japan. First, Japan had made concessions and moved some of its high seas salmon fisheries to the west. Furthermore, many of the world's fishing nations were blaming the depletion of fish stocks off their coasts on distant-water fishery operations. This prompted an international cry for changes in the long-held tradition of narrow territorial seas, and several Latin American states had extended their jurisdiction over the ocean space to 200 nautical miles from shore. The U.S. negotiation position vis-à-vis Japan was improving, and thus we decided to remain in the INPFC. On the other hand, the foreign fleets operating off Alaska gradually had made their way south, first fishing off the coast of British Columbia and by 1966 establishing ground fisheries from California to Washington. Local fishermen complained that the Soviet fleet, which was comprised of factory and freezer trawlers and mother-ships, was taking large incidental catches of salmon and focusing on ocean perch, a fishery that had been pioneered by Oregon fishermen. The U.S. government's position was that they were largely taking Pacific hake, which were not harvested by Canadian or American fishers.

Our offices received a stream of telephone calls from fishermen who said they were fishing in the same localities as the Soviets. The fishermen said while the Russians might be catching hake, they were also aggressively fishing for Pacific Ocean perch. These reports contradicted those made by the U.S. Coast Guard, which formed the basis of the official U.S. position. I decided it was time to go and have a look for myself. I chartered a plane from Kenmore Air, and a single-engine, combination float-and-wheel plane picked me up on the lake behind my Montlake office. We took off and headed for the ocean between Astoria and Newport Beach, Oregon. It was a relatively long flight south and, as we flew out over the ocean, we were buffeted by winds blowing between thirty and forty miles per hour. The seat belt pulled and tore at my gut.

It was perhaps a little over an hour before we spotted the Soviet fleet. They were fishing about thirty to forty miles off shore, just north of Newport. I asked the pilot to drop down and circle one of the factory trawlers that seemed to be hauling in its gear. After two passes, we closed in on his port side, and I could see a load of rockfish on the deck. The pilot located our position on his navigation chart and determined that the vessel was right in the middle of the perch grounds, some of which I had surveyed onboard the *Cobb* more than a decade earlier. We checked out several other vessels in the region, and they, too, were fishing the perch grounds. It was easy to identify the piles of red rockfish on the vessel decks. We spent about thirty minutes looking over the fleet, but the wind was throwing the small plane around, making it increasingly hard to continue.

I suggested to the pilot that we head to Newport, so that I could give a first-hand report to Gordon White, a fisherman who befriended me during my first adventure in marine biology. He had made one of the telephone complaints. The pilot nodded and said, "It's getting tough out here." We landed on a small airstrip just south of Newport and I drove to town and gave my report to Gordon and several other fishermen. They seemed pleased that someone had bothered to verify their story. The growing number of Soviet and other foreign fishing vessels off the Pacific coast and Alaska generated a lot of political interest in initiating bilateral talks with Japan, the USSR, Poland, and South Korea.

Large Factory Trawler, ARtic Trawler, 1980's

CHAPTER X
GLOBE TROTTING

The mid- and late 1960s were a struggle for the Alversons. The character of my job was changing. The government was under pressure to monitor the Soviet, Japanese, Korean, and other distant-water fishing operations that were harvesting fish resources off the northeastern United States and along the West Coast from Alaska to central California. This pressure called for replacing the historical explorations of living marine resources with an assessment of the impact of the foreign fisheries on stocks of interest to U.S. fishermen and conservationists. I was not thrilled with the change in priorities, but the political demand to deal with the problem and the growing frustration of coastal fishing communities was obvious. More and more of my time was being relegated to the international scene, which meant a great deal of travel.

It couldn't have come at a worse time for my family. At the urging of some of my peers, I had returned to the University of Washington for classes in oceanography, advanced statistics, and ecology. I wasn't eager to sign up for an advanced degree but nevertheless enrolled in the Ph.D. program. Even with one course per quarter, I still struggled to keep up with the educational requirements and the demands of the job. Bob had entered high school, and Sue was in junior high. Both were on the Normandy Park swim team, and Bob was also in the Mt. Rainier High School swim club. Both were struggling with the emotions and physical changes of adolescence. Ruby and I attempted to accompany them to their various swim meets and also to continue camping in the backwoods of the Cascades and fishing the rivers and lakes of the region.

Often, Ruby provided the family support. Meanwhile I scampered across one ocean and then another in seemingly frantic efforts to find Band-Aid solutions to the growing economic and overfishing problems caused by a global society that sought to exploit every possible ocean resource. Between trips, we would renew our social ties with neighbors and other friends. We stayed close to Vern Benedict and his wife. The kids had their own circle of friends. Ruby had the garden club and PTA to keep her busy. Of course, Bob and Sue had participated in Cub Scouts, Boy Scouts, and Girl Scouts for twelve years, and their mother led one or two scout groups.

Our trips to southern California became less frequent, but we still tried to make it down to see my family in San Diego and Ruby's folks in Los Angeles. My parents had moved to a property northeast of Escondido. They were building a new home and putting in an avocado orchard on seventeen acres they had purchased for $400 an acre. The new homestead was off what was then Highway 395, which ran north to Riverside and sat on top of a hill. The view was marvelous, with Mount Palomar to the east, the San Jacinto Mountains to the north, and, on a clear day, the ocean to the northwest. There were a few avocado ranches and the homes of orange growers on the surrounding hills.

Ruby's folks had moved from Maywood to Montebello, ten or so miles to the east. Her father built eight apartments near the center of town, and they lived in the largest one. His real-estate activities included the construction of an office building on Whittier Boulevard, an opportune move because the Los Angeles suburbs were spreading rapidly across the coastal plain, east toward the mountains. My brother first settled in Long Beach, where he worked as a consultant to the tuna industry. Later, he joined the Inter-American Tropical Tuna Commission as a scientist working on the exploitation of the yellow fin and skipjack tuna resources in the eastern tropical Pacific. Frank had done well there, publishing a number of articles in the scientific literature. The Alverson brothers covered the living ocean resources from Peru to Alaska, and the fisheries journals found room to document the results of both of our efforts.

In the late 1960s and early 1970s, business trips took me to Moscow and various other areas of the Soviet Union as well as Poland, Japan,

Korea, Taiwan, the Hawaiian Islands, East Africa, Rome, and Norway. Most of these travels are now a jumbled mosaic in my memory, but a few hold some details worth passing on. I flew mostly Pan American, an airline that I truly enjoyed, but unfortunately it eventually ran into economic trouble and disappeared from the international scene. Abroad, I found myself aboard an array of European, Asian, and African airlines, including Swiss Air, KLM, SAS, JAL, KAL, and East African Airlines. Most of the airlines used Boeing aircraft, except in the Soviet Union where Aeroflot flew its own craft.

Back to Paradise

Sometime in 1966, I was asked to participate in a conference on the development of the fishery resources around the Hawaiian Islands. The meeting was scheduled for Hilo, where I had played out a number of childhood dreams. The thought of returning to "paradise" brought back many pleasant memories. I decided it was time for Ruby to visit the land that had captivated my brother and me during much of our adolescence. The Bureau of Commercial Fisheries had a research laboratory on the campus of the University of Hawaii, and our local lab director suggested that I stop by the lab and meet Kenji Ego, a BCF scientist who was also going to the conference. Including the trip to Honolulu, the stay in the islands would last about ten days. Ruby asked her parents if they would come up from Montebello and take care of the kids. They were always eager to be with their grandchildren and were most agreeable. Thus, we made arrangements for a Pan American flight that got us into Honolulu in the early afternoon.

On the trip over, I kept slipping back in time and reliving my days in Hilo. I wanted to go back and walk on the breakwater and peer down into the clear tropical water and see angel, parrot, and butterfly fish darting in and out of the crevices between the rocks. I wanted Ruby to see the house I lived in near the breakwater, Coconut Island, Rainbow Falls, the Volcano House, Kilauea Crater, and a hundred other sights that flooded my mind. Most of all, I wanted to know if any of the Breakwater Gang were still in Hilo, especially Buddy, our close friend and constant companion during our stay in Hawaii. It wasn't at all clear to me how one calls an old acquaintance after a thirty-year absence. Would he remember our childhood friendship with the same sense of warmth?

After landing, Ruby and I took a taxi to our hotel just behind the Halikalani Hotel; I think it was called the Edgewater. The room, about five stories up, had a nice open lanai overlooking the swimming pool. After unpacking, we took a walk and found an access path next to the Halikalani that led to the beach next to a small restaurant, which was open on three sides and faced the water. At a sandy beach between the restaurant and the hotel a number of swimmers were enjoying the incoming surf. We walked toward Diamond Head until we reached the Royal Hawaiian, where we had mai tais at the beach bar. We then returned along Kalakaua Boulevard to our hotel, where we rested for an hour or two and chatted about dinner. Both of us had overindulged on the plane, so we decided to go down to the Halikalani waterfront bar for a drink and watch the sun fall into the Pacific Ocean. It was the beginning of a second honeymoon.

Early the next morning, we swam and played in the surf for an hour or more. We dried ourselves in the warm morning air and had breakfast in the open-air restaurant on the beach. It was Sunday, and we were scheduled to leave the next morning for Hilo. Since it was Ruby's first trip to the islands, I rented a jeep and took her around Oahu. From the hotel we drove south along Waikiki Beach, around the ocean face of Diamond Head, and then continued on toward Koko Head. On the way we drove into Wailupe Circle, where I showed Ruby our old house, the entrance to the cave and lava tube my brother and I had explored, the on-base movie theater, the swimming pool, and so forth. We drove to Hanama Bay, where we walked down into the crater and swam in the "keyhole." Ruby fell in love with the bay, and we visited it on each of our subsequent trips to Hawaii. We headed around the island, stopping at the Crouching Lion restaurant for lunch, and up to Waimea Falls. From there we drove north, swam in a small bay, and then went beyond the paved road onto lava beds and beach areas not open to normal traffic. We came down the west side of the island, past Pearl Harbor and back to our hotel. The hotel maid, a Chinese-Hawaiian woman, became friends with Ruby and brought her an orchid each morning, which Ruby wore in her hair.

On Monday morning, we met Kenji at the airport. He was a little over five feet tall, but what he lacked in size, he offset with a great sense of humor. He also had a terrific grasp of the fisheries scene in the Hawaiian

Islands and the western Pacific. On the short flight from Honolulu to Hilo, I got a good perspective of the fisheries and environmental problems of concern to the state's governor, who was sponsoring the conference. Distant-water fishing operations were increasing in the waters surrounding the islands, just as they were off of many other coastal states. Long-line tuna fishing vessels operating out of Taiwan, mainland China, South Korea, and Japan were extending their operations eastward across the south and north Pacific. The foreign fleets concentrated on harvesting yellow fin, skipjack and big-eyed tunas throughout much of the tropical Pacific, but also sought albacore and blue fin tuna from the more temperate areas of the world's oceans. The governor was aware of the commercial and recreational value of the living marine resources surrounding the islands. He was interested in a more informed documentation of the importance of these resources—including the precious corals, marine mammals, and pelagic oceanic and reef fish resources—to the people of the islands.

During the flight, Kenji filled me in on the conference details and our task to file a report on the opportunities for expanding fishing and aquaculture activities in the islands. As we approached Hilo, we flew down the Hamakua coast. I saw the Laupahoehoe Peninsula jutting out from the coast, and suddenly I was thirty years younger, remembering how Frank and I had swum off its outer point. A lump came to my throat as I remembered the large tidal wave that had engulfed the peninsula just after World War II and drowned all those children and teachers in the local school. Soon I saw the breakwater that helped form and protect Hilo Bay, and within minutes we were on the ground. The airport, once little more than a dirt runway, along which we had picked guavas in the 1930s, had been lengthened and a nice open-air passenger facility now sat along the east side of the landing strip.

We were put up in the Hilo Bay Hotel, and our room had a sweeping view of the bay and Coconut Island. I was uncertain about when, and even if, I should call Buddy and decided to wait until we had some spare time. The meeting kept me busy for three days and then we went with the other participants on a bus trip to the Volcano National Park, where we visited the Volcano House, Kilauea Crater, and a host of surrounding craters. On the way back, the bus driver, who had done a great job describing the sites, said, "wiki wiki," which translates as "quickly" or "fast." I immediately repeated a poem I had learned as a child in Hilo: Wiki wiki he no stop. Yesterday he go. Wiki wiki he no stop. Pretty soon he go.

The driver stopped the bus, looked around, and said, "You local boy?" We all laughed, and I didn't think much about it at the time. Later, I found out that the driver was "Chicken Little," one of my childhood friends. He had become a guide for one of the many tours operating out of the Hilo hotel complex. Later, I rented a car and took Ruby to the breakwater. I also showed her a row of coconut trees that Buddy, Frank, and I planted some thirty years earlier. They were about two and a half feet tall when we set them in the ground, but had matured into beautiful, gracious, forty-foot palms.

I had put Buddy's number next to the phone at the hotel but concerned about the response I might get, couldn't get up the nerve to call. Buddy's war and life experiences might have left him with only a flitting memory of what had been one of the great experiences of my life. I kept putting the call off, telling Ruby that Buddy might feel obliged to show us around or do something else for us. Besides, he was chief of the Hilo Fire Department and probably had a number of commitments. Our plane was scheduled to leave Hilo late on a Friday afternoon, and we had packed all our belongings for our return to Seattle. Before we left the hotel room, I got up the nerve to call and Buddy answered the phone after a few rings. After I told him who I was, he seemed very pleased and wanted to know where he could meet us. I told him our situation, and he said, "I will meet you at the airport."

As we got out of the cab, I saw someone standing next to the baggage area. Buddy was about five foot ten, muscular and heavyset. He walked over and said, "You Lee?" I said, "Yes," and we hugged each other. After I introduced Ruby, he asked, "How yore broder Frank?" Buddy was disappointed that we had not gotten in touch earlier, but the meeting rekindled a childhood friendship that would last throughout our lives. Sometime later, Buddy's grandchildren and all four of mine would play in his backyard, struggling to get the husk off a coconut and sitting down to a Hawaiian feast at his table. Each subsequent year, Buddy sent us a large box filled with candies, homemade jam, macadamia nuts, and brittle, and I mailed him a large frozen salmon.

Across the Soviet Union

I went to Moscow in 1966 to meet with Soviet scientists and fisheries managers, attempting to establish a formal exchange of the statistical and biological information needed to evaluate the trends in exploited fish stocks. I met with Professor Peter Moiseev, who was Deputy Director for VNIRO (The All-Union Scientific Research Institute of Marine Fisheries and Oceanography). He was an elegant gray-haired man, about six feet two inches tall, and he spoke quite good English. He looked like he came right out of the courts of the tsars. He told me that his family had been shipped off to Siberia when he was about ten years old and that he had grown up and been educated in Vladivostok. His work on the ground-fish resources off the east coast of Siberia and south to Japan, which had been translated into English, was the single most important work on the demersal fish fauna of the region. I had read its approximately three hundred pages several times. Moiseev had headed the Soviet scientific team during my visit to Moscow. His grasp of the marine sciences was excellent, but there was little doubt that he was a committed communist and steeped in party ideology.

In preparation for our talks with the Soviets, our group of scientists met at the U.S. embassy, holding our confidential talks in the famous "Hanging Glass" room. The room was suspended with cables and air blew around the outer walls of the glass "cage." I never knew what was so secret about our talks, but this cloak-and-dagger routine kept our tactics hidden from our Soviet counterparts. The evening after our preparatory talks, the ambassador invited us to a cocktail party. I thought that this was top-of-the-line treatment for our negotiating team, but it turned out that Senator Hubert Humphrey was in town and our group was lucky enough to join his party. I talked with the senator, who spoke elegantly for five or ten minutes, but in the end I sensed that he transmitted better than he received.

The first meeting with the Soviets was friendly, and there was a good exchange of views. Our delegation saw *Swan Lake* at the Bolshoi Theatre, another ballet at the Palace of Congress within the Kremlin walls, and the Moscow Circus. However, we never received the essential information regarding the level of catches caught off various parts of the United States. We were promised the information, but each day there was some delay and we left empty-handed. Needless to say, our scientists submitted detailed catch and fishing-effort data to the Soviet delegation. I was never quite sure whether the data was withheld on purpose or if the Soviet bureaucracy just couldn't get its act together. The Soviet fishing operations were under one government entity, and the fisheries science was under another. Thus, the statistics had to be accessed from different sectors of the fisheries ministry.

In June 1967, I received a letter from the FAO, asking me to join a study tour in the Soviet Union, scheduled from mid-September until late October. It was more time than I could spare from my job in Seattle, but the head office in D.C. thought it might improve our access to information about fisheries in the USSR. I flew to Rome in mid-September to be briefed on the mission by UN staff and meet with Professors Koroki from Japan and Blaxter from the United Kingdom. The three of us were to give lectures to a group of students selected by UN and Soviet officials to participate in a tour of marine and freshwater scientific institutions throughout the USSR. The students were drawn exclusively from Soviet bloc countries or those with leftist leanings; most were from Cuba, Yugoslavia, and Poland and a few were from South America. The tour group was made up of about thirty-five individuals, including several Soviet interpreters, a member of the KGB, about twenty-five students, a secretary, and the three of us from the FAO. We flew from Rome to Moscow on an Aeroflot flight on September 20 and stayed an Intourist hotel on the outskirts of the city.

The next morning, we were taken to the All-Union Institute of Marine Fisheries (VINIRO), where we were briefed on our tour and also attended several lectures by members of the Soviet Academy of Sciences. I again had the opportunity to meet with Moiseev, who assured me that, at our next formal scientific meeting, we would get the fisheries statistics we wanted. The opening session included a bit of Soviet propaganda, although not as much as I had anticipated, along with several good talks on fish-behavior studies that were underway in various parts of the country. The session also allowed me to become better acquainted with the two other guest lecturers. Professor Blaxter's work had been published in a number of European marine science journals, and he was well known for his work on fish behavior. I had read a number of his papers concerning the behavior of herring and other pelagic fish species. He had a good sense of humor and spoke in the "King's English." Koroki spoke good English, but with a moderate slight Japanese accent. Both men were witty, intelligent, and intellectually stimulating and turned out to be excellent traveling companions.

We stayed in Moscow for two days and were scheduled to leave for Batumi on the Black Sea, where we would visit the Georgia Scientific Research Institute. On the second night, I was rather tired and went to bed early, about 10:00 p.m. I had fallen into a deep sleep when the phone rang. It was about 2:00 a.m. and the operator said "Emergency, Emergency" in clear English and then lapsed into Russian. The operator repeated the message several times, and I said, rather intensely, "I don't understand Russian." She then said, "You don't understand Russian? Never mind, there is no problem." The next morning I mentioned the peculiar call to Blaxter, who said that he'd had the same experience. We asked one of the women interpreters about the matter. She laughed and said quietly, "Don't worry, it's just a standard KGB check." She then pointed out the KGB person in the tour group.

The next morning, we boarded an Aeroflot flight to Sokhumi, located at the far northeast end of the Black Sea. At the time, we were not sure how we would be transported to Batumi, which was located about 150 miles south near the Turkish border. The flight took about three hours, during which we were given an apple and an open-faced sandwich. We landed about noon, boarded a bus, and started up into the Caucasus Mountains. After several hours, some of the students asked if we would stop for lunch. We hadn't had anything to eat except the snack on the plane and everyone, including the interpreters, was expecting a pit stop and food.

The bus driver was listening to a soccer game blasting out on his radio and seemed oblivious to the complaints of his passengers. He just continued driving along the narrow road, which snaked through the hills and mountains. We had been on the road for almost three hours, and it didn't appear that we would stop to eat or for any other bodily functions. This driver has a big bladder, I thought to myself, and he is mesmerized by the soccer game. I pulled out three cans of mixed nuts that I had picked up at the U.S. embassy in Moscow and passed them around the bus. Everyone took part in the mini-feast, even the bus driver, but he drove on for another hour before stopping at a roadside stand with a public restroom. Although no one made an issue of the matter to the driver, everyone in the bus made a mad rush for the toilet.

Our tour guide had said that it would take a little over five hours to get to Batumi, but we continued on until after dark. After seven hours, Blaxter asked one of the Russians when we would get to our destination. There was a subsequent discussion with the driver, and we got a vague explanation about traffic problems. When we arrived around 9:00 in the evening, we found that the drive had been extended because our rooms had not been vacated until late in the afternoon. Located in a semitropical environment, Batumi was a popular tourist town. It had a long sandy beach, and several cold-water streams, inhabited by trout, entered the sea near the city. Since it was a warm night and it had been a long, hot trip, Blaxter, several of the Russians, and I decided to have a late evening swim. We were about waist deep in the water when searchlights shined on us, and military personnel with machine guns motioned us to get out of the water. We quickly complied. It was tense for a while, but after several minutes of discussion between one of our interpreters and the army officers, we were sent back to the hotel. We were informed that the beach was closed after dark because citizens occasionally tried to swim across the border into Turkey. Our Russian companions were as surprised by the event as Blaxter and I and had not been aware of the restrictions.

We visited the local fisheries institute the next day, where we were given a briefing on research into the region's fisheries resources, including electro-fishing and acoustical herding. The Soviets were exploring various acoustical frequencies in attempts to herd fish into the paths of oncoming trawls. They were also working hard to develop methods of differentiating fish near the seabed to improve fish-counting methods. Acoustical enumeration of pelagic fish populations had made major strides in western Europe and better estimates of fish populations near the seabed would be an important step in verifying fish-population enumerations. Koroki, Blaxter, and I each made a presentation on different aspects of fish behavior with regard to fishing gear. The three of us were becoming comfortable with each other, although our relationship with the students remained formal. The interpreters were friendly and eager to discuss both contemporary events and cultural issues.

We finished our day's work early in the afternoon and Blaxter and I encouraged a number of the students, interpreters, and our Soviet associates to join us in an afternoon swim. As everyone headed for the beach, I ran back to the hotel to get a towel. When I opened my door, there were two men standing in the middle of the room. They were caught off-guard, mentioned something about fixing the TV, and left. It was obvious that they had opened my briefcase and several of the drawers were open. But there was nothing that I could do and I thought, "Well, if they expected to find something, they must have been disappointed." I picked up my towel and joined the group at the beach, where I saw the KBG tour participant. Blaxter and I questioned him about the men in my room. He asked if anything was missing, and when I replied, "Not that I know of," shrugged his shoulders, conveying the message, so what's your problem? I knew better than to push the matter, so I entered the warm waters of the Black Sea and enjoyed a good swim.

When we returned to the hotel, several of the Cuban students were in the lobby, along with Steven Jukich, a handsome young biologist from Yugoslavia. He invited me to join him and other students for dinner. The young female interpreter, who appeared to be of Armenian descent, also would join us. On the way to the restaurant, the interpreter told us that the Georgians were not particularly fond of the Russians. Stalin had come from Georgia and, although he was no longer in vogue, there was still a large statue of him in the waterfront park across from the hotel. She noted that Georgia was only area within the Soviet Union where he was still honored. We walked for about half a mile to a roadside restaurant. The staff put together several tables for our group, and we ordered some local wine, dark bread, and chicken. The wine was very good, the bread wholesome, and the chicken excellent but it had relatively little meat on its bones. After the owners found out that there was an American at the table, several of the waiters began to ask questions. The conversation was friendly but took a good deal of time as we communicated through the interpreter. By the time the evening was over, I had broken the ice with the communist students from Cuba, and Stephen was rapidly becoming a friend.

We were up early the next morning to take a tour aboard one of the institute's research vessels. We saw their sonar and echo-sounding gear and listened to lectures on the fish fauna of the region and the history of the Soviet republic of Georgia. If my recollection is correct, the region had been occupied by the Persians, Greeks, and Romans and had ethnic ties to the Armenians to the south. We spent several hours cruising north along the coast. The cruise provided a casual atmosphere, which enhanced the dialogue between the lecturers, students, and interpreters. Stephen confided that the FAO secretary was spending her nights with various members of the student group and that she had become their main topic of conversation. Before we departed the vessel, I talked to the KBG agent but did not mention the search of my room. It was a time when the relationship between our two countries was strained, so we talked instead about the Soviet space program, sports, and the rapidly expanding Soviet fisheries activities.

Back in Batumi, we ate at the hotel. After dinner, there was a scheduled trip to a museum about a half hour away, in the hills behind the city. When we arrived, the guides formed us into several groups and closely chaperoned us through the various displays. Blaxter and I got into a conversation and soon found ourselves alone; the tour had moved on. We quickened our pace, looking for the group, and entered a large room. No one was present, and it was clear that we had strayed from the flock. I suggested that we try to find our group, but Blaxter had found something that attracted his attention. "Lee, come here and look at this convoluted stack of metal," he said. The display was enclosed behind a clear plastic shield. Blaxter was able to decipher enough of the writing on the plaque to conclude that we were looking at a piece of Gary Powers's U-2 spy plane. We surmised that the plane probably had been shot down somewhere near Batumi and that the Soviets had put the success of their anti-aircraft technology on display.

After a few minutes, one of the interpreters came up and urged us to return to the tour group. By that time, however, several of the students had found us and began to look at the showcase. Soon the entire group surrounded the remnants of Powers' plane. When the KBG agent arrived, I decided to have some fun and chided the Russians for shooting down a poor, unarmed plane that got lost over the USSR. But he was on to me and began to laugh, and soon everyone was laughing. He then told the group a joke about some Russian generals who went to Egypt after the Six-Day War and asked President Nasser, "Why did you lose the war so quickly? We gave you excellent tanks, rockets, small arms, and other weapons and you allowed the Jews to chase you back into your country." Nasser and his generals responded, "You don't understand, we were using Soviet strategy." The Russian general was indignant and replied,

"What Soviet strategy?" The Egyptian responded, "We have lured them deep into our country, and now we are waiting for the snows of winter."

From that point on, it was apparent that our chaperone had a good sense of humor and the tour group had jelled into a good team. We left Batumi after about four days and flew back to Moscow. The flight surprised us, since we had taken a nine-hour bus ride on the incoming trip from Sokhumi. But by that time, I had ceased trying to understand the Soviet bureaucracy. Besides, our hosts treated everyone kindly. We were to stay in Moscow for one night and then proceed to Murmansk, but the schedule was delayed for a day. I took the time to purchase a few sundries at our embassy. During my visit, I ran into the science attaché who asked me to dinner. He wrote out an address and told me it would be about a twenty- to thirty-minute taxi drive, depending on the traffic. He scheduled the dinner for 8:00 p.m., so I left the hotel around 7:30 and told the driver the address. We arrived at a large apartment complex just a few minutes before 8:00. I knocked and the door opened, but to my surprise, a heavy-set Russian woman stood in the entrance. She spoke no English, and I had no idea where I had gone wrong.

I returned to the street, checked the sign, and was horrified to find it read Leninski. The driver had misinterpreted my pronunciation of Leningradski, and I did not know where I was. In Moscow, it was difficult, if not impossible, to wave down a taxi. It was necessary to find a taxi stand, and they could be difficult to locate. It had started to rain, so I walked toward the center of town, watching for a taxi stand. After about a half hour, I was getting drenched and hadn't seen a taxi stand or met anyone who spoke English. The thought that I might not get back to the hotel in time for the tour departure for Murmansk spiked my adrenaline and I increased my pace, hoping to find a store or hotel or to flag down a taxi on the road.

Panic was beginning to set in when I spotted a policeman standing in a small shelter near the center of an intersection. I walked to the center of the street, looked straight at the policeman, and said, "taxi." He looked me over rather slowly and did not answer. I repeated, "taxi" and then, in a stoic manner, the policeman pointed to a hotel about two blocks down the avenue. I hurried down the street, entered the access road to the hotel, and spotted a taxi waiting at the entrance. I felt a sense of relief after I got into the vehicle, gave the name of my hotel, and the driver nodded his head. We arrived at the Intourist facility just fifteen minutes before the tour group left for the train station.

At the train station, we boarded about forty-five minutes before departure time. Blaxter and I were booked into the same compartment, which was furnished with a long seat across one side and two bunk beds on the other side. A shelf above the seat accommodated our baggage. We settled in, and soon we were on our way to Leningrad. Of course, it was dark and there was nothing for us to see except the lights of Moscow and later the occasional lights of other cities. The bunks had one padded mattress about three inches thick, but they looked comfortable, a lot like the bunk beds aboard the *John N. Cobb*. The ride must have been reasonably smooth, because I slept well and woke just after sunup. Shortly after we dressed, a large, bulky Russian woman entered the room and brought us some freshly brewed tea. She never smiled and gave the impression she was there to keep an eye on us. After we finished our tea, one of the interpreters took us to a dining car. Breakfast consisted of chewy black bread, butter, an orange, and more tea. The heavy bread had a good taste, was filling, and went down well with the tea. We entered the outskirts of Leningrad, but we didn't get much of a view as we were soon traveling on a track about fifteen feet below ground level. We stopped at the Leningrad station for about an hour before departing for Murmansk.

The train headed north, up the Kola Peninsula east of Finland. Outside of Leningrad, we entered a forest of largely deciduous trees that appeared to be a type of silver birch with a whitish outer bark. The forest seemed endless, covering a flat terrain. There were few cities or villages along the track, only endless stretches of trees, which after a time gave me a forlorn feeling. Every once in a while, we passed a small station with a dozen or more people and a dirt-road crossing. It all looked bleak and lonely. As the day went on, the birch trees got smaller and the leaves became a variety of yellows and reds. The tea lady made her way through our car three times during the day, serving all of the compartments. At one end of the car, there was a small room where she prepared the tea in a large container, which Blaxter called a "giant samovar." The rest of the time, the tea lady just sat and watched her passengers and, I suppose, helped those who had some special needs.

Shortly after dark, we ate in the dining car and then Blaxter and I returned to our compartment, where we chatted with the interpreters and some of the students. The major topic of discussion was the sexual activity of our traveling secretary, an attractive blonde with a very pleasant personality, and the question seemed to be who hadn't slept with her. Steven noted that he thought a lot of the chatter was wishful thinking. He did, however, admit to spending part of the night with the young lady and thought that she was on the prowl for a husband. We turned in around midnight and were up early for our morning tea and another bread-and-jam breakfast. When I looked through the windows, the birch trees were still there, but they were shorter, no more than fifteen feet tall. Half an hour before we arrived in the city within the Arctic Circle, the trees gave way to tundra. The birch trees could not survive the long, harsh winter climates of the coastal area. We moved up some coastal hills and then slid down into Murmansk.

The city was nestled in a bay off the Barents Sea and was built along the waterfront and on the surrounding hills. The waterfront was bustling with ship traffic, mostly fishing vessels. At a number of docks, various types of trawlers were being unloaded or preparing to go to sea, including several of the modern "Pushkin" type factory trawlers. My recollection is that they were just under three hundred feet in length. A number of larger mother-ships, which served the smaller Soviet catcher vessels, were taking on supplies or offloading their processed fish. Murmansk was the northern hub for the extensive USSR fishing operations that spanned the Atlantic, including the Barents, Norwegian, and North seas, the waters off Labrador, Newfoundland, the eastern seaboard of the United States, and the south Atlantic off Africa and South America. In the years following World War II, the Soviets who had built a variety of catcher and support vessels that operated for extended periods in remote areas of the world. The explosive expansion of their fishing enterprise had come as a surprise to much of the world's maritime community, including the United States.

Many speculated that the Soviet vessels' primary purpose was to gain information about U.S. and other noncommunist naval and military developments. Over time, it became obvious that some of the fleet indeed was used for intelligence gathering. Nonetheless, the Soviet commitment to fishing as a means of supplying animal protein to a growing population was genuine. If one examines a map of the world and compares the U.S. geographic position with that of the former Soviet Union, it is quickly apparent that most of the USSR is situated completely north of the United States. The climate and terrain placed limits on animal husbandry, so the Soviet Union developed a comprehensive plan to obtain food from the sea. Along with its ship construction program, it had committed funds to expand fishery science and oceanographic programs that would provide the technical information for guiding its expanding ocean harvest.

Our tour group was housed in a hotel on the hill just above the harbor. The rooms were spartan, but clean and comfortable. Our first afternoon, we were treated to a tour of the city, including some fishing vessels, a fur farm that raised sable and mink (the leftovers of fish processing comprised a significant component of the diet for the farmed animals), and a number of landmarks and memorials. Back at the hotel, Blaxter, Koroki, and I learned that we had been invited to have dinner on one of the larger mother-ships, which supported fishing activities in the North Atlantic. The captain expected us around 6:30 p.m. and would entertain us for the evening. I suspected that we were in for a long evening of spirits and immediately went to the hotel restaurant and treated my stomach to some Russian brown bread and lots of butter, hoping to soften the impact of the expected river of vodka.

We arrived at the mother-ship at exactly 6:30 p.m. and went up the gangway to the main deck. The vessel was about 450 feet long, with the bridge and main accommodations aft. The mess hall was furnished with a long wooden table set with fine china and linens, something I had not expected to see aboard a fish-processing ship. A few moments later, we were introduced to the captain. He was a handsome man, blue-eyed and slim, about six feet two inches tall. He gave us a short rundown on the ship's facilities and responsibilities. The vessel had recently returned from the Norwegian and North seas where the crew had worked with a fleet of trawlers and drift netters that were fishing herring. During our tour of Murmansk, we had seen thousands of barrels stashed on docks, next to buildings, and in vacant lots around the city. Blaxter supposed that they were most likely used to store salted fish, and indeed they were. We saw so many barrels around the area that we labeled Murmansk the "Barrel City."

Next, we toured the ship. Everything was spotless, and the metal railings shined as if they were making a statement. There was little doubt that the ship had been scrubbed and polished for our visit. I had been aboard and had observed a number of Soviet fishing vessels in Copenhagen and in north Pacific, and they were far less impressive. But then again, many of the U.S. fishing vessels I had sailed on left a lot to be desired.

After touring the ship, we returned to the dining area, where the table had been set with flowers, bottles of champagne, and several types of liquor, along with bread and butter. Blaxter, Koroki, and I sat across from the captain and several of his officers. The captain started off the evening with a toast of vodka to the three of us and, of course, to the USSR. Vodka was served straight in small glasses and, at the end of each toast, we were obliged to turn our glasses down to show that we had consumed the entire contents. The dinner then began with a series of courses, including king crab salad, soup, caviar, smoked herring, salmon, borscht, various cheeses, and lots of Russian brown bread. Two pretty young women, identified by the captain as members of his crew, brought the food to us. It was not unusual for women to be a part of the crew, and they often worked on the larger Russian fishing vessels. Throughout the evening, the captain raised his glass in a toast to friendship, peace, the sustainable use of the world's oceans, and so forth. After each toast he showed his empty glass and then looked directly at his three guests to ensure that we had joined him. We then showed him our empty vodka glasses. As might be expected, to balance out the evening celebration, Blaxter, Koroki and I felt obliged to make our own toasts.

Having been a guest at many Chinese Gom Bay (bottoms up) events, I was well aware that the party would be considered a great success if the guests were carried off the ship in a drunken stupor. During my tour in China, I had been shown a sneaky drinker's trick, which entailed expelling the rice wine into one's water glass after a toast. Every once in a while I got up with my water (or other chaser) glass, talked with one of the guests, and then went to the bathroom and emptied the glass's contents.

I used this trick again on this occasion and would return to the table, pour some more champagne into my glass, and begin the process all over. The clear color of the vodka made it easy to hide in the champagne. We had been there for a little over an hour when Koroki slumped in his seat, and our host took him off to bed. Blaxter, on the other hand, was doing his best to uphold the honor of the Brits. He did not fill his vodka glass to the top but, even so, he was fighting a losing battle. He had the courage to tell the captain that he just couldn't stay the course and bowed out of the contest before he passed out.

By this time, we had completed our dinner and the party was coming to a close. Blaxter's departure left me alone, however, and I became the object of the captain's attention. While I had consumed perhaps one vodka drink for every five he had swallowed, I knew that my capacity had already been exceeded. I toasted the captain for the excellent dinner and entertainment, and told him that it was getting late and perhaps we had better retreat to the hotel. "We have a long day ahead of us tomorrow," I said. The captain smiled and answered, "Yes," but then picked up a new bottle of champagne, took out the cork, and proceeded to drink the entire bottle. I told him he was by far the champion drinker in Murmansk (and he was). With some difficulty, I managed to walk down the gangplank and to the car, my reputation intact. The next day our Russian interpreters told me that I had gained the reputation of a great drinker. What a fraud!

Our trip to Murmansk was an eye opener in that the city was considered a high-security area, and I was surprised that I had been allowed to participate in the tour. The extensive and impressive infrastructure that supported the Soviet fishing industry illustrated their commitment to "food from the seas." While in the Barrel City, we visited the Polar Research Institute of Marine Fisheries and Oceanography (PINRO), the national fish laboratory for the northern waters. Scientists there were working on various techniques to improve the capture of bottom and pelagic species, both over the continental shelf and in the open ocean areas. They showed us the results of experiments using various sound frequencies, which forced herring schools to move to greater depths near the seabed where they could be harvested more easily. Other experiments were designed to take advantage of particular ocular responses of fish to various net designs and web configurations.

During the train ride to Murmansk, we had been told that, for military reasons, we could not be flown into the city; thus we were caught off-guard when it was time to leave, and we were taken to the airport and put on a plane bound for Moscow. The windows were covered on the bus ride to the airport, however, and then we were taken to a "holding room" until the flight was called. The shades on the windows of the plane also remained closed until we were well on our way to Moscow. As for the secretary, reportedly she switched her attentions to one of the students from Chile.

We returned to Moscow on October 7, my birthday. I asked Blaxter, Koroki, one of the interpreters, and a visiting FAO scientist to join me for dinner. We went downtown to a well-known restaurant not too far from Red Square, where we were given a table next to the dance floor. I can't remember what I ordered, but the food was good. Before we finished eating, a dance band began to play and the dance floor soon became crowded. For the most part, the women on the floor were a mixture of heavyset young and middle-aged persons. Borema, the FAO scientist, pointed out a young couple sitting about four tables from us. The fellow was a young army officer, perhaps in his mid-thirties, and his female companion looked like she had just stepped out of a ballet. She was about five feet six, with short, blond hair and a beautiful, slim figure. She wore a knee-length silky white dress with a modest amount of glitter. She was the center of the crowd's attention.

Our interpreter suggested that she was likely a ballerina or a star in the Moscow Circus. "Well," said Borema, "we should take advantage of the fact she is here and ask her to dance." We all agreed, but then a debate ensued about which of us would undertake the mission. In the midst of the discussion, a young heavy-set woman came to our table. She said something in Russian, presumably asking me to dance, and pulled on my jacket sleeve. She looked like she had just come in off the farm and was the antithesis of the young lady who had caught our attention. My friends told me it would not be courteous to turn her down, so we went out on the dance floor. From the time we got onto the floor until I returned to our table, the young Muscovite talked to me continuously with short intervals for breathing. Of course, I didn't understand a word she was saying.

When I returned to the table, the Russian interpreter said, "If she wasn't so homely, I would have guessed she was a KGB plant." This was all very amusing to the table participants, but the conversation had never left the dazzling young woman and her military escort. While I was gone, a vote decided that, since Borema made the suggestion, he should be the knight of our round table. All eyes were on Borema as he maneuvered across the floor to her table. When he reached the couple, there was a short conversation, and within about thirty seconds he was on his way back. We were eager to hear what had happened. "I asked if they could speak English," Borema reported, and the officer replied, "Yes, goodbye!" We all had a good laugh, but never discovered the identity of the mystery woman.

The stay in Moscow was short, and the next day we took a train east to one of the great lakes on the Volga River, where the Soviets had a major fresh-water research facility. Two days later, we reached Borok, a village on the banks of the giant Rybinsk Reservoir. The village was unique in that its entire population was comprised of scientists, technicians, and support staff for the Soviet National Academy of Sciences Institute for Freshwater Research. There were about 2,000 people in Borok, which was not only known for its work on freshwater fishes and their habitat, but also for its leadership. Borok was established in 1959 and put under the direction of Ivan Papamin, two-time hero of the Soviet Union. Perhaps the name is as unknown to you as it was to me when we arrived in Borok.

Papamin became known for his early explorations of the North Pole. He and his men tried to reach the pole by floating with the Arctic ice across the northern Atlantic. They ultimately abandoned their vessel and moved onto the ice. Communications with the group were lost, and the explorers drifted for several months before they were found exhausted, without food, and half frozen. Papamin and a few of his men survived the ordeal, and his government subsequently recognized him as a "national hero." Later in his life, he led the defense of Murmansk during World War II and was again given the highest award of the Soviet Union.

During our stay in Borok, we were invited to sail with one of the research vessels, and Papamin came along. He seemed very proud of the scientists and workers at the institute. We sat on the bow of the vessel for over an hour while he talked about the work at Borok and how he was able to get "unbiased" scientists to work with him. He noted that the village had its own school system, stores, and infrastructure. We talked until a cold wind coming off the reservoir forced us into the cabin, where we had an excellent meal.

After we returned to the village, the tour group attended a number of lectures from the local scientists. A great deal of attention was paid to the sensory mechanisms the fish utilized in their migration. Some of the Borok scientists developed a hypothesis that fish may be able to sense and use the earth's electromagnetic field to orient themselves. They also were working on electro-fishing, a method of controlling the fish muscles so that the fish would move continuously toward a pole in the electrical field, where they could be captured.

On the third day of our stay, our hosts asked us to play a game of soccer against the local high school team. With absolutely no experience in soccer and not being in the greatest shape, I wasn't very enthusiastic, but the students, almost all from countries that played soccer, were overwhelmingly in favor of the idea. So the next day we took to the field. We were not without some talent. Stephen had played soccer for the national Yugoslavian team, and several of the students from Cuba and South America also played well. In deference to my age and inexperience and given the shortage of players, I was assigned to be the goalkeeper. The youth of Borok were too much for us, and most of the game was played at our end of the field, where the young Soviets shot five goals past me. I was able to stop only one shot. In courtesy, the Borok team applauded my one saving effort.

The following day, the tour group crammed into a yellow bus that looked much like an old U.S. school bus. We were given a royal send-off as Papamin and the entire village stood waving as we rode down the main street to the train station. Somehow the train that had taken more than two long days to get us to Borok returned us to Moscow in only one long day. By then no one, including the students, was trying to make sense out of the Russian bureaucracy. It was enough to know that, although things didn't always make sense, we were following a prearranged schedule and that our hosts were treating us as "special cargo."

We had finished our tour of the USSR and were slated to fly to Oslo to attend a UN conference on fish behavior and fishing gear. It was mid-October, and we had been on the tour for more than a month. Although it had been a wonderful experience, I was anxious to return to the West and get home. After all, it was fall and the U.W. was into conference football. During our extended tour, the students, lecturers, and interpreters had bonded, and each of us had made many new friends, so it was with a certain sadness that we bid farewell to our Russian hosts.

We left Moscow on an Aeroflot flight to Stockholm, transferred to an SAS flight to Oslo, and then changed to another SAS flight that took us to Bergen, Norway. We met in the conference room in the new Northward Hotel in the middle of town. The conference brought together most of the world's leading scientists concerned with fish behavior and, in particular, the behavior of fish toward mobile fishing gear. The scientific strength in this topical area was concentrated mainly in northwestern Europe and Japan. The work of the English, Scots, Norwegians, Dutch, and Japanese was well documented and, while the Americans and Russians were well represented, their work was only starting to become known. Shortly after arriving in Bergen, the FAO staff advised me that I would chair the conference, a surprise and honor I had not anticipated.

It was an interesting period in the evolution of fisheries science. There was a major effort to collect statistical data on the catches according to area, effort expended, and size and age of fish taken and develop the theoretical basis for managing the oceans' living resources. At the same time, other scientists were examining methods to improve the capture and utilization of the diverse marine life that was being discovered. The vast majority of the papers presented at the conference discussed how the knowledge of fish behavior could be utilized to increase the selectivity and amount of fish that could be captured with a particular type of gear. The participating scientists were largely from the developed countries, although there were a few from African and South American nations interested in developing fisheries in the waters off their coasts. The Russians, who were expanding their fishing activities throughout the world, had a large contingent at the meeting and sent one of their fishery research vessels from Murmansk to Bergen.

Several days into the meeting, the Russians hosted a dinner aboard their vessel. As the chairman of the conference, I was invited to go, joining perhaps two dozen scientists from various nations at the dinner. I expected our host to serve an excellent dinner, but also to supply ample amounts of vodka and other spirits. As in Murmansk, I prepared my stomach for the anticipated toasts. However, my plan for the evening was made more difficult because I was seated next to the wife of an FAO scientist who had her own devious way of eliminating the vodka from her glass. As soon as I emptied my vodka shot into my mouth, she quickly poured her drink into my glass, turned her glass over, and smiled sweetly to all those at the table. I finally had to tell our Russian host that I could not keep up the pace because I needed to be alert at the next day's session.

At the end of the conference, I said my farewells to the group of students from the Russian tour. Somehow, the political and cultural differences that had made communications difficult at the beginning had disappeared. Before I departed, the five members of the Cuban delegation presented me with a box of fine Havana cigars. I wasn't a smoker, but accepted the gift as a token of friendship and placed it in my luggage. By the time I arrived at immigrations and customs in New York, the cigars had been forgotten. However, the customs agents soon discovered them. The agents took the booty from me and gave me a serious lecture about bringing "contraband" into the United States. The loss was not of any great importance, but I wouldn't be surprised to learn some of the customs agents enjoyed the fine cigars over the next several days.

Back Behind the Iron Curtain

When foreign nations began fishing in the north Pacific, the U.S. Department of State initiated dialogues with them in an attempt to establish a formal exchange of data on the geographic distribution of catches by species. The effort required governments to note various biological information regarding the size and sex of the harvested species. The Poles, who had been fishing the Bering Sea since the 1960s, were utilizing large factory trawlers very similar to those used by the Soviets; their catches of bottom fish were increasing rapidly. It wasn't long before I was asked to go to Poland to organize a formal data exchange. Thus, I made plans to go to Warsaw and visit with the Polish minister of fisheries. The trip was organized rather quickly, and after waiting several days to receive a visa, I was back in the air heading across the Atlantic.

Prior to my departure, I made plans to stop in Bergen, the site of the International Council for the Exploration of the Seas. From there, I would fly to Copenhagen and then on to Warsaw. After a night of sleep, I would meet with members of the ministry. After the meeting in Bergen, I visited Steinar Olsen, one of the leading scientists at the local fisheries institute. He and I had become friends in Rome when he worked there with the FAO. I spent a night at his home and then I departed for Copenhagen, dressed in blue jeans and a sweatshirt.

When I arrived in Warsaw late in the afternoon, I was caught off-guard when a staff member from our embassy said, "The Polish minister of fisheries wishes to talk with you today, so we have set up a dinner meeting for 6:00 p.m." I looked at my watch and it was already a few minutes past 5:00. I told the gentleman from the Polish ministry that I needed to go to the hotel and change, but insisting that there was not enough time, he proposed an alternative solution. His apartment was on the way to the restaurant, and we made a short stop there so I could change clothes. His wife was busy making dinner for two children and her husband and didn't seem at all pleased to see me. I was hustled into a bedroom with my luggage; it took me about fifteen minutes to change and shave. I apologized to the wife, and we were out the door.

Dinner was staged in a very attractive restaurant, and the table set with fine linens. The minister welcomed me to Poland and stated that his country was eager to establish data and scientific exchanges with the United States. Their major scientific institute for fisheries was in Gdynia, along the Baltic, where I would visit the next day. He also said that the rapidly expanding Polish fishing fleet was comprised of a number of large factory trawlers along with support vessels, almost a carbon copy of the Soviet fleet. These vessels were being constructed in the large shipyards in Gdansk (called Danzig under German occupation during World War II), which was close to Gdynia. The minister recommended that I also visit those facilities. The evening went well, and I went to the Grand Hotel—there must be one in every large city in the world to spend the night. Before going to bed, I took a short walk. I was stopped three times by very attractive women of the street, which came as a surprise. In the Soviet Union, we knew that prostitutes worked certain hotels, but they were far less obvious. I suspect that they served a dual purpose—entertainment and intelligence gathering.

The next morning a government car picked me up at 7:00 a.m., and we headed for Gdansk and the Baltic Coast. I was pleased that we had not flown because it gave me the chance to see the Polish landscape. It was mostly farm country, with meadows and grazing cows, and large areas of yellow flowering rape (Brassica napus, the seed used to extract canola oil). En route, we stopped in a rather large town—Osterode,

I believe—and visited a school that taught about fisheries and other biological sciences.

Both Gdansk and Gdynia were located in what was once called the Gulf of Danzig, which was renamed after the Germans were forced from the area after the war. The Gdansk shipyards built everything from tankers to fishing vessels, including the large new factory trawlers that were fishing throughout much of the Atlantic and off the Alaskan coast in the Pacific. Poland had not been one of the world's fishing powers prior to the war, but following the Soviets, it now was committed to increased fish harvests to augment its supplies of animal protein. The construction of large factory trawlers, however, was not totally a domestic endeavor, because a significant component of the fishing vessel construction was exported to the USSR, as well as other Eastern bloc countries. The pool of skilled ship construction workers in the Gdansk area served the Soviets' maritime goals well. These workers were later at the forefront of the anti-communist movement that led to Poland's return to a democratic form of government.

After the tour of the shipyards, I was taken to a beach resort hotel in the town of Sopot (called Zoppot by the Germans). The accommodations were excellent, and I could look out over the Baltic from my room. Around 7:00 p.m. I went down to the dining room, which also had a view of the sea, and it was packed with young people, including attractive young ladies. Most of them seemed to be eating bakery goods and drinking tea or coffee. The crowd was talkative and the room was buzzing, much like one would expect in Rome, London, or Copenhagen. Based on my experience, the atmosphere seemed much more animated than in Russian restaurants. Communism has not dulled the Polish zest for life, I thought to myself. My impressions were somewhat tarnished the next morning, however, when my guide told me that the hotel was a favorite pickup location for the professional ladies of the night.

We took a short drive to Gdynia, another manufacturing city on the Gulf of Danzig. Both Gdansk and Gdynia looked dreary. Many apartment buildings in both cities looked like they had come out of a common mold, particularly in the newer areas. Built by the government, they looked like a series of giant beehives made of thousands of small cells, much like they did in Moscow. On the other hand, there was a notable difference in the number of church spires. Seemingly at odds with the national politics, on this Sunday morning, I could see families on the way to Mass from my hotel window. The Catholic Church remained active despite communism.

Gdynia offered the chance to visit a marine science center that housed a significant number of Poland's marine biologists. There, we began discussions on the extent of the Polish fishing efforts in the northeast Pacific and the country's future plans for expansion. I received considerable information on their general plans for ocean exploration and was surprised to learn that their catch for marine water was then the sixth largest in the world.

By coincidence, the trip to the coastal area put me back in touch with a young Polish fishing and recreational vessel architect I had met at the FAO fishing gear conference in Hamburg. He came to reception held in my honor, and we chatted for some time. I learned that he had a young daughter about the same age as mine. We put the two girls in touch with each other, and the communications between them continued for several years before coming to an abrupt stop. I later learned that my architect friend had been jailed on suspicion of collaborating with the West.

Before leaving Poland, I went to one of the Nazi death camps, east of Gdansk, where thousands of Jewish people had been gassed. The trip had successfully initiated an exchange of information and longer-term ties between marine scientists in our two countries. It also was a stark reminder of the political and ideological differences between western and eastern bloc nations. Following my trip, I wrote a lengthy travel report for the National Marine Fisheries Service (NMFS) and an article for *Fishermen's News* about the rapidly growing Polish fleet. Copies were sent to the Polish fish authorities. The report and news articles must have been reasonably accurate as there were no complaints from any of the recipients.

Into Africa

Toward the end of the decade, the FAO invited me to review the marine fisheries of Kenya and Tanzania, including Zanzibar. The invitation came soon after Ruby and I watched a T.V. special on the Serengeti. I believe it was called *Remember the Serengeti* or *Return to the Serengeti* or a similar title. We were both impressed, and I told Ruby that I would like to visit that part of the world some day. However, I hesitated to respond to the invitation because there was reported unrest in Zanzibar between some of the blacks and Arabs. But on reflection, I pointed out to Ruby that the current director of the Kenyan Department of Fisheries, Mr. Odero, had been a classmate of mine at U.W. and stayed at our house one night when he was visiting Seattle. I was also a friend of the head of the Department of Natural Resources in Tanzania and was sure that things would work out. So, with the blessings of my department, which was eager for its scientists to assist developing countries, I was on my way.

The schedule first took me to New York and then to Rome, where I met with UN officials regarding the goals and objectives of the trip. There was a general belief at the FAO and among other scientists who had evaluated the region that there were large underutilized fisheries resources in the Indian Ocean. I was to evaluate the harvesting and processing capacity within the two East African states as well as their ability to monitor and manage the resources. The trip plan called for me to fly to Nairobi, then to Mombassa and nearby fishing communities. From there I would fly to Dar es Salaam, Tanzania. After meeting with local scientists in the coastal areas, I was go to Arusha, Tanzania, to write the formal trip report, which would be submitted to local authorities for comment and edited. The final report would be given to the FAO Department of Fisheries. In Rome, I received a variety of shots to ward off potential diseases and anti-malaria pills, and also renewed my small pox vaccination

I left Rome in late April on an East African Airlines flight non-stop to Nairobi. The route took us across the Mediterranean, east of Cairo, and then down through the Sudan into Kenya. It seemed like we had been in the air for hours when the pilot announced that we were over Khartoum (Sudan). Immediately I thought of writings by Kipling, although I wasn't sure they were appropriate. A car and driver from the Kenya Department of Fisheries picked me up at the airport and drove me into town. I was put up at the Stanley Hotel, which had a famous thorn tree just outside the coffee shop. It was a Saturday night, but I was tired and, after a light meal, retired to my room.

The following morning was Easter Sunday and, although I was not a regular churchgoer at the time, I wanted to attend a local church to get a flavor of the participants and the proceedings. In the lobby, a desk attendant gave me directions to a Baptist church several blocks down the street. My intentions were to attend the 11:00 a.m. service, but upon arriving I found that the service had already started. I made my way to a seat in the rear of the crowded church. The congregation was about ninety-percent black. The sermon, given by a black preacher, was in English as were the songs, which for the most part were familiar. Except for its racial makeup, it could have been any church in Seattle. After the service was over, several people greeted me and asked where I was from. I moseyed on back toward the hotel, walking down Kenyatta Avenue (Kenyatta had been a major player in the freedom fight for independence during the British rule and also was Kenya's first president). Much to my surprise, I spotted a Kentucky Fried Chicken restaurant and decided to try it out. The food was good, but the chicken didn't have as much meat on its bones as it would back home. I spent a quiet Sunday afternoon, ate a light dinner in the hotel's Thorn Tree Restaurant, and went to bed early.

The next morning a car from the Department of Fisheries took me to their main offices on the outskirts of town. I was disappointed to find that Odero was on a business trip in Rome. But the deputy director, John, welcomed me and provided a rundown on the coastal fisheries off Kenya. The continental shelf adjacent to much of East Africa is relatively narrow, and coral reefs abound in much of the area. There was a small amount of trawling for a mixture of tropical bottom fish. Most of the fisheries in the region were based in coastal villages and used a variety of small boats. Fish were taken with hooks and lines, traps, gill nets, and set nets. Lobsters were taken both by traps and divers who collected them along the reefs. The overall fishery production was relatively small, and its use to feed the population was limited by the lack of available cold storage facilities and transportation to the larger population inland.

John was short and trim, with dark black hair. He was very friendly, but obviously wanted to convey a message to me without damaging my ego. He apparently had experiences with other foreign fishery experts who had come to tell the Kenyans how to modernize their fisheries and shore infrastructure. He lectured me for some time on the limitations on effective management of the harvested resources and increasing production from the ocean waters. To begin, there was limited refrigeration capacity outside of Mombassa. Many of the coastal fisheries had largely a subsistence character, and little or no cold storage facilities. Little of the catch found its way to inland markets. A fair amount of the catches from the larger fishing villages and coastal cities, however, was dried and sold in the outdoor markets of Nairobi. Increasingly, outboard motors were being used to power small craft, but this also was impeded by the lack of service and maintenance personnel.

Trained scientists, John said, "go to Europe or the United States, get an education, come back to Kenya and, because of the shortage of trained scientists and college educated personnel, they move quickly up the civil service system. They are absorbed into the government administration." The government had sent two students to the U.W. School of Fisheries, one who had quickly risen to the level of secretary of wildlife. The other, Odero, had hardly been back a year before he was selected to be the head of the Department of Fisheries. "Although we have a few well-trained people," John went on, "they are almost all administrators, and few are trained scientists actually working in the field." I nodded my head. I had known both of the students from Kenya who had attended the U.W. Other students from Ghana and Nigeria also had trained as fishery biologists, but most were quickly elevated to high-level civil service positions once they returned home.

After the lecture on the state of Kenya's fisheries and the problems confronting a relatively new nation, John told me about the revolution that had led to the formation of Kenya as a free sovereign nation. Many of the capable white people had left the country. The emigration of these trained professionals had left the country with a relatively small pool of well-educated persons. John noted, "We are trying to deal with that problem." There had been, and continued to be, some hostility against Arabs living in Kenya, and tribal discrimination was rampant. "It may remain that way for some time," John said. After a long morning, during which John informed me about the complexity of Kenya's problems, I was shown around the headquarters facilities and introduced to the staff.

We had a short lunch break of fruit and sandwiches, and then John said, "I want you to see the Nairobi open market." We drove in a jeep for a few minutes to the outskirts of town. The market must have covered three or four square blocks with open sheds, partial tents, and hundreds of stands selling a great variety of merchandise. John took me to the fish stands, which were about fifteen to twenty feet across, with tables filled with dried fish. A variety of tropical fish were represented, but one species about eight to ten inches in length dominated the displays. The fish were split open, cleaned, and then dried, and the flesh took on a yellowish color. All of the fish were covered with a layer of flies. John said nothing about the sanitation problems, nor did he make any excuses about the potential health risks for buyers. I think that he just wanted me to face the realities of life in Kenya, but my education was just beginning.

John arranged for me to go to Mombassa the following afternoon to see the new cold-storage facilities there. We then would travel up the coast and visit several small fishing villages. I spent the next morning talking to some of the staff at the Department of Fisheries, assimilating what information I could regarding catch information by species, types of fishing vessels, effort deployed in areas off Kenya, and biological information being collected to support fishery management. About 11:00 a.m., a driver came to pick me up. I jumped in the seat next to the driver, and we headed east toward the coast.

During the drive, we occasionally saw wildlife, including a large number of baboons toward the end of the trip. The two-lane road from Nairobi down to Mombassa was relatively good and fairly straight until we came to the escarpment where the high plains that make up much of Kenya drop sharply down into the coastal flatlands. The road wound down deep valleys, and the vegetation became much more tropical. The air also became warmer and more humid, which tired and drained me during the long drive to Mombassa.

The major port city of Kenya, Mombassa seemed much less organized than Nairobi. It was spread over an extensive area, with a mixture of old and new buildings and canvas-covered shops. Open markets and street shops were common, and the streets bustled with people walking, carrying goods, pushing carts, and fighting for space among the trucks and cars. I was taken to a small but good hotel situated right at the harbor entrance. From the window, I could see a large oceangoing oil platform being towed into the harbor. The hotel had a casino so I played the slots and a little blackjack. My luck was very consistent; I lost at both! Before retiring for the night, I sat on a ledge behind the hotel, watching the boat and ship traffic moving in and out of Mombassa harbor. The evening air was warm, but comfortable.

The next morning a government vehicle took me to the major cold-storage plant. We spent only a few minutes there, and it was obvious that they weren't interested in showing me the relatively modern facilities. In the manager's office, I was introduced to a driver who was to take me to two small coastal fishing villages south of Mombassa. A large man wearing a pair of shorts and blue tee-shirt, he, too, drove an open jeep, but this one had a canvas roof. We drove a short distance to a ferry, which crossed a body of water to the south of the port and took us to the mainland. We then drove along the coast through coastal forest and shrubs, including occasional coconut trees. The farther we traveled, the hotter it got, and I was sweating profusely.

Within an hour, I was drenched in sweat, thirsty, and uncomfortable. From time to time, I looked at the driver, but he seemed completely at ease and enjoying the ride. He was, nevertheless, more observant and sympathetic than I had suspected. After an hour, he pulled over to the side and said, "You look tired and beat, Sahib." Without another word, he took a machete out of the back of the jeep and climbed one of the coconut trees lining the road, knocked down several coconuts, and opened their tops. He then said, "Drink this, you will feel better." I downed the coconut milk and he was right, I did feel better. It was only about a half an hour farther down the road to the first fishing village. It was probably only thirty to forty miles from Mombassa, but the road was narrow and, at times, clogged with locals carrying goods on their heads.

We turned off the main road and wound down a dirt road that ended near the ocean among a cluster of huts and several wooden buildings. I was introduced to a young African who ran the village fishing operation. He purchased the fish that were in excess of the local needs and processed them for distribution elsewhere. Along with the driver, we went down to the beach area. Several small powered vessels were anchored offshore, and a large group of long (fifteen feet or so) dugout canoes were on the sandy beach. One dugout with a father and son on board came in as we looked over the village fleet. The canoe had about twenty to thirty pounds of a variety of reef fish on board. But there was no ice, and the fish were laid under a cloth to keep them out of the sun. I suspect the cloth was wetted down periodically. After looking at the local "artisanal" fleet, we went to a small building used as an icehouse where the dressed catch was stored. The fleet went out each day, brought in its catch, retained what was needed for subsistence, and sold what was left to the local fish buyer. He would have them cleaned and placed in the icehouse. From there, the fish were taken, most often by bicycle, to nearby villages and even the outskirts of Mombassa.

After the tour, the fish buyer asked me to come to his house and have a Coca-Cola. It was a welcome interlude because the hot and humid weather had drained my energy. He was not at all apologetic for the primitive character of the fishing fleet or the way the fish were handled. Instead he told me, with pride, of the progress they were making in improving the quality of the fish sold and the distribution of the catch. We entered his small house, and his wife came out of a back room in native dress and served us Coca-Cola with ice along with some Nabisco cookies. She smiled pleasantly and seemed delighted that she could provide us with such luxuries. I asked if I could take a picture of the two of them in their house. She nodded yes, went to the back room, and in a few minutes returned in western garb. I was disappointed but said nothing, seeing how pleased she was with her western dress.

We visited another village similar in character to the first and returned to Mombassa long after dark. The next morning, I was scheduled to leave for Tanzania. The airport at Mombassa was very crowded, and everyone seemed to be hauling bags, boxes and an assortment of carry-on luggage. The East African Airlines plane was quite comfortable and from my seat I could see the clear blue waters along the coast. From time to time, blue-green areas gave evidence of large coral reefs relatively close to the shoreline. After landing in Dar es Salaam, I was taken to my hotel, which was just across the street from the waterfront.

The city sat on the edge of a large crescent-shaped bay fringed with miles of coconut palms. The bay was filled with Arab dhows ranging from about thirty to seventy feet in length. They carried trade goods along the African coast but also across the Indian Ocean to Oman, Yemen, and even as far as India. The streets of Dar es Salaam were filled with people of black and Arabic descent, and I found it difficult to make my way among the crowds on the narrow sidewalks. Several vendors pressured me to buy some elephant hair bracelets and, after being hassled for a time, I gave in and bought two "guaranteed elephant hair bracelets." Later, I found they were plastic.

After I settled into my hotel, I called the fisheries officials. They had planned a lot for the several days I was to be in Tanzania. The next morning I would meet with the department at 8:00 a.m. and then

head north along the coast to a fisheries training village operated by **the** government. It would take several hours to get there, and we would **return** to "Dar" after dark. The second morning called for me to visit a **marine** research laboratory on the island of Zanzibar.

I was given a quick tour of the fisheries offices and a one-hour talk on the fisheries of the region the following morning. As in Kenya, I was informed about the primitive nature of many of the artisanal fisheries. There were only a few high-powered oceangoing craft, and the extensive canoe fisheries along the coast used hand lines, pots, and gill nets. There was a limited use of refrigeration or ice on most of the small boats that fished near shore, but there was an effort to upgrade the quality of the catch and build cold-storage facilities in a variety of ports. Immediately after the briefing, I was picked up by a driver in a vehicle that looked very much like those used to carry passengers in the game parks; perhaps it was a Land Rover. We headed north and drove along a two-lane road, through heavily forested tropical jungle and open areas with small farms. After a three-hour drive, we pulled into a government training camp that was dedicated to improving subsistence fishery skills, boat maintenance, handling and quality of fish, navigation, net construction, and conservation concepts. After several hours, we returned to Dar.

The following morning I caught a small ferry to Zanzibar. The crossing took several hours, and I was met by the director of the then named East African Fishery and Oceanographic Institute. The government officials in Dar told me that Zanzibar was recovering from a rebellion that had caused the death of several thousand islanders. The confrontation might have developed between people of Arab descent and black Africans, but I am not certain. At any rate, the officials told me not to leave my hotel without a government chaperone. The hotel was a wood building, several stories high, situated on the edge of the harbor. It had once been a club for British government officials. According to my host, it was constructed in the nineteenth century and suffered from lack of maintenance. Nevertheless, the rooms were spacious and clean with high ceilings and a mosquito net over the bed. I wondered why I had been put up at the club and not at the local hotel. I was told that the main tourist hotel was full, but the FAO reservation had been made at least three months earlier and was shown as confirmed on my travel documents. This mystery would become clearer during my stay.

It was early afternoon, and my first scheduled appointment was not until the next morning. Without a guide, I didn't feel free to walk into the center of town. I did walk out to the sea wall in front of the club and watched the boat traffic in the bay for more than an hour. Several small craft came up to one of the docks, and one offloaded what appeared to be about forty or so pounds of a herring-like fish. As I looked to the south and north of the city, I could see a ring of coconut trees along the water and several Arab dhows anchored out from the shore. The view reminded me of several stamps from Zanzibar I had collected as a boy that gave an almost surreal look to the island. I had never imagined that some day I would look out from the Zanzibar harbor and recapture the scene from the stamps. But I didn't know a great deal about the history of the island and its people, only that it had been a hub for the slave trade in the nineteenth century and was known for its export of certain spices. On a more contemporary note, in Rome I had been advised that Zanzibar had stopped using DDT to spray mosquito breeding areas because of the potential ecological damage. Given the absence of an effective alternative, there had been a devastating outbreak of malaria. As a result, I had religiously taken my malaria pills.

After I had returned to my room and rested, there was a knock on my door. A short heavy-set man introduced himself as a member of the local police and showed me an identification card. He asked me if I would join him and several of his coworkers for dinner downstairs. I agreed to meet them shortly. It was a warm evening, so I put on a pair of slacks and a short sleeve tee-shirt and went down to the dining room. The officer was waiting there with three of his friends. We sat down, ordered our food and the officer began to talk. He told me that he had been advised of my mission by the minister of wildlife. He assured me that I was in no danger, as things had settled down on the island, and he hoped that my visit would be fruitful. Then the group began pelting me with questions. From the tenor of the inquiries, it appeared that they doubted my assigned FAO tasks.

"Why was it important for Tanzania to modernize?" "Wouldn't modernization lead to the unemployment among a number of the small-scale coastal fisheries?" "Did I think that the U.S. fisheries science had anything positive to contribute to problems in Tanzania?" "If new conservation measures were put into law, wouldn't it impact the artisanal fishermen more than the industrial fisheries?" The questions were of legitimate concern, but they seemed to me too well orchestrated and crafted to have come from the men who sat around the table. Shortly after Tanzania had become an independent country, its politics had drifted to the left. By the time of my arrival, the country was receiving considerable aid from mainland China. The policemen may have thought I was working for the CIA.

I told them that I had not come to any conclusions and then answered their questions to the best of my ability, noting that I was not in Tanzania to impose U.S. science or policy. My assignment was to contribute, as an individual scientist, my expertise in evaluating national management and development efforts in light of problems that had been identified by the East African Community (a now disbanded working arrangement between Kenya, Tanzania, and Uganda). I ended by noting that the East African countries themselves ultimately would review and evaluate my recommendations; they could accept or reject them. My answers seemed to shift the discussion away from their concerns regarding local input, but the dialogue continued throughout the dinner. After dinner, we went to the bar for a drink. I assumed that they would attempt to loosen my tongue with alcohol, but as soon as we finished the first drink, the officer's three friends excused themselves.

The officer stayed, however, and we continued our talk. We moved from the bar and sat side by side on the stairs that led to my bedroom. The topic shifted to problems inherent in the growing number of young independent African states, the role of the United Nations in the future of Africa, and tribal and ethnic bias in the evolving civil service in the new states. The officer also gave me a short discourse on the people of Zanzibar and the historical involvement of the Arabic people in the region. At this juncture, I was doing most of the listening, occasionally asking him what I considered pertinent questions. We chatted well past midnight, and I later lay in bed wondering what it had all meant. At the time, I was unsettled by the evening's interchange, but several months later I received a note from the officer. He thanked me for my efforts and noted "the wonderful talks that took place at dinner." Maybe I had misinterpreted what had occurred and, then again, maybe my dinner companions had been satisfied with what they had heard.

Early the next morning, the director of the Zanzibar East African Marine Center picked me up and drove me to the laboratory. He was a tall, blond Englishman who had recently taken on the job at the center. He pointed out that funding had eroded as the result of problems in Uganda and the general deterioration of the East African Community. He also advised me that there were a few good scientists working at the laboratory, but for a number of reasons, it was difficult to find competent local scientists. The working conditions and national instability made it difficult to attract expatriates to the island. What a pity, since it was in such a beautiful setting. The lab had a rather large research vessel, but the costs of maintaining the ship and its crew was becoming more and more of a challenge. He noted that the fisheries of the region were largely artisanal in character and that data on catches by species and areas was, for the most part, not monitored. After lunch in town, we chatted with a few of the lab's scientists, who were studying the state of several of the local fish stocks.

About mid-afternoon, we took a tour around the island, which I can describe as an ideal tropical paradise. But the beauty masked the civil strife, health problems, and poverty confronting the local population. Just before sunset, we drove to the director's home, which had a great view of the deep-blue ocean to the south. The large house had spacious rooms, high ceilings, and many flowering plants in pots throughout the first floor. The living and dining rooms opened onto the surrounding tropical foliage. The whole area looked like a green mat descending down to the blue ocean, which stretched out and blended into the horizon. The director told me that many of the better homes were built in a similar manner, but that it caused a problem for some people because the rich insect population could not be kept out of the home. Only the upstairs bedrooms were windowed, screened, and air-conditioned. We had a wonderful dinner, and he spent some time filling me in on local politics. When I returned to my hotel, my officer friend was waiting. It was rather late, but we sat down for a drink, and I filled him in on my day's activities. He wondered what my report might include. I assured him that the local fisheries office would receive a copy.

In the morning I returned to Dar and reported to the ministry of wildlife. Officials had arranged for me to travel to Arusha the next morning, where I was to write my findings and recommendations. With ticket and documents in hand, I made my way through a horribly crowded airport lobby and stood in a somewhat unstructured line, waiting to get into the departure area. I was amazed that such a small and rundown airport served

Tanzania's then capital city. While I was in line, a dozen or more people crowded in front of me. I was concerned about making the flight, but everything worked out. It was a relatively short hop to Arusha and, I was shocked to find myself in a large, beautiful, air-conditioned airport with marble floors, panoramic windows, and spacious waiting areas—but it was almost empty. I surmised that the elegant airport had been built to attract tourists to the wonderful wildlife game parks of the area. Arusha sat on the southern slopes of Mount Kilimanjaro at an altitude of perhaps 6,000 feet and was noted to be the setting of a number of motion pictures based in Africa, including the Tarzan films. It was near the famous Serengeti plains and the Ngorongoro Conservation Area.

The drive to Arusha took an hour or so, but it was well worth the wait. The relatively modern hotel had a swimming pool next to the lobby and dining area. My room, which was in a building a hundred feet or so from the main hotel, was large, clean, modern and very comfortable. It included a couch, a small desk, and a queen-sized bed. It was a great place to unwind and write my report. I was given a week to look over my notes and the material that I had collected and organize my thoughts regarding UN and other possible aid projects. My FAO obligation was to evaluate the potential for making more effective and sustainable use of the region's available marine resources.

That afternoon, I walked to a cluster of buildings at the center of the mountain village. As I approached the main intersection, I saw a bar on the far side of the road that looked attractive. It was relatively warm, so I decided to join its patrons and have a cool beer. Pictures of Hollywood stars covered much of the wall space, evidence of Arusha's connection to the film industry. I spent an hour or so looking at the stars who had visited Arusha and the apparently well-known watering hole. I then visited several souvenir shops before returning to my hotel for a short rest. Dinner at the hotel was excellent, and by the time I returned to my room, the air temperature had dropped to a comfortable seventy degrees.

The next morning I woke early, had a good breakfast, and then returned to my room, where I wrote until early afternoon. I took a break for lunch, returned to my room, and wrote until late afternoon. I followed the same schedule for three days and made excellent progress on the report. On the evening of the third day, a young man from the wildlife ministry came to my room and informed me that I should take a one-day break and see some of the local sights. He joined me for dinner, and I told him about my progress on the FAO report.

The next day we traveled to the fringe of the Serengeti, where I observed an amazing abundance of zebra, Thomson's gazelle, wildebeest, buffalo, giraffe, etc. The trip brought back the mental images of the T.V. special about the Serengeti. It was a long day filled with great sights of African wildlife, and it inspired me to complete my report in the next several days. After I finished the rough draft, I decided not to polish it until I had discussed it with Kenyan and Tanzanian government officials, who arrived on the seventh day of my stay in Arusha. Both governments sent two members from their fisheries departments, and they spent the better part of the day reviewing and commenting on my report. They seemed to view the draft favorably, but pointed out certain interpretive errors that I had made. They also wanted stronger recommendations regarding the character and level of aid that was needed to develop latent fish resources in the Indian Ocean.

Various marine biologists had previously noted that the fishery resources off Kenya and Tanzania were underdeveloped. It was my impression that there wasn't a great opportunity for significant sustainable growth in the coastal fisheries. The continental shelf off the two states was relatively narrow, and the productivity of the coastal water was not high. Even so, there were some limited opportunities for expansion in the coastal region, although the likelihood of significant growth (particularly of tuna) was in offshore pelagic regions of the Indian Ocean. The development of the shore infrastructure and high-seas fleet necessary to exploit these resources would be costly and probably not in the best interest of the region.

During my session with the government officials, I made extensive notes on their observations and made changes that I felt were best for the region. By late afternoon, the report had evolved into an acceptable document, and we adjourned to the dining room for a lengthy dinner. I was driven back to Nairobi the following morning, circling to the west of Kilimanjaro, and then heading south. During the trip, we stopped at the border between Tanzania and Kenya, where I bought several carvings and some beadwork. Further on down the road, we had to stop for several minutes while a large herd of giraffe made their way across the narrow two-lane road. Occasionally I caught a glimpse of members of the tall and handsome Maasai tribe. The next morning, I left for home with some regret that my lengthy and exciting tour was over. I had made new friends and become intoxicated with the people, scenery, and the region's history.

The late 1960s were characterized by a heavy travel schedule; I also managed to crowd in visits to Taiwan, Korea, Japan, and Iceland. Because Ruby's parents were able to take care of Bob and Sue, Ruby and I had the opportunity to return to Hawaii, visit Buddy at his home, walk again on the Hilo breakwater, swim in Hilo Bay, and drive around the island, rekindling old memories. We had also had a wonderful chance to go to Rome, where I debriefed my clients on my work in Africa. It was one of those delightful late falls when the weather in Rome was at its best, and the streets were crowded with tourists and locals taking advantage of the warm temperatures. We arrived on a Friday morning, and I checked in with Jack Marr, one of the top American scientists working with the FAO. We had coffee with Steinar Olson, a Norwegian colleague, and discussed where we might go for a weekend retreat. We decided to visit Lake Nemi, south of Rome, where one of the emperors once entertained guests on a large warship he built to sail in the small lake.

We left the next morning, accompanied by Steinar's girlfriend, and stopped for lunch in the gorgeous valley where Lake Nemi is situated. We found a roadside restaurant, much of it covered by trellises. We sat to the side under grapevines, as the main portion of the restaurant was given over to the wedding party of a young Italian couple. We started our feast around 2:00 p.m., ordering several antipastos, dark bread, and some excellent Frascati wine. We took our time and it was an hour or so before we ordered our main meal of fresh lake trout. During our meal, the Italian wedding party became increasingly celebratory; we could hear toasts and music and many guests were dancing in the aisles.

By the time we finished our meal, it was late afternoon and the wedding party was still in full swing. I believe it was Steinar who suggested that we honor the party by giving them a half-dozen bottles of champagne. We ordered the drinks and toasted the wedding group. Then some of the young men in the party came to our table and asked the women to dance.

The party continued for another hour or so. Before departing, we all enjoyed some excellent Italian gelato. The entire afternoon cost less than $15 per couple, including the champagne. It was one of those moments in time that gets frozen into a corner of your brain and reappears from time to time.

Chapter XI
Climbing the Corporate Ladder

Moving to Washington, D.C.

The second half of the 1960s was a dynamic period in my life. I was in my forties and no longer considered myself a young man. My children were in high school, and my profile within the bureau was changing. I had published scientific papers and had made a large effort to communicate with the fishing community and industry by writing for popular fishing journals, trade papers, and the local news media. These included the *Seattle Times* and government publications targeting various fisheries groups. I had also been a guest on a Seattle radio program hosted by John Weiden who, at that time, was the editor of *Fishermen's News*. Probably, due to my involvement in bilateral and international fisheries negotiations, I had contact with congressmen and senators who dealt with fisheries and ocean issues in Washington, D.C.

As the end of the decade approached, I came under increasing pressure to take over one of the top positions within the Bureau of Commercial Fisheries in D.C. Web Chapman, perhaps one of the best-known fisheries scientists in the country, and his good friend Milner B. Schaffer, head of the Inter-American Tropical Tuna Commission (IATTC), dropped hints that it was time for me to do my part for U.S. fisheries. John Weiden, who had strong ties to Senators Jackson and Magnuson, also asked me to consider moving to Washington. Ruby and

I had discussed the possibility and had decided that we loved the Pacific Northwest. Neither of us was keen on getting into the politics of fisheries policy and administration, and we were unwilling to pull Bob and Sue out of their high school so I could climb the corporate ladder.

Instead of making these points clear to my supporters, however, I generally sidestepped the issue by noting that I needed to complete my Ph.D. However, this excuse was rapidly running its course. By 1966, despite my travel schedule, I had completed the course work I needed, taken and passed my French language exam, and was studying hard to complete my second language requirement, Spanish. Through much of that winter, I carried my pack of Spanish vocabulary cards around with me, studying during plane flights, in the evenings at home, and whenever I could find a spare moment. I was anxious to pass the Spanish test because I had failed it twice, which meant that I couldn't take the required written and oral exams and complete my thesis. By the end of March 1967, I was ready to have another go at it. The exam did not test oral skills, only the ability to translate several written pages in a given time frame.

On the appointed date, I was confident that I would ace the exam. As I scanned the material to be translated, I said to myself, "Well, there are several difficult sections, but in general it doesn't look too bad." Two weeks later, the test results came in the office mail and I opened them with a touch of apprehension. My eyes quickly caught the word "passed." It wasn't a great score, but my language requirements were behind me, and I had overcome the only real obstacle on the road to my doctoral degree. I could concentrate on my final tests and completing my thesis, which was titled "The Demersal Fishes and Fisheries of the Northeast Pacific Ocean."

About a week later, Abbey, my superb secretary, dropped the day's mail on my desk. I spotted a letter from the U.W. and opened it immediately. It was a notice conveying an apology from the Language Department stating that they had made an error; I had actually failed the Spanish test for the third time! It left me confused, deflated, somewhat angry, and uncertain about my future in academia. I had studied hard for the exam and wasn't sure that I wanted to make another effort to jump through the hoop.

Working full-time and acquiring an advanced degree challenged my fortitude and endurance. I told Abbey that I was having second thoughts about continuing at the U.W. and asked her to call my committee chairman. She began to laugh and said, "April Fool!" I had been caught completely off guard and, after joining her and the staff in several minutes of laughter, my spirits soared once again.

The scoundrels had acquired the U.W. School of Fisheries letterhead, drafted the letter, and timed it for arrival on April 1. I'd spent the morning on the telephone with D.C. and lost sight of what day it was.

Within several months, I had completed my written and oral exams and submitted my thesis; sometime in 1967, I received my Ph.D. I had started working on the advanced degree with some skepticism, believing that certain works were revered not because they were based on scholarly science, but because their authors had doctorates. The four to five years of schooling had reinforced my commitment to rely on good science not well-known scientists. The completion of my schooling, however, only left me more vulnerable to the pressure to move to Washington.

In the late 1960s, resentment against the distant-water fishing fleets grew into a movement against the historical legal concept of freedom of the seas, particularly as it related to the living resources near many coastal states. Several South American states, including Peru, Chile, and Ecuador, had unilaterally extended their national jurisdiction to 200 nautical miles, much to the disappointment of the U.S. Departments of State and Defense. Within the State Department, these countries often were referred to as "rogue states." Their jurisdictional extensions placed sectors of the U.S. tuna fleet operating out of southern California in some difficulty. The fleet traditionally had fished and caught considerable quantities off Central America, Peru, and Ecuador. Although the United States did not accept the claims of the "outlaw" states, they inadvertently supported them by paying the fines of American tuna fishermen caught fishing off their coasts.

While Peru, Ecuador, and Chile were seen as rogue states by the U.S. and many European governments, these countries were gaining increasing support in southeast Asia and Africa, where distant-water states were having major impact on fishery resources. My international experience only gave more ammunition to those pushing me to join the Washington hierarchy. In late 1968, Skip Crowther, deputy director for the Bureau of Fisheries, asked me to take a two-year assignment as the D.C.-based director for the bureau's science section. The opportunity did not really attract me. Nonetheless, I discussed the matter with Ruby and, for a variety of reasons involving career prospects, decided to accept the assignment. The one caveat was that the transfer not occur until Sue finished her junior year at Mt. Rainier High School. We thought we could endure a couple of years in Washington if we secured a promise that we could return to Seattle.

Skip approved our plan, but asked me to spend half of each month in Washington until after the end of the school year in June. This required a commute to Washington every other week, going to D.C. on Sunday and returning on Friday almost two weeks later. After several months, I became familiar with the passengers commuting into Dulles on United Airlines, including a number of the state's congressional delegation and Senators Jackson and Magnuson.

I started my training at the bureau's offices in the Department of Interior, which was just a few blocks up from the Reflecting Pool and close to the old Department of Navy building that I had frequented during World War II. My assignment was to take over as director of the science program, which I knew galled a number of old-timers. I was relatively young and often had expressed concern that the department's scientists usually produced results that supported national policy. That is, the scientists parroted the fishing industry's economic and social perspectives, as well as those of the Department of State. It was also apparent that many of the bureau's scientists had advanced quickly, without any substantial contribution to the peer-reviewed scientific literature. My popularity among the nation's regional fisheries' directors was questionable because they knew I believed that the placing of the nation's biological laboratory under the direction of regional directors selected by the governing party provided a mechanism to politicize science.

The spring months went by quickly as I flew from one coast to the other, juggling my regional and national responsibilities. In the Pacific Northwest, there was increasing political pressure from many sectors of the fishing industry to extend national jurisdiction over ocean fisheries to 200 nautical miles off the coast. This extension was strongly opposed by the departments of State and Defense. Within the bureau, there were mixed opinions. By the late 1960s, foreign fishing fleets were intensely harvesting the ground-fish resources in the Bering Sea, Gulf of Alaska, off British Columbia, and as far south as central California. These foreign fisheries had overfished some stocks, at times targeting species traditionally fished by North American fishermen. Thus, U.S. fishermen saw them as a threat to their access to resources and economic survival.

The situation in the capital was a mirror image, except that the complaints and growing political pressures came from almost all of the coastal states, particularly the New England and Mid-Atlantic regions. The nation was attempting to cope with the unraveling of international ocean policy, which allowed free access to fisheries resources beyond three nautical miles. Throughout much of the developing world, this policy was undergoing significant change. At home, a new administration was enlisting new leadership, reorganizing its departments concerned with the oceans, and struggling with the emerging global debate. In response to the White House-mandated Stratton Report on the nation's ocean policy, issued in 1969, the Bureau of Commercial Fisheries was slated to be transferred to the Department of Commerce. The new entity would be called the National Oceanographic and Atmospheric Administration (NOAA). At the same time, scientists with state and federal fisheries agencies and those in academia were feverishly attempting to monitor the status of fish stocks off our coasts through fishery data, which for most species was questionable at best.

In addition to the negative impact of large-scale foreign fisheries on the United States, the economic environment made it difficult for most American fishers harvesting relatively low-valued bottom-fish species to compete in the world market. The country was importing large quantities of frozen ground-fish filets, steaks, and blocks (from which these products were made) from Europe, Iceland, and Japan. Part of the problem was due to different labor costs, but some of the distant-water fisheries also were supported by significant government subsidies. These problems, along with the emergence of the extensive Soviet, Polish, East German, Japanese, Korean, and other fleets off our coasts, caused small-boat fishermen in the United States to feel disenfranchised. Many felt that the Bureau of Commercial Fisheries and the State Department had sold them out to achieve other national security and economic interests. As the result of the U.S. commitment to support the historical freedom of the seas, solutions to international disputes were being resolved by a series of bilateral agreements. These agreements provided some relief to the fishing parties in conflict, but the agreed-upon arrangements were often difficult to monitor and enforce. They did little to quiet the growing disenchantment with our nation's domestic fishing policies.

There were, however, exceptions to the economic dilemma facing American fishermen. The tuna industry operating out of southern California, which had traditionally fished off Mexico and Central America, modernized, using nylon or other synthetic nets and new net-loading technology and operational modes. Big, new tuna clippers used larger and larger nets, extending their fishing operations as far south as Peru. A valuable and viable tuna industry was operating out of San Diego and San Pedro, and their catches were largely secured off foreign countries south of California. These fishermen, along with the associated processing industry, were not sympathetic to the cries of many of their fellow fishermen for extended national jurisdiction, since it was not in their best interests.

To the north, off Alaska, a pot fishery targeting king crab was growing rapidly, and many Washington-based fishers took their vessels north to join in the late twentieth-century's version of the Alaska gold rush. Between the mid-1960s and 1980, Puget Sound fishermen would help build a modern fishing fleet for harvesting crabs off Alaska. Many who had toiled hard and scraped for years to make a living would become multimillionaires. Prominent among them were several Puget Sound and Oregon trawl fishers I had interviewed during my years with the Washington Department of Fisheries. Unlike the tuna industry, the crab fishermen were up in arms over foreign trawl fleets fishing on grounds where they traditionally harvested crab. These fleets were directly fishing or killing crab they took as bycatch while harvesting bottom fish.

Finally, there was a small but profitable distant-water, shrimp trawl fishery that operated out of the Gulf of Mexico. It fished off the northeast coast of South America. These fishermen also were not overly enamored by the extension of national jurisdictions.

The result in Washington, D.C., was a bureau struggling with: 1) trade and fishery-jurisdiction policies set by other government agencies; 2) a reorganization mandated by Congress; 3) restaffing of key positions in the Department of Interior following President Nixon's election; 4) growing criticism of foreign fisheries operating off the United States; and 5) concerns of fishermen and conservation groups regarding overfishing of stocks adjacent to the United States. With the exception of fisheries that were conducted under international agreements or off U.S. territories, coastal states managed adjacent fisheries. I couldn't have picked a worse time to join the Washington bureaucracy.

Shortly after the school term ended, Ruby and I packed our belongings, leased our home in Normandy Park for two years, and headed east. Our son Bob had just completed his first year at the University of Washington, and our daughter had just completed her junior year of high school. Both were accompanying us to D.C. Bob would return to Seattle in the fall and live in one of the U.W. dorms, while Susan would enter a school in Virginia, where we intended to live during our two years in the capital.

As we drove out of Normandy Park, I revisited almost twenty-five years of our family's life in a region with a wonderful living environment. We lived one block away from a community swimming pool and two blocks from a long stretch of beach on Puget Sound, and enjoyed quick access to two streams, where Bob and Sue had caught cutthroat trout, and to a myriad of camping sites in the Cascades. We were leaving a group of neighborhood friends, scientific peers with whom I felt comfortable, and a diverse group of fishermen with whom I had interfaced even before obtaining my B.A. We knew that the family soon would be divided for the first time, with Bob in Seattle and the rest of us in D.C. I was also acutely aware that my work would lead me away from day-to-day contact with science and into the science-policy arena.

To this day, I am not sure why I accepted the Washington assignment. After two years, I told myself, I would complete my assignment and return to the Pacific Northwest. Nevertheless, I cannot rule out that the desire for notoriety and power may have been present somewhere in the deep recesses of my subconscious.

The drive was rather quiet for the first fifty or so miles, as we were all in the same pensive mood. It soon faded away, however, as we began to enjoy the opportunity to see America. Not too far east of Missoula, where we spent the first night, we came across the Lewis and Clark Caverns. As I tended to have a touch of claustrophobia, I wasn't thrilled with the thought of visiting them, but democracy overruled my concerns as the vote was three to one. The caverns turned out to be an enjoyable experience, except for one section where the cave narrowed and we had to crawl under bats hanging from the ceiling and cross over a small stream. The trip from there to an expanded cement tunnel made me a little queasy, but then we rode a coal car back to the entrance. We also stopped in Deer Creek and made a casual visit to Montana state prison, so we only drove a few hundred miles that day.

We spent the second night in Miles City, Montana, and the third somewhere in Wisconsin. It was a relatively easy run on the fourth day around the east side of Lake Michigan to my cousin's home near Rochester, just north of Detroit.

By 6:00 p.m., we still had several hours to drive. We decided that it would be more convenient for my relatives to receive us in the late morning than at night, so we pulled over into a motel. I don't think Ruby or I looked carefully at the buildings or surroundings. We were tired and just wanted a place to sleep. A few moments after entering the room, Sue yelled, "There is a giant spider in the bathroom." I went in to kill it with one swat of a magazine, but that was just the beginning because Ruby spotted another good-sized spider on one of the drapes. It also met its end from a second swing of the magazine. Despite the arachnid infestation, we all got ready for bed, but when Ruby pulled back the blankets, another large culprit ran out. That did it! We packed up our belongings and drove another sixty miles north to a Holiday Inn, where we spent a comfortable night.

The next morning we drove to my Uncle Omer's home and got better acquainted with my father's brother's family. We spent the next two days visiting with the Michigan Alverson clan, including two aunts and uncles and half a dozen cousins, before heading on the last leg of our trip to Washington, D.C.

We found a motel in Virginia, just south of Seven Corners. The next day Ruby got in touch with the local school authorities and registered Sue for fall classes. While Ruby was attempting to find and establish a new home, my work at the Department of Interior started off with a barrage of major issues that needed to be addressed.

At the top of the list was the growing international interest in altering the traditional narrow jurisdiction over ocean waters adjacent to the coastal states. This involved developing policies within the department to help formulate a national position on freedom of science on the high seas. Another issue of great concern to me was the state of science within the bureau. There were some excellent scientists scattered around the country, but the program seemed to have little organization and direction. With the growing impact of expanded international fisheries off the United States, we needed not only to evaluate the consequences of the international fisheries on our marine resources, but also the impact of our domestic fisheries. There was also the question of how to foster a scientific institutional arrangement that would limit politicization of the information produced by the bureau's scientists.

The bureau organized its scientists under regional directors (RDs) who were political appointees. Some of the RDs did not attempt to promote preconceived conclusions and relied on the scientists to provide technical information on the effects of fishing on marine resources. Others, however, seemed to provide answers that supported the positions they thought best-reflected constituency interests. Finally, I wanted to fill the vacancies in my own staff. I needed a deputy who was well recognized in the field of marine fisheries, someone who could help improve the overall status of science within the bureau.

To set the wheels in motion, I got in touch with Jack Marr, a fishery scientist who had an excellent reputation. He had served as director of the Bureau of Commercial Fisheries Laboratory at Stanford University, directed research on oceanic tunas conducted by the Biological Laboratory at the University of Hawaii, and was well known in the international arena. After some discussion regarding his role in guiding the bureau's science program, Jack agreed to join me in Washington.

A Short Aside

I have now been working on my historical memoirs for over five years and the story has moved forward from 1924, the year of my birth, to almost 1970. Be patient, I am still crafting the story, which only has another thirty-three years to go. As a scientist, I followed with interest stories regarding extra sensory perception (ESP) and the transfer of thoughts from one person to another. Although TV commentaries and stories regarding ESP were interesting, they never seemed convincing and I have remained a cautious skeptic. But, while I had not heard or talked with Jack for some twenty years, seconds after I typed his name on the computer, the phone rang and Ruby told me, "Jack Marr is on the phone." He wanted to know the name of the office I had hired him for in D.C. We were both independently writing our memoirs and had arrived at the same point in our stories at exactly the same time. Those who are mathematically inclined might wish to calculate the fortuitous probability of that happening.

Back to Reality

Back to the original story at hand, I arranged for the directors of the bureau's major science labs to come to Washington and discuss the internal organizational problems. I also wanted to hear their views about how to monitor more effectively the impact of foreign fishing off our coasts. The meeting would provide an ideal forum for a debate on extended national jurisdiction and the future of scientific research on the high seas.

It was no surprise that many of the key scientists in the bureau strongly believed that science programs should not be under politically appointed RDs, even though some had excellent science backgrounds. Most agreed that the current arrangement allowed advocacy groups to pressure the scientists via the RDs. Thus, I set out to convince Skip Crowther that the bureau needed an organizational change that would establish a line of command from the field science laboratories to the central office. At the same time, we proposed that the science program be consolidated under four major science centers, located at Scripps Institution in La Jolla, California; Montlake in Seattle; Woods Hole, Massachusetts; and St. Petersburg, Florida. This would provide regional coordination of research and establish a critical mass of personnel who could more effectively carry out the bureau's expanding science mission. It meant consolidating a number of smaller labs into the major science centers and making government more aware of growing conservation needs.

As might be expected, the proposal was well received by the bureau's scientific community, but most of the regional directors were less than happy. The reorganization plan had been floated, and I was rapidly educated on the Washington political environment. Through their constituents, several RDs immediately moved to scuttle the proposal. It was amusing because our scientists had not yet signed off on the plan. It had been leaked to the field and a back-wave set in motion even before the plan was completed. I quickly learned that everything you say and write that might have consequences on the established system would be communicated instantly throughout the organization and to the industry and Congress. It was tantamount to working within a giant spider web; if the system was even slightly disturbed, it was instantly sensed throughout the web.

The simple laws of physics were at play in the daily Washington routine: every force has an opposing force. It has often been said that information and knowledge are power, and to many civil servants building a power base was achieved by passing on information to their friends, constituents, and the press. They raced to the phone as quickly as possible after a staff meeting to enhance their status. It didn't matter whether or not the information was considered confidential. This never gave me any great concern, however. I accepted that it was a part of the transparency that is necessary in a democracy, but it did fuel the rumor mill with a lot of nonsense, which at times could bite you back.

When we completed the proposed restructuring, it was clear that program priorities would be established by the RDs, but the hiring of scientists, the execution and supervision of the science, and the selection of individuals to testify on scientific issues would be under the direction of the center directors (CDs). The proposal would establish new National Science Centers, placing them under an administration science director in D.C., and give higher priority to monitoring and assessing the consequences of fishing on the living resources off the United States. I presented the plan to Skip and he endorsed it, pointing out that it made good sense, particularly since Congress was considering moving the Bureau of Commercial Fisheries and several other ocean-related departments into the newly proposed NOAA.

The plan moved ahead with seemingly little resistance; I wasn't sure why. The RDs had a lot of political clout, and I had expected a more heated debate and attempts to scuttle the plan. But then I had a peculiar call from one of the CDs, who filled me in on a fabulous distortion of fact, but one that worked in my favor throughout my government career. Apparently the CDs and RDs were convinced that I had a close working relationship with Senator Magnuson, who was a powerful force in the senate and had a personal interest in ocean fisheries.

I'm not sure who made up the rumor, but it was right on two accounts. The senator was a force to be reckoned with, and he did pay special attention to fisheries matters, particularly regarding the problems caused by foreign fisheries. The gossip mongers, however, were dead wrong about my so-called close relationship with the senator. I once had a very short discussion with his administrative assistant, Gerry Grinstein, regarding the incidental catch of salmon taken by Russian trawlers fishing off the coast of Washington. In addition, in 1966, I made a short presentation to a committee chaired by the senator on the consequences to U.S. fishermen of extending national jurisdiction from three to twelve miles. Up to that point, I had never even been introduced to Magnuson, much less communicated my ideas regarding restructuring science to him or his staff. In any case, the plan had yet to be cleared within the Department of Interior. Nevertheless, the myth was in place and remained, to my advantage, throughout my tenure with the government.

Sometime during the late summer of 1969, we moved into a town house south of Seven Corners. (I bought it before Ruby saw it—a big mistake.) Ruby was busy putting the new home together and spent some time with Bob and Sue visiting the various memorials, parks, and museums in the Washington area. I was just settling into my new job when the commissioner of fisheries, Charles (Chuck) Meacham, asked me to come up to his office. The new Nixon administration had decided to put its own man at the helm of the bureau, which meant the dismissal of Skip Crowther, my boss. After advising me about the intended action, the commissioner asked me to remain in Washington and become director of the bureau.

The offer presented my family with a whole new set of problems. First, I thought very highly of Skip and accepting the job would appear to be at the expense of a friend. Second, the job easily, and undoubtedly, would undermine my return to Seattle at the end of two years. The man who had made the commitment to allow me to return would be gone from the bureau. Third, it would put me into a high-level administrative position that I did not feel prepared for or competent to hold. And finally, it would take me out the marine-science arena that I had loved. Of course, all my friends and acquaintances had advice for me. Some were convinced that I would do a great job; some most likely felt their position would be enhanced by mine; and others believed it was politically correct to let me know they were on my side. Although I didn't want to believe it, there may have been an intrinsic desire on my part to be "number one."

In the end, however, the family became the deciding factor. Our daughter had developed horrible allergies and was in and out of the hospital with nosebleeds, hay fever, etc. She just wasn't doing well in the D.C. environment. I went to the commissioner, gave him my regrets, and told him that, in the interests of my family, I would have to return to the Pacific Northwest. I did commit to stay on as acting director until a new director could be nominated and put in place. Chuck was very sensitive about the problem and told me, "If I were in your shoes, I would have made the same choice."

Several days later, Ruby packed what she could in a couple of bags, pulled Sue out of school, climbed onto a Northwest flight, and returned Seattle. She had to find a place to live, get Sue back in school, and start up our life again in Normandy Park. Luckily, after renting an apartment about a mile from our home, she discovered that our tenants were unhappy with the neighborhood. The lawn had not been mowed in several months and the inside of our home looked like it had not been cleaned since we left. Their son's room had peanut butter on the walls, and the stove was so greasy it looked ready to burst into flames. With a little bargaining, Ruby released them from the contract. Ruby had her home back, and Sue was in her home school for senior year. Both were happier than clams at high tide.

Counting the time I spent commuting to Washington, D.C., I was there for only six months. My future in Seattle wasn't even certain because my prior position might have been taken by someone else. I felt a sense of remorse for Jack, who would be left to carry on the changes we had initiated, but he was an intelligent man, had more longevity than I did, and knew the political workings of Washington. Within several months, however, he, too, left Washington for a job with FAO in Rome.

I had not grown up in a religious environment. My parents believed in God, but they never said much about theology to my brother or me. We were never encouraged to go to church, Sunday school, or any religious camps. The Boy Scouts probably was the place where God was formally introduced into our lives. Yet, as far back as I can remember, I have always had a deep-rooted spirituality. From the moment I told Chuck what we had decided, I felt a sense of relief and something inside me said, "Don't worry, things will work out." I knew the decision was right for our family.

Within a week, I moved into the director's office and took over the helm of the Bureau of Commercial Fisheries. There was a lot to deal with, including, the request for 1970 funds to the Office of Management and Budget; input into the president's state of the union address; public relations efforts for the "Man in the Sea" project; and the role science would play in the new ocean agency Congress established to deal with ocean fisheries, weather, oceanography, etc.

The first lesson I learned was how to work within the "spider web." Not long after I first sat in the director's chair, I met with key staff members. One of the topics was the interior department's input into the state of the union address. Walter Hickle had just been appointed interior secretary and apparently was reviewing the department's responsibilities and actions over the past year to evaluate whether there was something of value to contribute. Within the halls of Congress, there was a widespread, ill-conceived assumption that U.S. fishermen could not compete with foreign fisheries. A number of Congress members believed they were technologically inefficient compared with the high-seas, catcher-processor fleets operating out of western Europe, the Soviet bloc nations, and several Asian countries. Thus, someone in Hickle's office had suggested the president propose a program to subsidize the construction of a number of high-seas factory trawlers.

Indeed, such technology did not exist in American fisheries but not because our processors and fishermen were unaware of it. They had carefully examined the cost of construction, operations, labor, foreign subsidies, etc., and found that, even with the new technology, they could not compete. The willingness to adopt technological innovations was apparent in the high-seas tuna seiners and the Alaska crab fleets; they were among the most technologically competitive fishing fleets in the world, but the harvested species supported the investment. I suggested that we draft a memo to Secretary Hickle pointing out the potential problems with the proposal. However, we never had the time to draft the memo, because later the same afternoon I was asked me to the secretary's office. When I arrived, I was surprised to find that someone had alerted the office that I thought the idea was "ridiculous." I had never used that word, but it closely echoed my view.

It wasn't very comforting to know that someone on my staff found gratification in communicating information to attain power. On the other hand, I recognized that I should be able to defend whatever was said in a staff meeting. If the spider web were disturbed, someone would seek to find out why. I explained the basis of my concern to one of the staff members, and the matter was resolved. The proposal was dropped. I have no idea if the idea ever reached Hickle, but, as an Alaskan, he was no doubt aware of fisheries matters and would have rejected the proposal anyway.

The government budget process always seemed strange to me and designed to ensure that operational efficiency would never be rewarded. At the onset of each fiscal year, each unit of government is given a sum of money that the entity is expected to spend over the year to achieve its obligations. If the entity lives within its budget, it is fiscally responsible. If you are a good manager and end the year with money left over, the funds will most likely be taken back and given to some unit of the government that has been less responsible and overspent its budget. Furthermore, if a manager does not spend all the funds allocated for the year, the government may cut the budget for the coming year. Of course, all the government managers dealing with annual budgets were aware of this and worked feverishly to make sure that all the allowed funds were spent by the end of the fiscal year. There are no incentives in the federal government to do your job effectively with less.

Each year the heads of the major departments or bureaus present a budget supporting their activities for the coming year to the Office of Management and Budget (OMB). The amount that a department head might request is fairly well set by administrative guidelines and the attitude of Congress toward the department's responsibilities. I became the BCF's acting director only a few weeks before the bureau was scheduled to make its presentation to OMB. It had been brought to my attention that our bureau frequently had been criticized for not having economic information to support proposed funding needs for meeting the ever-increasing fishery conservation and development obligations imposed by Congress and the administration. Although I had some strong doubts about beefing up our economic justification, I nevertheless proceeded down this path. In fact, I asked James Crutchfield at the U.W., perhaps the best-known and most capable fishery economist in the country, to review, evaluate, and strengthen (providing he agreed) our economic justifications. I thought that this would improve our justification for additional funds.

"A stroke of genius," we thought at the time, but in reality it was a bad call. Jim came to D.C. for our presentation to the OMB officers. We brought to their attention the growing problems of monitoring the effects of foreign fisheries, including overfishing and competition with our domestic small-boat fisheries. Our agenda also included the shifting global attitude regarding coastal states' stewardship responsibilities and priority of access to the resources off their coasts. If international efforts to change the status quo from narrow to extended jurisdiction moved ahead, the United States suddenly would find itself with stewardship responsibilities for living marine resources in more than a million nautical square miles. We would be ill prepared to take on those obligations.

From the onset, the words seemed to fall on deaf ears, perhaps due to conservative guidelines from the new administration, but most likely because Jim was there to respond to questions. From the moment Jim started providing the economic justification for the programs, the hair on the necks of the hearing officers seemed to stiffen. They may have wanted better justification, but it was clear that they didn't want it from an academic outside the system who could respond effectively to their questions and was more conversant with the economic theory and jargon than they were. Needless to say, we didn't get any increase in funding. We actually lost a few hundred thousand dollars—not a good start for the new acting director.

During my short stay as the head of the bureau, there was a lot of discussion internally and among the public regarding the "Man in the Sea" project, in which scuba divers lived for several weeks in a small, underwater shelter in the Caribbean. There was a great deal of interest in the project, which generated positive media reports regarding the opportunities that it offered for increasing our understanding of the aquatic environment. I chatted with Conrad Mahnken, one of the aquanauts and an acquaintance from Seattle, about the nature of the scientific observations that were made. On one hand, they didn't add much to ocean science. On the other hand, any extension of underwater observation modes—such as submersibles, remote sensors, underwater cameras, or scuba and other diving techniques—could make a considerable contribution to our understanding of the life history of fishes, their behavior with respect to fishing gears, ocean habitats, etc.

As a result of the issue's high-profile character, I was asked to review some of the results of the project and consider future activities with Assistant Secretary Leslie Glasgow. The meeting went well, and I became better acquainted with the assistant secretary, who had taken a particular interest in the "Man in the Sea" project. I did not extol the scientific results that had been achieved in the first effort, but carefully noted Conrad's words about the need to improve our ability to observe what was going on under the sea surface. They served as the major theme in my presentation. I sensed that my comments pleased Glasgow, which would help during a conflict that subsequently engulfed me.

I had been in the acting director's seat for about three months when Chuck Meacham informed me that the BCF was attempting to recruit Phil Rodell, who had directed the California Department of Fish and Game. Phil was approved for the job and was expected to arrive in D.C. sometime in mid-December, which meant I might be home for Christmas. Since Ruby's departure for Seattle, I had lived with Ed Schaefer, who had come from Seattle to be director of the Exploratory Fishing operation. When I knew the definite date of Phil's arrival, I arranged to remain in D.C. for a little more than a week for transition talks and prepared to return to Seattle. Ruby and I sold the townhouse we'd purchased only six months earlier, and I had retained the car. I called Ruby and asked if she would like to come to Washington and drive home with me. She made the appropriate arrangements and was in D.C. the day before my assigned departure date.

We left Washington in mid-December, following a morning discussion I had with Phil. Although I was well aware of his work and the contributions he had made to marine science during his tenure in California, I had no idea about his views concerning the bureau. We chatted for a few hours regarding the proposals for science, the establishment of national centers for marine science, and the impending move to the National Oceanic and Atmospheric Administration. Our views were similar, so I felt that the few policies I had pushed during my six-month stay in Washington had a reasonable chance of moving forward.

From the moment I left the Interior building, I felt a great sense of relief. When I took over as acting director, I was concerned that politics would somehow prevent my return to Seattle, and I would be stuck in an administrative position and a working environment for which I had no zeal. Minutes after leaving Ed Schaefer's home, Ruby and I were on the Washington beltway heading west. By the time we crossed into Ohio, it had started to snow, and we pulled into a motel sometime after 8:00 p.m. It continued to snow all night, but the next morning, the freeway traffic was still moving, so we hit the road around 5 a.m. and continued our trek. Snow piled up on each side of the freeway and continued to fall, but we still made relatively good time. I believe we stopped somewhere in Wisconsin that night.

I woke up about 4:30 a.m., unable to sleep. It was still snowing, and we were concerned that the storm might trap us, up so we got up, quickly put our belongings in our '64 Impala, and resumed our trip. The radio reported snow all the way from North Dakota to Pennsylvania, but it became lighter as we moved across North Dakota and ended completely somewhere east of Montana. We spent the third night in a small town in western North Dakota and the next day made it to Missoula. From there it was a relatively easy drive back into Seattle. We drove up our driveway in Normandy Park just after dark and were back in the home we had built in 1951. It seemed that we had spent a lifetime in six months, and the world never would be quite the same.

I was reassigned to my post as regional director of science. I hadn't been back in Seattle for a month when Don Johnson, regional director for the Pacific Northwest, asked me to accompany him to the airport to pick up Under Secretary Dunn. I had never met Dunn during my stay in Washington, and all I knew about him was that he was a strong Nixon supporter and that he or his wife was related to the first lady. Dunn was en route to Seattle for a celebration of the first U.S. fish-protein concentrate plant to operate in the Pacific Northwest.

The plant was established in Aberdeen, Washington, following some of the early hake assessment work conducted from the *John N. Cobb*. The surveys, along with egg and larva work of the bureau's lab in La Jolla, delineated a significant population of hake, or whiting, inhabiting the waters from Mexico to northern British Columbia. Experimental tows, using the *Cobb* mid-water trawl, demonstrated that it was possible to achieve catches in excess of twenty tons per set. The success of the plant, therefore, was contingent not on whether there was an adequate hake population to support a major fishery, but on whether fishermen were willing to harvest large volumes of fish for less than two cents per pound. There was also the risk that global fishmeal prices, which were very volatile, could decline, taking away the relatively thin profit margin forecast by the plant owners.

The celebration would include several state dignitaries, Senator Magnuson, and the under secretary. Don and I were to escort Dunn to Aberdeen and brief him on the history of the plant and the government research that had persuaded local investors to build it. We greeted the under secretary, a tall thin man with gray hair and a reddish face, at the airport gate. When we arrived at the baggage area, Dunn looked at Don and said, "Here are my baggage claim slips, go pick up my luggage." It took some time for Don to find the luggage, since he had no idea what it looked like and was given no help.

On the way to the hotel, some roadwork required us to wait several minutes before proceeding. The under secretary noted that we should have made a "dry run" before picking him up. He also wasn't pleased with the government car we were using and stated that we should have rented a limousine. After all, he was used to riding with the president. By now, the smell of liquor on his breath had filled the car, and it was obvious that he wasn't in a good mood. He made several other curt and demeaning comments but Don, who was driving, said nothing. When we arrived at the hotel, Dunn told Don to check him in, so I waited with the under secretary. He was acting like a sergeant in the marines giving orders to new recruits in boot camp. When Don got back a with his room keys, Dunn looked at me, held out the keys, and said, "Here, take my luggage up to my room."

I had become increasingly riled about the way the under secretary was treating Don. When he handed me the keys, I handed them back and informed him that I was a professional civil servant, not a bellhop, and that the hotel personnel would be pleased to take care of his belongings. His face turned several shades redder and, after some foul comments, he let me know that I was unwanted in his presence. I returned home, told Ruby the story, and let her know that there was a reasonable possibility that my career with the government would be terminated. I would gladly have helped Dunn, or anyone else, with his luggage if he had asked for help politely. But his arrogant manner and dictatorial attitude had pushed the wrong button.

The situation became very sticky. Dunn obviously didn't want me in his company, but Ruby and several other wives of bureau employees had organized a welcoming party for him at our home, which was only ten minutes from the Hyatt Hotel. Don and I decided to cancel the reception, since the under secretary wasn't aware of the party and wouldn't know what he was missing.

Because of my words to the under secretary, I was asked not to participate in the ceremony in Aberdeen. Senator Magnuson previously had asked me to be present because of the extensive work conducted by our exploratory group to delineate the distribution and abundance of Pacific hake along the coasts from California to British Columbia. Nevertheless, I remained in Seattle and waited for the reaction that I was sure the Dunn incident would not have initiated when he returned to Washington. I didn't have to wait long, and about a week later friends in D.C. reported that he had called for an extensive review of my career.

I had not mentioned the incident to any of my coworkers, but news of the event had spread quickly, not only within the department, but also among the bureau's clientele. To add to my dilemma, Dr. Web Chapman, the state department's first ambassador for fisheries, wrote a scathing letter to Secretary Hickle about the under secretary's behavior toward Don Johnson and me. I had seen a copy of the letter and was sure that when it reached Dunn's hands, it would curl his gray hair, but not help me. From that point on, things moved quickly. Several weeks later, I received a communication from D.C. stating that, since I had returned to Seattle because of my daughter's health, I would be required to pay all the moving expenses, which amounted to well over $7,000.

I wondered if anyone in D.C. had intervened in my defense. Phil Rodell, the bureau's new director, called to say that he wasn't privy to all the facts, but felt that he shouldn't get involved since he was new on the scene. It was a disappointment, but not an unreasonable move on his part. Don had been supportive, but he was in Seattle and all the action was taking place in D.C. under the direction of a powerful political figure. Even though I was never sorry about the words I had said, it was becoming increasingly clear that I might have shortened my career with the Department of Interior.

Still, several key figures within the department had taken issue with the Dunn's attempts to punish my "insubordination." From what I was later told, Chuck Meacham went to Assistant Secretary Glasgow and told him about the SeaTac Airport incident. Glasgow, in turn, apparently went to Secretary Hickle. Then about three weeks after the incident, I got a call from one of Hickle's aides asking me for a detailed accounting of the confrontation. I spent a few minutes recounting the event to him, and he said, "Thanks, don't worry about the matter, it's all being settled."

I have no idea what Hickle and Dunn may have discussed, but a few days later friends in D.C. called to say that Dunn had left the department. The reasons for his departure were never made clear. I don't think it had anything to do with the event in Seattle, although that may have added some icing to the cake. There were rumors that Hickle and Dunn had strong ideological differences and that Hickle was upset that Dunn had his office re-carpeted at great expense—a matter that had been exposed by the press.

My twenty years with the service had outlasted the under secretary's short-term vendetta. Hickle had his own problems with the Nixon administration and left the Department of Interior several months later. In less than a year, I had climbed to the top of the civil service ladder and slid back down to deputy regional director of science for the Pacific Northwest region. Several months after my return to Seattle, the Bureau of Commercial Fisheries and other ocean-oriented agencies were transferred to the newly created National Oceanographic and Atmospheric Administration under the Department of Commerce. Shortly after its creation, Dr. Robert White, its first administrator, reorganized the bureau, which was renamed the National Marine Fisheries Service. He also implemented the proposal to establish the four national science centers, and I became the first director of the Northwest and Alaska Fisheries Center.

CHAPTER XII
BUILDING NATIONAL FISHERIES POLICY

Finding a Path for Science

The Northwest and Alaska Fisheries Center (NWAFC) constituted an amalgamation of the facilities at the Montlake Laboratory, several small labs around the state of Washington, the major National Marine Mammal Laboratory located at Sandpoint in Seattle, and a relatively large fisheries laboratory in Auke Bay, Alaska. One director supervised research programs to study the life history of commercially important fishes, shellfish, and marine mammals, and the general ecology and population trends of marine fishes, invertebrates, and marine mammals inhabiting the waters from central California to the Bering Sea. A significant component of the work also included a major program designed to protect and rebuild the great runs of salmon that once entered the Columbia River.

Many of the salmon runs had suffered primarily due to overfishing and the environmental degradation resulting from the construction of dams and poor agriculture and forest practices over the last seventy-five years. Finally, there was also a deleterious impact from the use of sometimes scarce water resources by a variety of industries that relied on low-cost electric energy. The biggest issue, perhaps, was society's unwillingness to acknowledge that maintenance of large salmon runs in the Columbia, in combination with multipurpose use of its waters, would require compromise from all the river's users.

The NWAFC had a mandate to carry out a number of research programs to help the fishing industry develop new products as well as more effective ways to preserve and maintain quality. It also was charged with providing information to user groups on biological findings as well as the general market. As I recall, the budget at that time was somewhat in excess of $20 million, which seemed like a lot of money. However, the center's research responsibilities encompassed about forty commercially harvested fish and fifteen species of shellfish (many of which were targeted by foreign fishers). Added to this mandate was the need to accumulate a great array of data on marine mammals that were distributed over millions of miles of continental shelf and slope as well as open ocean waters. The budget also included the funds for operating the major research vessels, the *John N. Cobb* and a newer, larger vessel, the *Miller Freeman*.

My first major endeavor was to staff the office, which entailed naming a deputy director, secretary, and office manager. These were drawn from the existing staff at Montlake and included Al Pruter, who had taken over my job as director of the Exploratory Fishing unit when I left for Washington, and Heater Heyamoto, a biologist from the Washington Department of Fisheries. Al had a great ability to work in harmony with people and was liked by all. He moderated my own style of management, which tended to place considerable emphasis on merit and performance and perhaps lacked sensitivity to personal matters. Heater was also a gem; he had worked with the exploratory group and had a keen understanding of government bureaucracy. In addition, I assigned Charles Gill, also from the exploratory group, to ensure that our work and hiring policies were fair and equitable to minorities. Charles was African-American, had a good work ethic, and was well known at the center. Finally, I hired Linda, a young woman who could take shorthand at an amazing speed, to be my secretary. She had a great personality and managed to correct my poor spelling and unravel my miserable handwriting without any great problem.

It didn't take too long before I recognized that, despite its advantages, the job allowed little time for personal research and required dealing with an abundance of unpleasant personnel problems associated with a staff of 400 or more. Other matters included dealing with a growing number of equal-opportunity matters, the territorialism inherited from the previous organizational structure, and ways to streamline the excess paperwork. In this regard, I was particularly disturbed by the 400- to 500-page quarterly progress reports we had to produce, which I was convinced no one in Washington ever read. To test this hypothesis, I had the Exploratory group withhold its quarterly reports for more a year and never heard a word from Washington. Later, when we began resubmitting these documents, it was noted that "some seemed to be missing."

After the first year, these matters began to sort themselves out. We managed to upgrade our computer facilities to some of the best in the nation, purchase an electronic microscope, and hire a number of top scientists out of the University of Washington and other Pacific Northwest universities. I had been named an affiliated professor at the U.W. School of Fisheries and so could keep track of superior graduate students. In a matter of a few years, we added to the staff about a half-dozen Ph.D.s, who excelled in the quantitative aspects of population dynamics, including Mike Tillman, Jim Balsiger, Brian Rothschild, and Loh Lee Low.

The first three eventually would become center directors in various areas of the country, and Low would head one of the major divisions at the Alaska Fisheries Center. On the environmental side, we hired Usha Varanasi, a specialist in toxin in the marine environment and sea life. She later became the director of the Northwest Fisheries Center.

We established staff-performance expectations, including the publication of scientific findings in peer-reviewed literature. Within several years, the center became a cohesive research unit with capabilities in multiple disciplines and positive interactions with other federal and state agencies and academic research institutions.

The Growing National and International Conflict

My job as director of the NWAFC became increasingly difficult because of the growing national concern with the foreign fisheries activities off the U.S. coast, including those of the Soviets, Poles, East and West Germans, Japanese, Taiwanese, South Koreans, Portuguese, Spanish,

Chinese, and probably several others. The United States was attempting to resolve, or at least address, some of these problems through a series of bilateral meetings that required frequent travel to Japan, Korea, Poland, and the Soviet Union. These bilateral efforts improved communications and collaborative research between the nations that were the major harvesters of fish off the U.S. coast. The commercial and sports fishing industries and environmental groups, however, did not see the meetings as solutions to the perceived overfishing problems or the social and economic struggles facing the nation's fishing groups. Although several international efforts partially resolved some of the most contentious issues, they only served as Band-Aid remedies. By the early 1970s, a political tide demanding national policy changes was sweeping across the entire country. It was also highly volatile on the international front.

Opponents of the extended jurisdiction desired by most U.S. coastal fishermen included the American distant-water operators (fishing tuna and shrimp) concerned about being shut out of their historical fishing grounds. Some of the salmon processors feared losing the protection they had achieved under the Abstention Principle. These factions were quite satisfied with the status quo. To place themselves on a more solid political footing, the industry staged a meeting in San Francisco to address the differences among the major players from the tuna, menhaden, salmon, shrimp, and ground fisheries. Perhaps because of my international experience, I was the only non-industry person invited to participate. I had to get approval to attend from headquarters, which authorized my attendance as long as I did not propose any strategies inconsistent with the U.S. commitment to retain narrow coastal-state jurisdiction. For all practical purposes, this meant that I could answer only technical questions.

The "Golden Gate Meeting" took place sometime in the early 1970s, with participants from Alaska, the Pacific Northwest, California, New England, the Mid-Atlantic States, and the Gulf of Mexico. Given the radical differences of opinion within the various sectors, I thought there would be a lot of acrimonious debate and discussion, but I was wrong. At the start of the meeting, one of the participants turned to me and asked, "Is there any different approach to managing ocean fisheries that might accommodate U.S. national interest?" I answered, "Well, you might wish to consider using some sort of biological and ecological considerations." Within short order, the group drafted what was called the "Three Species

Approach." In essence, it gave the coastal states jurisdiction over species that inhabited the continental shelf and slope (cod, hake, flounder, and rockfishes), as well as over the anadromous species (salmon and trout) in all ocean areas. Tunas and other highly migratory species were left to the community of nations that harvested or had a special interest in such fisheries.

The speed with which the group came to a consensus caught me off guard and, within a few hours, a subcommittee was writing up the proposed concept. The rest of the group debated its acceptance by the international community and how to present the issue to the U.S. government. Someone asked if I thought the approach would fly. "It makes sense in terms of the biological and ecological aspects of the fishery resources," I replied, "but I suspect it will also be seen as a strategy that reeks of national interest. Developing states, except for the few that have distant-water ocean fisheries, will not see it as a management concept in their best interest. Nevertheless, it perhaps stands a better chance than the current U.S. position." In the end, the group suggested that I reveal the proposal to the NMFS and the Department of State. At least within the Department of Commerce, the Three Species Approach was seen as a backup position, if the proposals to retain narrow areas of coastal jurisdiction failed.

Even prior to the 1970s, issues regarding the use of ocean space and its living resources had become extremely controversial among the world's coastal states and even other nations. Some of the more important issues included use and management of ocean fisheries, ownership of mineral and living resources of the continental shelf and slope regions, resources of the deep-ocean seabed, freedom of science, and freedom to cross the straits and other open ocean areas. Two UN conferences on the law of the sea had been held since World War II, but they failed to resolve these issues to most nations' satisfaction. Thus, the United Nations was planning a third conference. In preparation, the Department of State called a planning session of its own sometime in the early 1970's; the NMFS asked me to attend.

Participants included members of the departments of Interior, Commerce, and Defense, as well as State. I was impressed by a number of the representatives. They were articulate and spoke in a knowledgeable manner. I was startled, however, by their rather Pollyanna perception that the U.S. fishery position would easily carry the day at the third United Nations Convention on the Law of the Sea, referred to in short as UNCLOS III. They may have been parroting words of their superiors, but it sounded like they believed what they were preaching.

As a newcomer who was largely unknown to the players, I listened patiently throughout the meeting. Near the end, however, I made the mistake of noting that, "the overwhelming position of most of the world's developing countries was leaning toward extending their right to manage living and non-living resources of the oceans and the seabed to 200 nautical miles." To this, I added, "At home, the nation's environmentalists, conservation groups, sports fishermen and most of the commercial fishers are not supportive of the status quo, or of extending some minimal rights to placate coastal states." Because this strongly contradicted the position of the State Department staff, the looks I got from the supporting cast were tantamount to "Who is this poor uneducated character?"

Nevertheless, the participants were mostly pleasant to me, perhaps because it was all an in-house affair. Later that year (1971), the first formal UN planning session was held in Geneva, and I was selected to be part of the team providing fisheries scientific input to the U.S. negotiators. The meeting was very educational, giving me a glimpse of the complex process of an international bureaucracy. Each nation participant took turns making its views known to the international community. There were a variety of statements—some good, some excellent, and others that were overdone and preachy. Listening to the endless commentaries, it quickly became obvious that most nations favored extending jurisdiction over ocean space to 200 nautical miles. The support for the U.S. position was limited largely to western Europe, Japan, and the Soviet Union. These nations all had significant distant-water fisheries, ocean commerce, and/ or naval power. South America, Africa, and the Arab states primarily pushed for extended jurisdiction.

At times the meeting seemed to repetitious and boring, but there were entertaining moments. I was sitting behind the U.S. negotiator when the delegate from Nepal, a landlocked nation, began his oratory on the problems to be handled at the conference. Three hours later, he still held the floor. This fellow was from way up in the Himalayas, but he had a better grasp of the issues than many his colleagues from nations sitting on the edge of the sea. There was also the issue of "freedom of science," which meant the ability of scientific research ships to go wherever they wanted in the world's oceans. In general, there seemed to be a lot of support for not interfering with biological and oceanographic sciences, but the "Pueblo Incident"—in which the *Pueblo*, a U.S. electronic spy ship, was captured by the North Koreans, who claimed the vessel was within their territorial waters—made a lot of nations suspicious about what was going on under the auspices of "freedom of science."

The director of the Scripps Institute of Oceanography, a member of the U.S. advisory group, wanted to present an appeal supporting the freedom of science. I suggested that it would be far better if such an intervention came from Canada, which was seen as more supportive to the developing countries. The advice fell on deaf ears, and the director made the appeal to the conference.

It was a good talk and was treated with respect, but it convinced none of the skeptics. After he returned from the meeting, the director and I went to dinner in the hotel restaurant. When we began to talk, it was apparent that he had consumed a fair quantity of alcohol and had little interest in discussing how the paper was received. In fact, he was not in the mood to chat about anything having to do with the conference, but proceeded to outline his theory of continental drift. He continued to fortify himself with red wine and the discussion of ocean science drifted away from reality. We moved on to a discussion of the possibilities of continental convergence. Two hours into dinner, he became so overcome by alcohol that he went to sleep with his face in the dinner plate. It was clear that the director's speech had gained little support from countries wary about any research activities in their backyards. I suspect he knew that ocean science was a likely loser at UNCLOS III.

At times I felt that the United States and its few allies were talking to themselves, failing to hear the repeated statements of the many countries that were pushing for extended national jurisdiction over ocean resources. During the hours of listening to the open sessions, I often sat with Jake Dykstra, a trawl fisherman from the New England area, and at one meeting of the U.S. staff, we looked at each other and said, "Is anyone listening?" When we pointed out that the United States had little support in the developing world, we got looks that said we were "out of the loop." Because Jake and I continued to debate the issue with our U.S. peers, one of the state department aides took us aside and said that we could be arrested if we promoted a philosophy inconsistent with our national policy. We had only discussed the issue internally, but we got the distinct feeling that we had pushed our case about as far as we could.

In addition to the official U.S. delegation, there were a number of American academics at the international debate. Edward Miles was an internationally known expert on law of the sea. He had been raised in Trinidad and did his graduate studies at the University of Denver. Ed had a great sense of humor and was one of those individuals whom you quickly recognize as more intelligent than everyone else. His use of the English language was elegant, his vocabulary unlimited, and his grasp of the array of ocean issues, including fisheries, superb. When Ed talked, I frequently found myself looking at my shoes and feeling dwarfed by his intellect. Within a few minutes, Ed cleared up a lot of my concerns about the U.S. negotiating position. As I recall, he saw that the United States was holding on to its "no-win" fisheries position to secure other goals at the conference. He expected that the United States would move closer to the scenarios suggested by the developing countries at an appropriate time, noting that the U.S. positions on freedom of passage and the deep seabed were far from resolved. Ed remained my link to reality and helped me to get a better handle on the international ocean political arena. Later, he joined the faculty at the University of Washington, becoming my mentor and sounding board on ocean policy matters.

From my vantage point, little substantive progress was made at the planning conference. On the other hand, it did shape the organizational and operational aspects of UNCLOS III. Furthermore, many of the positions of the key ocean states had been clarified, and the players had a much better understanding of their strengths and weaknesses. The opening of the next LOS meeting, to be held in Caracas, Venezuela, was just months away. I enjoyed visiting nearby cities in Switzerland, the contacts made at numerous cocktail parties, and the company of many interesting politicians and statesmen. But resolution of the myriad of fisheries issues facing the United States and other countries seemed a long way down a road filled with potholes.

Family Matters

I returned to Seattle somewhat skeptical about the international process, but it was the only game in town where global solution could be achieved by consensus. Ruby, as usual, kept the family life going. Bob completed his third year in college, and Sue graduated from high school and enrolled in a junior college. Several friends in the neighborhood purchased property on Case Inlet, located on the arm of the long fjord that extends north from the southern end of Puget Sound. Our family took part in numerous picnics and overnight stays at the Brights, friends who had bought property on the water near Grapeview, which consisted of a gasoline and fire station. Ruby loved the area and had a sense of investment that was more optimistic than mine. She urged that we also buy a lot on the water, where the family could gather in the future. It was a good time to plan a second home because the Boeing plant had undergone a major reduction in its work force, which led to a local economic slump and a decrease in labor and material costs. We also had paid off the loan on our house and had a few dollars in the bank. We purchased a lot with a low bank on the water about two miles from the head of the inlet and planned our summer home.

The lot faced east across Case Inlet. We looked at the northern point of Treasure Island and Mount Rainier stood up majestically in the distance, that is, when it was not raining. The beach in front was covered with a mixture of sand, gravel, and small rocks. There were some wild oysters on the beach, a plentiful supply of steamer clams, and a good number of sea-run cutthroat trout inhabiting the shallow tidal waters along the shoreline. The beach sloped off gently and, during low tide, we had several hundred feet of beachfront. On a high tide, the water swept against our shoreline, and we were forced to put in a bulkhead to prevent its erosion. This was accomplished before the passage of numerous environmental acts, so we only had to get a permit from the U.S. Corps of Engineers. Shortly after we told the contractor to proceed with the cement bulkhead, we got a call from the corps telling us that we had built a bulkhead without a permit. In panic, we called our contractor to see if he had obtained the proper papers. He said that he had, but he hadn't even started the work. Apparently the corps made a survey by plane and incorrectly identified the location.

About a year after we bought the lot, we celebrated our twenty-fifth wedding anniversary at the beach. Our Normandy Park friends erected a pre-built outhouse and hung a silver moon on the door. A year or two later, we built a cabin that remains our favorite hideaway to this day. It has been variously improved throughout the years, but we always have refused to install a phone. It has caused some problems, but avoided a great many more. The property was an excellent investment; even more, it was a place to relax, fish, eat clams, water ski, and swim. When I was at home, particularly during the late spring months and summer, the family headed for the cabin almost every weekend. The kids, who were now young adults, sometimes brought friends with them. Their passions were water skiing, inner tubing, riding a single board, and occasionally fishing. My order of priorities was fishing, clamming, some water skiing, and walking with Ruby in the fir forest that surrounded the cabin. Ruby's loves were boating, blackberry picking, pie making, and watching the local wildlife. Numerous seals lounged on the floats or dock, and bald eagles, kingfishers, great blue herons, and a number of other seabirds frequented the inlet. Deer occasionally walked on the beach.

One afternoon Ruby and I decided that it would be great to have a berry pie. We went off to the local blackberry patch about fifty yards from our cabin. The blackberries grew in a meadow, which we entered through a gate that was always open. Just before we reached the gate, we observed a doe with two fawns walking directly toward us. She apparently saw us at the exact same time, and we all came to a full stop. We stood still for a few seconds, and then Ruby and I retreated a few steps. The doe proceeded down the path and entered the meadow with her two fawns. We followed and started picking blackberries. We were quietly picking some plump berries when we both heard a noise directly behind us. Standing no more than three feet away was one of the fawns. When we turned, it jumped probably four feet straight up and bounded off in great leaps back toward his mother before turning to watch us again. He was curious about what we were doing.

At the office, Al Pruter carried a greater and greater share of the load of running the center, and I was increasingly on the road dealing with evolving national fisheries policies. Sometime during the early 1970s, I was honored to receive the Fisherman's Highliner Award. It was a very auspicious award because it honored the best of the nation's fishermen based on their fishing productivity, commitment to conservation, ingenuity, and recognition by their fellow fishermen. Recipients were the so-called "leaders of the pack." Occasionally the award was given to someone outside the industry who had made major contributions to the benefit of U.S. fishers. Over the years I received a number of other awards from the government, elements of the commercial and recreational fishing industries, as well as scientific groups. The Highliner Award, however, was always my favorite because it reflected my interest in the welfare of those who used the ocean's living resources.

On to Caracas

The opening session of UNCLOS III took place in Caracas in 1974. Many of the participants questioned the location since there had been a lot of instability in Venezuela. Rebel factions periodically attacked government forces, individuals had been kidnapped for ransom, and there seemed to be considerable unrest among certain political elements. Nevertheless, there had been a strong push to open the conference in a South American country since the unilaterally extended jurisdiction over adjacent ocean waters to 200 nautical miles originated in Peru, Ecuador, and Chile. During the late 1960s some South American countries had been designated as "rogue" states, but by mid-1970s they had gained considerable support from a large number of African, Asian, and island countries. Thus, Caracas had been selected as the opening site for the conference.

Upon arrival at the Caracas airport, which is on the coast below the city, we boarded a bus that was preceded and followed by a military escort. The convoy made its way up the steep winding road to the city, which is at about 5,000 feet, without any trouble. As we approached the hotel, we were stopped by guards with submachine guns, and each of us was taken to a security room, where officials checked our papers and issued special photo IDs. We were required to wear them at all times within the hotel campus and wherever we traveled in the country.

After Jake Dykstra and I got situated in the room we shared, we checked out the hotel and found armed soldiers at every entrance. To get to the meeting site, which was several blocks away, we had to go through an underpass. Several men, not in uniform, but all carrying automatic submachine guns guarded each end. The armed encampment character of the facilities didn't do much to make us feel at ease. The downtown area is situated in a valley with hills and mountains on two sides. Portions of the city looked rather affluent with nice buildings, parks, and a number of high-rise structures. But from our seventh-floor window, Jake and I could see small mud-stucco structures along the adjacent hills, where thousands of poor people lived without indoor plumbing. These living facilities had only sparse access to water and were reached by many steps that wound their way up between the buildings.

During the meeting, it quickly became apparent that the status-quo view that limited control of narrow territorial seas would give way to those who sought much broader national jurisdiction. The 200-mile Fishery Conservation Zones (FCZ) and Exclusive Economic Zones (EEZ) were gaining support from a broad number of the participants. Their manifestation as international law, however, would have to wait for the building of a broad agreement on a number of ocean-related issues before its ratification by the UN. To many, this was tantamount to an international debate without foreseeable end.

The issues that needed resolution included freedom of transit through important straits; freedom of movement of cargo and military ships through FCZs or EEZs if, or when, they came into being; freedom of access and use of the deep seabed; and free movement of scientific research vessels throughout the world's oceans. Shortly after the start of the Caracas meeting, the U.S. strategy seemed to change from "we will overcome" to "how to acquiesce to the 200-mile steamroller, find some formula to deal with the interests of our tuna and salmon fishermen, and retain the freedom of transit through important straits." The salmon industry saw the extension as a retreat from the Abstention Principle that gave protection to the country where the salmon originated regardless of its distribution on the high seas. The tuna fishermen saw it as closing off of ocean areas they had fished for several decades. The question was how to integrate these concerns into the extension movement and not yield on the freedom of passage issue.

The U.S. team composition was much the same as it had been for the opening session in Geneva. However, I had been told that Shirley Temple Black's husband had been added to the list of fisheries advisors and that Shirley herself would be attending. I never quite knew why Charles Black was on the U.S. advisory team but discussions with him showed that he was very conversant with the issues. Furthermore, he had a son who was somehow involved with fisheries. Black soon became an important part of the daily discussions related to fisheries. One day he approached me and asked if I would join him, his wife, and several other members of the fishery delegation at the apartment they were renting on the outskirts of town. As I recall, the dinner guests included Don McKernan, the ambassador for fisheries and oceans at the Department of State; Bob Schoning, director of the NMFS, myself, and several others whose names I have forgotten. I think all of the invitees were rather excited to meet the great child movie star Shirley Temple.

It was about a fifteen-minute taxi ride to their abode on a hillside overlooking the city. Charles introduced each of us to his wife, Shirley, and we soon sat down to a delicious dinner. Shirley had a marvelous personality and made everyone feel at home. When she laughed, anyone who had seen one of her movies would have immediately recognized the Shirley Temple from the screen, despite the passage of time. During the evening we discussed a number of fishery issues, including the removal of some rocks on the Rogue River in Oregon that reportedly blocking salmon migrations. Shirley was not only conversant on the issue, but also debated with Bob Schoning, who had been Oregon's Director of Fisheries at the time of the action, about the wisdom of removing the rock barrier.

Sometime during the Caracas meeting, members of the Venezuela Fishery Agency asked if I would give talks at a university and a field research station. I agreed, and they arranged to fly me to a city about one hundred miles to the south and then on to the research station located on a nearby island. On the day of departure, I was taken back down the winding road to the airport. However, something had gone wrong and the plane was overbooked. I couldn't get out until the next day. As fate would have it, the first plane crashed into a mountain, and all aboard were killed. Ruby, who knew I was supposed to be on that flight, was in a panic, not knowing that God, or good luck, had spared me. The next day, when we landed on the island, we saw the remnants of the crash.

As Jake and I listened to an endless stream of interventions from countries all over the world, it became increasingly apparent that a new legal regime would be established to control access to, and management of, the ocean's living resources. The nature of changes to come was a part of the formula being forged at the UNCLOS conference. Limiting coastal states to a narrow zone adjacent to their coasts, a policy that had prevailed for several centuries, was being jettisoned in favor of a new 200-mile exclusive economic zone. Coastal states would have exclusive rights to manage the fishery and resources within these zones. The preferential right to utilize the living and nonliving resources would be coupled with a responsibility for stewardship of fisheries and other living resources within the FCZ/EEZ. The question was no longer whether extended jurisdiction would occur, but how long the community of nations would wait while the conference debated the remaining unresolved matters, such as how such a zone might limit freedom of passage for merchant and military ships. At this point, the industry advisors to the U.S. delegation began a more serious discussion of the character and management authority that might implement fisheries management under a legislated extended jurisdiction. The conference closed with attendees taking no action on the coastal state jurisdiction issue over distant-water fisheries; a number of U.S. participants went home looking for more results from Congress.

The Indian Wars

National and international resource policy consumed a great deal of my time in the mid-1970s. But the work situation became even more stressful when I was unwillingly swept up into long-simmering debates between state, interstate, and federal authorities and Washington Native American groups. This debate entered its final stages in the historic case, *Washington v. the U.S.* A number of Washington tribes, which had been a part of an overall U.S.-Indian treaty, the Stevens Treaty of 1855, claimed that state fishing regulations favored non-Indian fishers. This is an oversimplification of a rather complex legal issue. From the standpoint of most fisheries professionals, the heart of the debate was that the tribes were not allowed to catch their share of the salmon returning to state waters nor were they being given their treaty rights to manage fish stocks returning to reservations and/or traditional fishing grounds.

The court decision, frequently referred to as the "Boldt Decision" after the presiding judge in the case, found the tribal claims to be justified and awarded the tribes fifty percent of the allowable salmon harvest for U.S. fishers that was bound for Washington streams and rivers. The court also reaffirmed tribal rights to manage, subject to certain provisions, fisheries on their reservations and in traditional fishing areas. The court ruling came as a shock to non-Indian fishers who saw it as unfair, disastrous, and a misinterpretation of the Stevens Treaty. The ruling left many non-Indian fishers in Washington and adjacent states bitter and disenfranchised. Some were on the verge of anarchy.

At the time, I felt disconnected from the ruling because I was unfamiliar with the literature and legal findings that were part of the court record. I also did not possess the legal knowledge for understanding the basis for the court's allocation between the two factions. It was obvious that tribal options and access to the state's salmon runs had been eroded by the extensive growth of non-Indian gill nets, seines and troll fisheries in waters outside of, and frequently far from, reservations. But, in my option, non-Indian fishers should not have been held responsible for this evolution. The growth of non-Indian fisheries had been allowed and fostered by both state and federal authorities for more than fifty years. Now, suddenly, many non-Indian families who had fished Puget Sound and other areas of the state for generations were being told to stop salmon fishing. On reflection, the tribes, after a long period of neglect, inherited what was theirs to begin with while the non-Indian fishers were never fairly compensated by governments that had allowed the inequity to evolve and persist.

Given these circumstances, the rebellious nature of the non-Indian fishers at the time of the Boldt Decision becomes more understandable. Nevertheless, the hate that built up should have been directed at poor governance rather than at the Indians who benefited from the ruling. The conflict became increasingly hostile with threats of violence from both sides and, contrary to state and federal regulations, some non-Indians continued fishing. To address this open conflict, the president formed an independent committee comprised of one of his advisors and high-ranking officials from the departments of Justice and Interior. The committee was to evaluate the consequences of the Boldt Decision and to recommend actions that would mitigate the impact on both Indian and non-Indian fishers. Basically, its goal was to find a solution that would be reasonable and acceptable to all parties concerned. To provide local input to the committee's investigation and develop recommendations, three regional individuals were named to assist the Presidential Committee. Unfortunately, they included John Howe from Interior, John Merkel from the Justice, and me.

I say "unfortunately" because for the local participants, and probably for those at the national level, it was a "no-win" situation. The future of the national appointees was tied to the parties in power, but the local members were all civil servants whose careers in the region would last a lot longer than those of their bosses in Washington. All the participants must have known that the two sides were so far apart that there was no common ground or space for negotiation. Whatever the character of the proposed compromise, it would be rejected by both Indian and non-Indian participants. Furthermore, the work of putting together the historical catch records, operation patterns, and other relevant data and arranging meetings with the disparate participants would be a time-consuming process. The participants included a range of recreational and commercial fishing groups, the key processors of salmon caught in Puget Sound, and all of the tribes associated with the Stevens Treaty.

Over a period of months, we attempted to meet with all of the important stakeholders, sometimes jointly with Indian and non-Indian participants, and discuss possibilities for solutions. This may not have been the wisest strategy, because we generally were not seen as impartial by either side, and at several meetings the two sides came close to physical confrontation. For a very short period, we had an agreement on the part of both Indian and non-Indian fishers "to live and fish according to the recently established Boldt ground rules." When it broke down, both sides berated the local committee, and some of its members were threatened verbally and by mail.

The level of emotions and the commitment of the non-Indian fishers to escape the consequences of the Boldt Decision were reflected during a meeting of the local and presidential committees at the FBI offices in Washington, D.C. Half way through our discussion, we were interrupted by loud voices outside the meeting room. Within minutes, a group of about twenty fishermen's wives, led by "Tink" Mossness of Seattle, burst into the room and began to lecture the committee on the inequities of the Boldt Decision; some had tears in their eyes. The committee members listened and sympathized with a number of their concerns. The real question, however, was how this group had managed to pass through the various security points, find our meeting room, and enter it without being challenged.

Jim Waldo was hired to help develop background technical information and assist the local committee by doing everything he could to find the "middle ground." With his help, we finally put together a report with recommendations supported by all three local committee members. This had not been an easy task, as there were turbulent debates between the local participants and, in one instance, one of the members was sent packing from the home of the meeting's host. By the time we were all in sync and ready to present our findings to the Presidential Committee in Washington, we were fairly washed out. In fact, we presented copies of the song *Take This Job and Shove It* to our Washington counterparts, who received them with good humor.

The Presidential Committee findings were presented to the Indian and non-Indian stakeholders and, as expected, were castigated by all. The report recommended a new allocation formula and established a government salmon-mitigation program, which was expected to improve the quantities of salmon for both sides. It altered, however, the proposed court allocation, which was unacceptable to the tribes, and failed to recommend a Supreme Court review of the case, which was fundamental to the non-Indians. Shortly after the proposal was unanimously rejected, the Supreme Court sustained the Boldt Decision and, although many non-Indians have never accepted it as equitable, the hostility has diminished over time. In retrospect, it was probably a good thing that the compromise proposal was rejected. Changing ocean environmental conditions during the late 1970s and early 1980s would have derailed the enhancement components of the proposal, and the failure of the Supreme Court to review the case would have left questions of legality among the critics of the Boldt Decision.

Drafting of National Legislation

During all of this hometown commotion, the political wave for extended national jurisdiction continued. Between 1974 and 1976, several more UNCLOS III sessions were held in Geneva and New York, and progress continued, but at a snail's pace. However, Canada and the United States were making headway on gaining greater protection for salmon, regardless of their oceanic meanderings. Canada sought to gain support for the view that the salmon should largely benefit the nation of origin because its rivers had to be protected and maintained for the runs to be sustainable. During one session, the Canadian delegation passed out hundreds of wooden boxes containing the life history of and Indian legends about Pacific and Atlantic salmon as well as beautiful paintings of all the major species of the trout and salmon families. These gifts were presented to key delegates, including the fisheries representatives on each national delegation. I have never been certain of the political success of this event, but the issue concerned was one of the few agreements between the two countries.

The slow pace of the international debate was wearing thin on the U.S. supporters of extended jurisdiction. As a result, two key U.S. Senators, Magnuson from Washington and Stevens from Alaska, began to listen more carefully to their constituents. As the tedious discussions proceeded during open sessions in Geneva in late 1975 and early 1976, many of the fishery advisors were chatting about how to frame legislation that would mandate unilateral extended jurisdiction. These talks included consideration of how a federal management might be regionalized; the character of authority to be vested in the federal government; the basic goals and objectives that would guide any newly established authority; and what conservation and allocation principles would guide the utilization of the fishery resources under U.S. authority. Most of the basic concepts already had been set in place during the UNCLOS session. Although no notes were taken, bits and pieces of these chats found their way to aides in Magnuson's and, I suspect, Stevens' offices.

In Seattle, Harold Lokken, head of the Halibut Line Fishermen's Association, believed it was time to put together a comprehensive legislative draft for requiring unilateral extension of national jurisdiction to 200 nautical miles. Sometime in early 1976 he asked me to review his efforts. The draft contained most of the important requirements, but the boundaries of the regional management were lacking and the proposed entities lacked any technical or scientific support. Within an hour we concluded that six major advisory bodies (councils) should be established based on the existing NMFS administrative regions: New England, Mid-Atlantic, Gulf of Mexico and Southeast Atlantic, southern California, Pacific Northwest (northern California, Oregon, and Washington), and Alaska. The number of council members and the exact process of selection was left for further discussion. Lokken and I did agree, however, that an Industry and Scientific Advisory Group should support each council and that a set of principles should guide the councils' activities, but the drafting of these was left to a later date. This advice ultimately was used to craft the 200-mile legislation.

A group of scientific and industry advisors spent an afternoon discussing guiding principles at the next UNCLOS meeting in Geneva.

As a strong proponent of principles, I was asked to keep a record of the suggestions put before the group. My recollection is that Jake Dykstra, Lowell Wakefield, Lucy Sloan (all members of or advisors to the U.S. delegation), and several others were at the drafting session. All I had was a large manila envelope, so I jotted down the suggestions, which we named the "National Standards," on the back of it. Later, a redrafted copy was sent to Bud Walsh, who was Senator Magnuson's administrative assistant. Lokken also sent his preliminary draft of the bill to Bud, and no doubt numerous suggestions were submitted from New England and conservation and sports groups. As far as I know, Bud and his staff were the main architects of the extended jurisdiction bill, originally titled the Magnuson Fishery Conservation and Management Act (FCMA).

When the draft legislation ultimately was submitted to the Congress, it was not very popular within many of the key segments of government concerned with ocean policy. I believe that it had no official support from any sector, and many believed that President Ford would veto it even if it were passed by Congress. Despite the internal government resistance, the bill found great support among most coastal states, recreation and commercial fishing groups, and conservationists. In addition, a number of high-ranking senators and congressmen were pushing for the legislation.

While the bill was under consideration, Senator Magnuson convened a group of industry representatives, academics, conservationists, and members of NMFS to discuss the proposed act. Close to a hundred people attended the meeting in Washington, D.C. Magnuson, who chaired the session, made a few opening remarks and then began to discuss key elements of the proposed act. After about a half an hour, someone handed him a note. He then looked around the room, caught my eye, and said, "Lee, will you take over the chair and continue the meeting?" I nodded my head to say "yes," but I was really caught off guard. The senator quickly left the room. I had come to the meeting wearing Levis and a polo shirt and was in the back of the room. I walked to the front of the room, past attendees dressed in more formal attire, sat at the head of the table, and reconvened the meeting. Feeling somewhat ill at ease, I managed to get through the afternoon, wondering how many other people had chaired such a meeting in Levis.

To make a long story short, the legislation came before Congress in September 1976 and passed, much to the chagrin of the administration. Despite continued rumors that President Ford would veto the act, he did not do so. According to a story that has persisted over time, Senator Stevens told the President that, if he vetoed the act, he would not be reelected in November. The act constituted the first comprehensive statement of U.S. fishery policy since independence. It provided for U.S. sovereignty to manage the living resources within the FCZ, with special treatment for salmon and highly migratory species. It gave preferential rights to U.S. fishers over foreigners and established national regional councils to advise on fishery management and allocation.

The act also included requirements for Scientific and Industry Advisory Teams for the councils, and incorporated national standards that set forth the legal basis to enforce the laws and regulations put into effect by the secretary of commerce. Much of its content and structure had been taken from ideas developed within the UNCLOS framework. Although there were flaws in the act, it did provide the foundation for natural-resource governance of our nation's living ocean resources in a sustainable manner. Whether or not this goal would be achieved would depend upon subsequent council advice, political manipulations, and the actions of the federal government.

Chapter XIII
The Implementation of Policy

While I attended a sequence of UNCLOS III meetings, as well as national discussions on emerging ocean policy, Al Pruter took charge at the center. By the time Congress had passed the FCMA, Al was looking for a reprieve from the daily administrative duties. He asked if he could take a special assignment to look at the potential ecological consequences of the expansive trawl fisheries in the Bering Sea. He had more than earned the opportunity, but I hated to lose him at the center. Nevertheless, there was a plus side to my approval of his request.

Frank Fukahara, who headed our Fisheries Assessment Divisions, had been of great help to the center in dealings with our Japanese counterparts. He was not only a good scientist, but he also spoke Japanese relatively well. He had been invaluable to the American delegation at various U.S./Japanese negotiations and the next International Pacific Fisheries Commission (INPFC) would convene in Japan in another month. The unilateral extension of the U.S. jurisdiction to 200 nautical miles was bound to have horrific effects on the Japanese fisheries in the northeastern Pacific Ocean. Thus, after scanning possible candidates for the job, I selected Frank to move into the deputy director's slot. Al Pruter took on the task of summarizing all the data we could find on the impact of trawling on the ocean seabed.

Preaching the Gospel

The INPFC meeting was scheduled for late October and early November 1976, and the session was expected to be lively. Frank and I took off from SeaTac and headed north to Anchorage where we would refuel and head on to Tokyo. The flight up the coast of British Columbia and southeast Alaska was smooth and without incident, but when we began to descend over Turnagain Arm, just south of Anchorage, the plane hit rough air. The turbulence was as bad as I had ever experienced, and the plane was jumping and twisting at the same time. This was before doors were installed on overhead storage bins, and a lot of what was stored in the racks came tumbling out onto the seats and floor. When we landed in Anchorage, several of the passengers refused to go on to Tokyo until the following day. However, there was very little turbulence when we took off an hour later, and the flight to Tokyo was calm and incident free.

We stayed at the Hotel Acura, my favorite hotel in Tokyo. Its class and service was unmatched by any other hotel that I had stayed in during trips abroad. It was right across the street from the American embassy, so our delegation had easy access to a meeting room. The commission meeting was held several blocks away at the Japanese equivalent of the state department. The walk took us past the Gaimusho, the Japanese senate, and close to the walled grounds of the Imperial Palace.

The commission meetings generally had two major components. For several days, the scientists compiled the statistical aspects (landings by species and the effort expended in taking the catches) of the fish harvested off the United States and Canada by those two countries, Japan, and other nations. From this information, we generated an analysis of the status of exploited stocks. This data, plus the results of any special research such as tagging or genetic information, formed the backbone of the annual report published by the commission. The formal sessions of the INPFC followed introductory remarks made by the heads of each delegation. These comments normally set the tone of the remainder of the sessions, which followed a formal agenda. Of course, the U.S. unilateral extension of jurisdiction was the major theme driving the conference talks in 1976. In a way, it was rather sad because we all knew that the U.S. act would erode the need for the commission and that its long-term survival was dubious. The commission and the friendships that had evolved over a quarter of a century among scientists, industry advisors, and negotiators would be altered, diminished, and replaced by other arrangements.

From the beginning, there was a sobering tenor to the meeting. The chairmen of each national section praised the commission's historical accomplishments and its contributions to science and the rational management of the ocean's living resources. They expressed their hopes that the institution or its missions would be absorbed into a new scientific organization for the north Pacific. I expect that a vast majority of the U.S., Canadian, and perhaps even the Japanese scientists would not have given the commission high marks for its contributions to sustainable use of the north Pacific's living resources. Nevertheless, it did serve as a forum for debate and partial resolution of conservation issues confronting the resources of the region. On the plus side, it had stimulated the expansion of productive ocean science in the north Pacific and kindled bonds of friendship between the scientific communities in three countries. It also had established friendships among industry advisors that would, in time, influence the development of U.S. fishing ventures in the north Pacific.

I have not forgotten the great concern the Japanese had regarding the U.S. unilateral extension of jurisdiction. Many Japanese thought that when the law went into effect the following year, it would terminate their extensive trawl, line, and other fisheries in the eastern Bering Sea, which landed billions of pounds each year in ports of northern Japan. During the Tokyo meetings there were a number of organized street protests against the U.S. action.

Just prior to the meeting's formal closing ceremony, Frank and I were invited to the embassy to discuss a public-relations mission for the new legislation. The ambassador was concerned that the act would generate anti-American feelings among the Japanese, who were the world's leading consumers of fish products. The Department of State had decided that the act should be explained and justified to the Japanese fishing communities that depended on fish caught off the U.S. coast. This was undoubtedly a desirable mission, but one that was not likely to be easy. The audience, I thought to myself, might not be familiar with the adage "Don't shoot the messenger."

The plan called for discussions in Sendai and Hachinohe, northern Honshu and Hakodate, and Sapporo and Nemuro on the northern island of Hokkaido. These were main fishing ports or processing centers that relied on fish supplies caught by distant-water fishing vessels, much of it in the waters off Alaska. Frank and I thought that the local diplomats should undertake the task, but they insisted that they needed someone familiar with the fisheries and the FCMA. We were given our marching orders, so we informed our wives that our return would be delayed and began to plan the nature of our presentation.

Although the act gave preferential rights to the United States, these rights were contingent on the fishing industry demonstrating its capacity to harvest and process the identified sustainable fish yields. Many U.S. fishermen thought that passage of the act would lead to a rapid departure of distant-water fleets off the Pacific coast, but it wouldn't be that easy. The world market for cod, hake, and flounder (whitefish) remained soft. To a large extent, the higher-priced markets for the species of fish harvested off the contiguous Pacific coastal states and Alaska were predominately in Japan. There, a major market existed for pollack surimi, a fish paste used to formulate a number of products called Komoboko, such as artificial crab, shrimp, and hundreds of local Japanese confections. In addition, there was a high Japanese demand for red and other types of salmon, slope rockfishes, black cod and some flounders, all of which were plentiful in the Bering Sea, the Gulf of Alaska and, to some extent, along the coasts of Washington, Oregon, and California. However, the Japanese markets were not easily accessed by U.S. fishermen and were protected by a number of policies and fees for conducting in-country business.

Frank and I decided that our best course of action was not to build any expectations that the Japanese distant-water fishery off the U.S. coast would survive in the long term. Nevertheless, it was likely that it would be phased out over a period that could last from a few years to a decade. It all would depend on how quickly the U.S. fishing industry rallied and whether or not it could compete economically on the fishing grounds and in the world markets. We planned to present a general summary of the act, including any elements that related to continued foreign fishing off the United States, and answer any questions to the best of our ability.

We left Tokyo by train for our first stop in Sendai, along with an advisor from the Department of State and two interpreters. Both Frank and I were nervous. There had been a great deal of criticism of the U.S. action in the Japanese press. Sendai was not significantly dependent on distant-water fish catches, but it did process some fish taken off Alaska.

The meeting seemed to go quite well, and there was only a touch of animosity from the Japanese. For the most part, they were polite, asked a number of questions, and pointed out the importance of fish to Japanese society. We left with several traditional gifts from the region and headed north to Hachinohe, situated on the northwest side of Honshu in Aomori Prefecture. The governor of the prefecture, the local heads of fishing organizations, and fish processors met us there. We gave the same pitch that we did in Sendai, but here there was a strong and long grilling on the possible long-term consequences of the act on the people of northern Japan. Hachinohe received considerable pollack from the Bering Sea and was extremely dependent on fish taken by their distant-water fleets. Still, the session ended on a friendly note. Frank and I began to think that our mission would be a "piece of cake."

Late that evening we left for Sapporo, a large and seemingly prosperous city on the southern-central portion of Hokkaido Island. On the way in from the airport, we noted that the streets were wide, as were the sidewalks. Someone told us that a western engineer had plotted the city early in the twentieth century, but I can't confirm this. Our group was booked into a business hotel. The room was small but clean, and more than adequate for our needs. We turned in early, to rest up for the next day's meeting with the island's fishermen and processors.

The next morning, our hosts said that a crowd was expected to protest outside the meeting room. We also learned that a very large Japanese delegation, made up of fish processors, members of fishermen's organizations, and local government officials, would be in attendance. The meeting started around 10:00 a.m. in a government building a short ride away from the hotel. As we approached the site, we saw that the streets were swarming with protestors, and we were quickly ushered into the building. The meeting room was quite large and the center filled with a long rectangular table. Behind one side of the table were seats for observers, press, and other interested parties. Representatives filled every seat on the Japanese side of the table, while all the chairs behind them held their constituents. The room was literally buzzing with people talking among themselves and members of the news media. Equipment, including a dozen or more television cameras, could be seen around the entire room. Our group of six sat with our backs to the wall, occupying only a small length of the massive table.

As we sat down to start our presentation, an embassy staff member told us that an estimated 10,000 people were protesting in the streets. Our entire contingent was nervous, and the confidence we felt the day before vanished completely. The meeting started with someone from the governor's office making a few remarks regarding the recent U.S. unilateral extension of jurisdiction over ocean space and the importance of marine food to the people of Japan. He also expected our presentation to help the Japanese people understand the reasons for the U.S. action. It was all presented in a serious but courteous manner. Following these introductory remarks, a member of the embassy staff introduced Frank and me, and then we were on our own.

I started the session with a description of the FCMA and noted the large number of species in the northeast Pacific that appeared to have been overfished by distant-water fishery operations. I did not name Japan as a participant in the overfishing, but it was implicit. I then turned to the details of preferences and allocations under the act's provisions. I noted the constraint on the U.S. industry; to displace a foreign distant-water fishery, it had to demonstrate a capacity to catch, process, and market the fish in question. As I spoke, I watched the faces of those across the table; most were stoic, while several directly across from me just kept shaking their heads. It was going to be a long day.

Frank then provided more details, using specific examples of how the phase-out might occur for one or two major fisheries of the region. He noted that the U.S. fishing capacity and markets for crab and shrimp were strong both in the United States and abroad. Of course, the Japanese had been purchasing large quantities of crab from U.S. companies (many of them owned by the Japanese) for years. He noted these fisheries were likely to be quickly "Americanized," a term coined by our fishermen for the transfer of fishing activities to the United States. He concluded by stressing that the U.S. ground fisheries were not well developed off Alaska, and thus we did not expect the Japanese fleet targeting these species to be phased out rapidly.

It was a strategic move for Frank to close out the presentation. He was Japanese American and spoke passable Japanese. He had grown up in a Japanese-speaking family and understood their culture and behavior patterns. He was bound to be taken more seriously and could respond well to their questions. I am not sure, however, that it helped much with the Sapporo contingent. As I continued to watch the Japanese across the table, their facial expressions remained unchanged and those directly across from me continued to shake their heads. When Frank concluded, there was a short recess for coffee or tea, and then we returned to the table for questions, clarifications, comments, and general reflections from the Japanese delegation.

They wasted no time. A gentleman named Abe, who was the president of one of the largest Japanese associations and cooperatives, began with a stern lecture regarding the unorthodox manner in which the United States had extended its jurisdiction. He strongly criticized the unilateral action and opined that the we should have worked within the framework of the UN law of the sea. He then related how the Japanese had undertaken extensive fisheries explorations off the Aleutians and in the Bering Sea area in the 1920s, long before the United States was even interested in the ground fish of the region. In the early 1960s, he emphasized, the Japanese spent great sums of money to develop a process to make surimi out of pollack and subsequently built an extensive industry dependent on distant-water pollack catches. These catches supplied much needed protein to the Japanese people. He outlined the expected impact on Japan if its fleets were forced out of the Bering Sea and Gulf of Alaska. He continued for the better part of thirty minutes. His tone was firm, at times harsh, and he constantly looked at both Frank and me. Abe was obviously an angry, unhappy man, and his comments and questions were not easy to handle.

Both Frank and I realized that much of what he said was founded in fact and that his comments dealt with Japanese developments over the past several decades. Neither of us knew the magnitude of the infrastructure that formed the foundation of the Japanese surimi industry. We did know that his comments regarding the history of Japanese fisheries off Alaska were correct. We acknowledged his recollections about the development of fisheries ff Alaska and said we were well aware of the Japanese dependence on ocean fisheries, including those off North America. We also pointed out that most of the countries attending the UNCLOS III were opposed to the massive distant-water fleets fishing off their shores and were promoting a new regime that would give coastal nations greater authority over the fisheries in adjacent waters. We added that the extended jurisdiction act was generally in accord with principles and conservation concepts promoted within the UNCLOS III.

We finished by saying that the United States would have preferred to wait until the conference concluded in a comprehensive agreement, but that the aggressive overfishing of ground-fish stocks and economic dislocation of our local fisheries, and others in the world, required more immediate actions. Furthermore, the United States could not wait for the conclusion of a conference that was mired in details and issues unrelated to fisheries. We also reiterated that the act did not require an immediate withdrawal of Japanese fishing activities, but permitted a level of foreign fishing, referred to as the "Total Allowable Level of Foreign Fishing" (TALFF), which depended on the rate of development of the U.S. catcher/processing industry. We expected that it would be some years before the Japanese were phased out of the area.

Abe did not appear to be aware of the overfishing problems, so we used an overhead projector to show the trends in abundance of some of the Bering Sea flounders and rockfishes. He still seemed unsatisfied with our explanation and asked how many years the Japanese fleet would be allowed to fish within the U.S. zone. Both Frank and I refused to conjecture on that point since so many variables could influence U.S. fisheries development.

When Abe completed his comments, all of the Japanese representatives stood, one after the other, and made similar remarks, almost all of them forceful but stoic. I'm sure Frank sensed the true tone of their presentations. I was lucky, however, because their remarks had to be translated into English and the interpreter spoke in a much softer voice. The morning and a good part of the afternoon had passed by the time the session ended. Throughout the five-hour meeting, T.V. journalists covered the conference and photographers took pictures from a variety of angles. Frank and I had been put through the proverbial "ringer" and were not sure that we had eased any of the Japanese concerns. Their contingent left the table without further comment.

Our Japanese host asked if we could wait for a few minutes. I asked our embassy escort why we were waiting, and he didn't seem to have a clue. We remained alone at the table for about fifteen minutes, and then Abe asked us to join him in another room. I could see a beautifully set banquet table decorated with flowers and crossed Japanese and U.S.

flags. The Japanese delegation was standing around the table, smiling and clapping as we entered the room. I was in total shock. The people who had appeared shaken by the U.S. extended jurisdiction, and had questioned us without mercy, were welcoming us to a special meal and a friendly discussion. The testy meeting participants became kind and gracious hosts. It was time for a festive dinner. I was not sure about the implications of the seating protocol, but Abe sat directly across the table from Frank and me, and some of those who had engaged us at the meeting sat on either side of him.

The conversation included comments about Vice President Mondale, who was in Tokyo. Several of our hosts noted that our meeting in Sapporo was receiving greater press coverage than Mondale. It was probably a frivolous comment to set us at ease, but there was little doubt that the fisheries issue was high on the Japanese agenda. Abe explained in detail his association's activities in Hokkaido and the economic importance of fisheries in the region. When the third or fourth course was being served, he looked at me and said in a quiet and serious tone, "I was blessed with a granddaughter yesterday, and I just hope she will have adequate food to grow up healthy." It was a rather surprising statement as well as a coincidence, since I had received a cable message notifying me of the birth of a granddaughter on the same day. I mentioned the fortuitous event and said, "I hope they both have adequate food and that they may grow up to be friends." Abe smiled, shook my hand, and we toasted our new grandchildren.

Later that evening, Abe asked Frank and me to visit his office the following morning and we accepted, as we were not leaving until the afternoon. I told Frank, "I'm just starting to get a slight feeling for Japanese culture." The dinner had not changed their minds, but it had allowed both sides to develop closer personal relationships. The next morning Frank and I arrived at Abe's place of business right on schedule. The building was a large four-story structure with a stucco-like exterior. When we entered the large room on the first floor, we saw forty to sixty employees standing and bowing down to their waists. We walked up through a corridor between desks, chairs, and other furniture to Abe's desk, which sat on a raised platform. We were asked to sit, and then tea and an assortment of cakes were presented to us.

Abe reiterated his concerns about the U.S. extended jurisdiction act and then said, "I am glad you have a new grandchild in your family and that she will share the same birthday as my new granddaughter. To celebrate this occasion I would like to present you with a little gift." Then several of his staff brought us a glass-encased, beautifully dressed Geisha doll that was about three feet high. I was somewhat stunned and did not know if I could accept the doll. Abe's graciousness had caught me off-guard, and I was caught up in the emotion of the moment. Later I was told that it would be impolite for me not to accept the doll and that I should take it home. It remains in our family today as a memento of a kind man and a special moment in my life.

From Sapporo, we flew to Kushiro, which sat on the northern shore of Hokkaido facing the Okhotsk Sea. There, Frank and I were presented with the traditional carved wooden bears and met with two groups on the same night. First, we sat down to a long dinner meeting with the Japanese Trawlers Association, which was calm and without difficulties. The participants were mostly owners of distant-water vessels that operated in the Bering Sea and Gulf of Alaska. For some reason, they seemed to have a better understanding of the U.S. act than the politicians did. They were in the heartland of the Japanese pollack industry, however, and recognized that over time, they would be the hardest hit by the extended jurisdiction. They also noted that passage of the act in the United States had given the green light to other countries to take similar actions. This was correct, but these actions were stimulated by conservation and economic concerns and fueled by strong emotions.

Our dinner with the trawl group ended just after 9:00 p.m., and we were asked to join the Longline Association for a second dinner. The two groups did not get along well and preferred to meet separately. The Longline group was perhaps the only fishery group in Japan that supported the U.S. extended jurisdiction. Most likely, they believed there was little they could do to derail the act and would fare better in a "phase-out" program by supporting it. The meeting was over in less than an hour, and then it was time to celebrate. It was difficult for our group to partake eagerly in a second dinner after dining royally only an hour or so earlier. We did our best, however, and the day in Kushiro went well. After traveling the next morning to Hakodate to visit the local university, which was well known for training Japanese fishery scientists, our mission was complete.

We returned to Tokyo; both Frank and I carried an array of souvenirs that had been given us at each location. They included a samurai warrior from Sendai, a carved wooden horse from Hachinohe, carved wooden bears from Kushiro, and, of course, the famous doll from Sapporo. Both of us had to obtain extra bags to get the gifts home.

Frank and I were asked to return to Japan the following year to provide an update of what had happened since the act had taken effect in 1977. In fact, very little had changed for Japan beyond a reduction in its crab catches. Our meetings went without a hitch. Before I left for Japan, I asked Ruby to acquire a Scarlet O'Hara doll, which she placed in a glass display case. In Tokyo, I asked the embassy to set up a meeting with Abe. When Frank and I arrived in Sapporo, we returned to Abe's office building. As in the previous visit, the office staff all stood bowing as we went down the same aisle to Abe's desk and gave him the doll for his granddaughter. He had one of the staff take several photos, and he seemed to have tears in his eyes

Establishing a Database

When I returned to Seattle in November, the NMFS was busy organizing the newly established fisheries management councils. The Northwest and Alaska Fishery Center provided scientific information to the Pacific and North Pacific, and Alaska councils. As a result of the new policy, the research centers were swept into management business. They became the federal entities responsible for the preparation and analysis of information regarding the resources under exploitation within the EEZ. For most centers, this constituted a major shift in research priorities and, at NWAFC, it required some reorganization and reallocation of personnel and funds in order to respond to the NMFS/NOAA stewardship responsibilities. Much work along these lines had been done within the INPFC and in bilateral arrangements. Still, the centers and state agencies became the entities responsible for providing the councils with scientific and technical information regarding conservation and allocation of the resources off the U.S. coast

The fish stocks and fisheries off Alaska were by far the largest in the United States. The historical database on harvest levels and trends in abundance were based on information supplied by the distant water-fishing operators. Our first task was to attempt to improve the quality of the catch data by requiring more specific information on the catch, the effort expended to take the catch, and the time and area of fishing activities. By placing an increased number of observers on board foreign vessels, we obtained a clearer picture of both target and non-target catches and were able to sample the size and age of the fish being harvested.

An expected fallout from the observer program tracking the foreign hake fishery off the Pacific coast states was a significant overnight improvement in the Soviet catch rate, as I remember, about twenty-five percent. This only confirmed our previous suspicions that distant-water operators had been underreporting their catches. At the onset of extended jurisdiction, we relied largely on data acquired from foreign sources and our observer program. In the next several years, the center received financial support that enabled it to carry out its own research surveys. This allowed estimates of stock trends independent of data collected from fisheries operators. It was several years before our research effort developed enough credibility to allow us to be reasonably confident about the trends of the major exploited species. Even then, we had just begun to scratch the surface in terms of collecting the range of data needed for effective fisheries management. In science, the more you learn about matters you are attempting to understand, the more you realize that there are other sets of data that can shed additional light on your questions—in this case, how to better manage ocean resources.

First, how what was the impact of fishing on the abundance of the target species? And could we measure this effect independent of nature's hand, which was always stirring and shifting environment parameters? The impact of fishing on non-target species also had to be taken into account. That is, all fishing activities were known to generate death of unwanted species caught along with the target species. This bycatch included a broad range of fish and invertebrate species, as well as birds, turtles, and marine mammals, that were returned to the sea. Understanding their fate in the ecosystem constituted an important goal of fisheries managers. Added to these two difficult questions was the impact of different fishing gear on the seabed and other environments. Thus, the complexity of the issues confronting the marine scientist required the collection and interpretation of varied new data sets, which often generated greater uncertainty and a desire for more conservative approaches to management.

Many of the data requirements needed to support the conservation goals of the MFCA were put in place in the northeast Pacific within a few years of the act's passage. Underlying science and information needs, however, are always in a state of metamorphosis. From the formation of the North Pacific Fishery Management Council (NPFMC), its council members tended to be cautious and conservative in establishing harvest levels. This was somewhat atypical of what was going on in other areas of the United States and the world. Quotas were set, leading to relatively low harvest rates and, hence, a low mortality among the fished populations.

Throughout my career, I often was asked why the North Pacific council had an excellent conservation record. Many factors may have contributed, but I believe that it was largely because of the excellent reputations of the Pacific Halibut Commission and the International Pacific Salmon Commission. The two organizations, which managed halibut and salmon stocks fished by U.S. and Canadian fishermen, had resulted in a level of respect and support for fisheries science unparalleled in most other regions of our country. The council members inevitably stood by the findings of its scientists. Furthermore, for years the Alaska Department of Fish and Game had promoted the policy of erring "on the side of caution."

After the passage of the FCMA in 1976, I was elected to serve on the Scientific and Statistical Committees (SSC) of both the Pacific and North Pacific councils. These responsibilities, coupled with running an expanding fish center, kept my plate full. Added to these tasks was the burden of continued travel to Japan, Korea, and Taiwan. After several years on the two SSCs, I resigned because I could not continue serving on the committees and also deal effectively with the center management. Furthermore, the administration of the center was becoming increasingly encumbered by the increased paperwork required by the Washington bureaucracy, which unfortunately left little time to keep pace with the expanding field of ocean sciences. Sometime in the late 1970s, I found myself becoming more and more jaded with the manner in which the centers were operated. It was not a welcome feeling because for most of my career, I had enjoyed my work. Suddenly, I found myself looking forward to retiring from the government. I could potentially retire by 1980, only a few years downstream.

Dancing with the Canadians

I had no real aspirations to retire, only to get out of the service and work in the private sector, but at that point I had only given the matter passive consideration. These thoughts, however, were put in the deep freeze in the late 1970s when Don McKernan, ambassador for fisheries and oceans, died during a mission to China. For several years, Don had been negotiating with Canada about a range of salmon problems extending from California to Alaska. Don was only one of a number of U.S. negotiators who had been searching for solutions to the fish controversies between the two countries. The difficulties were apparent. A series of meetings had been underway for almost two decades with little progress.

The debates with Canada over Pacific salmon actually extended well back into the 1800s, when both U.S. and Canadian citizens began fishing the extensive runs of sockeye and pink salmon that originated in the Fraser River. The river and its tributaries were completely within the province of British Columbia, but most of the salmon entering the Fraser at that time migrated east through the Strait of Juan de Fuca. Some of the salmon swam up the strait on the U.S. side, and others were within Canadian waters. Thus, U.S. fishermen were able to harvest significant quantities of Fraser-bound salmon in the strait and in northern Puget Sound.

From the late nineteenth to the early twentieth century, the competition between American and Canadian fishermen for the Fraser runs of salmon became ferocious, as did the fight over legal access to these fish. Efforts to establish joint management of the Fraser salmon stocks began in the 1890s. Unfortunately, no agreement was reached, and there were signs of overfishing of the Fraser runs early in the twentieth century. The heat of the debate decreased in 1913 after railroad construction dumped tremendous quantities of rock into the river at Hell's Gate, a narrow and restricted portion of the river about a hundred miles upstream of the river mouth. The discarded rock caused a major block in the river, which threatened most of the significant spawning grounds.

The catastrophic event galvanized the two countries into action, and both the United States and Canada provided funding to remedy the Hell's Gate blockage. To solve the problem, they constructed fish ladders that allowed the salmon to access the upper-river spawning areas. Although the talks continued, it was not until 1937 that the two countries established the International Pacific Salmon Fisheries Commission. The formation of the commission did not quell the contentious issues, but it did provide an international management and scientific institution with the authority to manage the river for sustainable production and allocate its harvest equally between the two fishing nations. For the most part, both sides remained reasonably happy until the post-World War II years. Following the war, both sides built up their fleets and began a major expansion of their fisheries into the ocean waters off Canada and the United States. Although the Salmon Commission effectively regulated the harvest and provided for the migration of fish into the Fraser, a growing sense of nationalism in Canada led to questions about why a river totally within British Columbia's borders was being studied by U.S. scientists and managed by an international commission.

The debate was fueled further, perhaps, after the early UNCLOS III discussions, with the enactment of extended jurisdiction legislation, first by the United States and then Canada (1976–77). During the UNCLOS meetings, the two nations united in a major effort to establish that the nation in whose rivers the salmon spawned and the young were reared should reap the major benefits (i.e., fishing rights). This principle strengthened the Canadian claim that the U.S. harvest of fifty percent of the Fraser-bound fish was inordinate even in light of the United States' contributions to the commission and the Hell's Gate blockage. Furthermore, by this time, the nature of the debate had become much more complex and extended into areas far beyond the convention.

Canadian fishers operating off the British Columbia coast and in the waters inside of Vancouver Island were intercepting considerable quantities of silver and king salmon of U.S. origin during their northward feeding and southerly spawning migrations through Canadian waters. To the north, off southeast Alaska, U.S. fishermen were intercepting salmon bound for major rivers in British Columbia.

Complicating the matter further, there were heated arguments over the allocation of salmon in the rivers that crossed the Alaska Panhandle into Canada. In the north, there was the added issue of dividing the catch in the Yukon River, which originated in Canada but flowed through extensive areas of Alaska into the Bering Sea. In short, there was a major brouhaha over whose ox was being gored. The Canadians felt that the American fishermen were intercepting many more Canadian fish than the fish of U.S. origin that Canadian fishermen were intercepting. Conversely, the U.S. salmon industry was convinced that the value of the fish of U.S. origin taken by the Canadian fishermen exceeded the value of fish originating in Canadian rivers that was intercepted by U.S. fishermen. Resolution of the matter was complicated, in part, by the fact that the "impartial" scientists on each side of the border supported national advocacy positions.

At the time of Don's death, the negotiations were almost stagnant and the major issue stalling progress was the concept of "equity." This was the term used to promote the principle that each nation should reap the benefits of the salmon spawned and reared in its rivers. Although the United States had helped to foster this principle, its application to the fisheries off the Canadian West Coast was difficult to swallow. Many on our negotiation team argued that the equity concept did not account for American investments. In addition, U.S. fishermen had fished the Fraser and other Canadian fish stocks in southeast Alaska and the Yukon for generations, and the equity concept made no provision for historical participation in fisheries development.

Don's death forced the Department of State to search for a new leader for the U.S. negotiating team. The job held a certain amount of prestige but also a number of persistent and intrinsic problems. First, the federal governments of both nations sought regional solutions that would satisfy the diverse constituents on each side of the border. That is, they wished to avoid the political heat of an agreement that was seen as unfair to any major constituent group. Second, the U.S. advisory group was fraught with internal discord on how to reach an agreement. Shortly after Don's death, the Department of State named John Negroponte, a career diplomat, to head the next scheduled negotiation with the Canadians. At the time, John was not well known to the Pacific Northwest industry and had a limited understanding of the complex West Coast salmon issues. He also had to deal with the fact that the lack of progress in the negotiations had soured some delegation members about the likelihood of achieving an accord with Canada. Hence, there was reluctance on the part of many on the U.S. team to pass the leadership to an individual outside of the Pacific Northwest and Alaska. Regardless, he took over the next negotiation, which took place in Vancouver, Canada, in late 1978 or early 1979. I was asked to go to the meeting and assist John, whom I found to be pleasant and intelligent.

Unfortunately John's initial meeting with the U.S. delegation did not go well. He had not had the opportunity to become familiar with the range of issues, so he seemed uninformed, and there were outbursts of sarcasm from some disenchanted members. John sat through the morning session and quickly sensed how the complexity and difficulty of the issues that stood in the way of a treaty that would satisfy both the U.S. and Canadian delegations. Late in the morning, he told me of his uneasiness with the tenor of the proceedings. After a short discussion, he asked me to take over the remainder of the session.

I'll never know all the factors Negroponte considered when he stepped aside but suspect he saw a certain futility to the negotiations. Later, I would better understand the wisdom of his decision. Within a week or two after the Vancouver meeting, I was named the new chief negotiator for the U.S. section. John went on to have a distinguished career. Many years later, he was appointed ambassador to the United Nations and then ambassador to Iraq in the early part of the twenty-first century.

My first task was to review the history of earlier negotiations and become acquainted with the U.S. advisory group members and their positions regarding a new treaty. The historical review was not a major challenge since I had been involved to some extent in the discussions over the contentious fishery issues between Canada and the United States. The group's composition and diversity of views was a much more difficult issue. Included were representatives from a number of Indian tribes from Washington State, state agencies, commercial fishing groups, recreational fishing interests, and the departments of Interior, Commerce, and State.

There were at least sixteen different Indian tribes with strong concerns about the nature of any new treaty, divided among three major geographic regions of Washington State. First, there were the coastal tribes living along the northern coast. Then there were the tribes that inhabited Puget Sound and portions of the Strait of Juan de Fuca, and finally the tribes that depended on the salmon migrating up the Columbia River and its tributaries. Among the tribes and even within geographic groups, there were differences about how to resolve the conflict with Canada.

The tribes living along the coast and the Columbia River largely were concerned about Canadian interception of the silver and king salmon bound for Washington's coastal rivers. Tribes in northern Puget Sound were more interested in how the fish bound for the Fraser would be dealt with, while those in southern Puget Sound focused on the interception of silver and king salmon bound for the lower Sound. Several tribes depended on Fraser River salmon and stocks bound for other areas within Washington. Despite the range of views within the tribes, however, they all agreed that the treaty should not alter any aspect of their recent victories in the federal courts.

The family of commercial fishing interests influencing the negotiations had perhaps even more disparate views than the tribes. They included commercial troll, purse seine, gillnet and set-net fishermen and salmon processors from California to Alaska. The Puget Sound fishermen and processors did not want a treaty that in any way eroded their historical access to fish bound for the Fraser River. Trollers fishing off the contiguous coastal states wanted to decrease the interception of salmon originating in their states and caught off Canada and Alaska. On the other hand, the Alaskan trollers did not want a resolution that reduced their access to the king and silver salmon off their coast even if the fish had originated in Canada or the southern West Coast.

Recreational fishermen, who targeted king and silver salmon from California to Washington, wanted a reduction in the interception of salmon stocks from their respective states that were taken off Canada and Alaska. The Alaska sports groups, on the other hand, were not interested in a solution that would reduce their access to fish off their coast.

The states with the greatest interest in the outcome of the negotiations were Alaska, California, Oregon, Washington, and Idaho. For the most part, the southern states sought a reduction in interceptions of their stocks of salmon that migrated north along the Canadian and Alaskan coasts. However, the state of Washington had to deal with a major split between fishing interests that depended on catches of pink and sockeye salmon bound for the Fraser and those that wanted to reduce the quantity of silver and king salmon taken off Canada and Alaska.

The federal government sought a resolution that incorporated the principle that the state of origin should reap the benefits of the harvest of the mature salmon runs. But it also wanted the historical participation of U.S. fisheries in northern waters to be taken into consideration. Like our neighbors to the north, the federal government wanted a solution that would give immediate attention to conservation and overfishing issues. However, it was noticeably silent on how to reconcile established international policy and the historical access various U.S. fishing groups had to salmon runs of Canadian origin.

As might be expected, all the U.S. parties professed to be committed to the conservation and sustainability of the salmon resources. Put simply, conservation was the number one priority, but its implementation should place the burden on those responsible for the conservation dilemma in the first place. In this game, most participants pointed their fingers at other players or said that habitat degradation was outside of their control. Deep down every member of the U.S. team knew that if an agreement was reached, some elements of our fisheries would have to be reduced considerably. With the exception of the more enlightened state officials and scientists, who clearly recognized that compromise was the only way to an agreement, the topic of conservation generally was deliberately avoided. On one hand, no interest group wanted to be the sacrificial lamb; on the other hand, no one wanted to foster internal discord by offering up their competitors' fish.

The Canadian advisory group included many of the same constituent groups as the U.S. delegation, but not as many diverse Indian groups. Furthermore, their fishing interests seemed to be more consolidated, perhaps making their head negotiator's job somewhat easier. But that might have been a perception from south of the border. Strong advocates of minimizing the interception of salmon originating in another country, the Canadians also were very aware that many areas had a mixture of salmon stocks originating in both countries. There was no practical way of achieving the goal of no interceptions other than to eliminate all the salmon fisheries that took place outside or in close proximity to the rivers. Many of the long-established fisheries off the United States and Canada took salmon in waters that were a considerable distance from spawning rivers. It was just not a politically viable fix. In addition, the Canadians insisted on gaining control over the Fraser River, which entailed their

461

exclusive right to study and manage the river. This meant an end to the highly successful International Pacific Salmon Commission, the caretaker of the river since the late 1930s.

In reviewing my assignment, I found it fortunate that the chief negotiator from Canada, Mike Shepard, was a scientist I had known for a number of years. I think it was the first time in the debates that the leadership on both sides of the border was in the hands of scientists. But Mike and I knew that this fact alone would not lead to a successful conclusion to the long and exhausting conflict between our nations. Still, the scientific leadership did establish a more positive negotiating environment. Mike and I had a mutual respect for each other, and we were both concerned about science tainted with nationalism. We also both had been associated with the UNCLOS process and knew that no agreement could lead to the overfishing of salmon stocks of mutual interest to our respective countries. We were well versed in the anadromous fish principles promoted by Canada and the United States at the international conference.

During the first full planning session with the U.S. advisors, I asked the participants whether we were willing to accept, as a delegation, the concept that we had fearlessly promoted at the UNCLOS. Should the major rights to the salmon belong to the nations in whose streams, rivers, and lakes the fish were spawned and reared? The principle served the interest of the United States, particularly as it related to the nations fishing salmon species on the high seas, e.g., Japan, Taiwan, and mainland China. It was more difficult, however, for a number of the U.S. interest groups to apply the principle to Canada.

The joint U.S.-Canadian participation in the Fraser and the U.S. financial contributions to mitigation, research, and the management of the river had to be woven into a solution. Also requiring our attention was the long history of U.S. fishermen harvesting the vast majority of salmon that originated in British Columbia and migrated into the rivers of the Alaska Panhandle. I argued that these concerns were legitimate and would have to be negotiated with the Canadians. Nevertheless, they were not a reason to abandon the basic allocation and management principles for salmon advanced within UNCLOS III. We discussed the issue for the better part of two hours and finally agreed to accept the principle as a basis for further dialogue with Canada.

The second and more difficult issue was the concept of equity, or a balance in the quantity or value of fish taken by fishermen in ocean waters where salmon of the two nations intermingled. This would require the two countries to find operational modes that would lead to a balancing of interceptions. Although most participants agreed that a balance was necessary, almost no one agreed on how. There was the longstanding question on how equity was to be measured—that is, in numbers of fish or in their value—considering the range of prices paid fishermen for the different species of commercial salmon.

In the early 1980s, many U.S. scientists believed that if value was the measure of equity, we were close to a balance, and the Canadians possibly were harvesting a higher value of American fish than we were of Canadian fish. Thus, although there were concerns about accepting the equity principle, the numbers didn't suggest that we were indebted to our Canadian brethren. As a result, we moved to endorse the two concepts at our next negotiation. We felt that this might set a new tone that appealed to the Canadians, particularly since under the leadership of McKernan, the United States already had agreed to the Canadians' right to manage the Fraser, which was paramount to reaching an accord.

By the time the next U.S.-Canada negotiation rolled around, I believed that if our scientists supported a balance of interceptions, the equity issue would not create a formidable obstacle. There was one major "if" standing in the way. In the early 1980s, there had been an increase in the Fraser salmon runs moving down the inside waters east of Vancouver Island. As a result, the fifty-fifty split that took place in the convention area was becoming a smaller portion of the overall run to the river. This change appeared to be the result of a gradual warming trend in the north Pacific and the influence of strong El Niño years, which moved much of the migration of the sockeye toward Vancouver Island, playing into the hands of the Canadians. In effect, this development forced the United States to negotiate for a percentage of the overall pink and sockeye salmon runs that would expand the convention area.

Nevertheless, we made relatively good progress in developing treaty language for the annexes that would govern the harvests of each nation in four regions. These included northern Puget Sound (Fraser River fish), the west coast of Vancouver Island, southeast Alaska and northern British Columbia, and the panhandle of Alaska, which had several rivers that originated in British Columbia. Although we moved closer, we could not close the gap to an acceptable level in any of these regions. We had agreed to accept the equity principle, but there remained scientific differences in each nation's accounting of interceptions. The numbers game couldn't be resolved, so we set a date to resume the talks several months later.

Although we met about twice a year, the negotiation process with the Canadians was an endless effort that required constant contacts between the various advisory groups: the Department of State, state agencies, and involved scientific groups. Progress was painfully slow. Some of my most vivid memories were of the meetings to obtain input from the various Indian groups bring them up-to-date on the U.S. negotiating strategy. Sometime in 1980, I met in Portland with the Indian tribes that fished the upper Columbia River, including the Warm Springs, Yakima, Shoshone, and several other tribes. The Indian advisors were dressed in their formal native dress, including head feathers, beadwork, buckskin, etc. I had become friendly with one of them, Levi George, and I whispered into his ear, "George, if you people keep coming to the meeting in this formal Indian garb, I am going to dress up like John Wayne at the next meeting." He laughed and then passed my remark around to the other participants. From that point on, the tribal members dressed like the rest of the delegation.

Later at a reception hosted by the Canadians, Levi and I spent the better part of the evening discussing the importance of establishing an agreement that would allow both sides to attend to the conservation issues that had taken a backseat to the allocation topic. Around midnight, Levi took off his elegant beadwork and presented it to me. The gesture was one of friendship and possibly due, in part, to the influence of the cumulative toasts we had made to each other. I did not want to jeopardize the friendship we had established or my position of impartiality among the other advisors, so I took the gift for the night and returned it the next day. Nevertheless, I was fond of Levi, who wanted desperately to find a solution that would reduce the Canadian interceptions of salmon bound for the Columbia River.

At another meeting held near Westport, Washington, which included a group of gillnet fishermen and representatives from the Quinault tribe, the debate between the two factions got out of hand. A gillnet fisherman and a tribal advisor were on the verge of a fist fight. I intervened by saying that we would have a better chance of settling the matter with our neighbors to the north if our delegations remained unscathed. The two potential combatants cooled down, and the session continued.

Perhaps, the strangest and most interesting interaction with the native Indian tribes occurred at a meeting in Anchorage. The Alaskan Indian groups that fished the Yukon had gathered to consider the Canadian concerns that the salmon runs bound for their waters were being decimated by Alaskans fishing the lower river areas. The U.S. tribal groups were not particularly concerned about the plight of their brothers in Canada, but were worried about interception of salmon by Japanese and other nations fishing with gill nets on the high seas. To characterize the nature of the debate with Canada, I noted that the Yukon River originated in Canadian territory. However, the Alaskan native groups would not acknowledge this; they simply remained silent on the matter. Perhaps history had made them cautious, and they did not want to acknowledge anything that might reduce their access to fish they had caught for centuries.

The first negotiation at which the two sides came close to an agreement was held in Lynnwood, just north of Seattle. We were close to agreement on most of the regional matters, but still had not settled on the division of fish runs entering the Fraser River. Since there was an increase in the number of salmon migrating into the river using the Inside Passage east of Vancouver Island, the United States was attempting to find a split that would apply to the overall Fraser run. Our delegation had agreed on a thirty-five percent U.S. take, but the Canadians were holding firm on a lower U.S. percentage. In an informal chat with Mike Shepard, I asked if he thought their team would go for a thirty-three percent U.S. take of the Fraser River pink and sockeye runs. He answered that this might be acceptable. Unfortunately, I could not sell my delegation on the split, largely because the northern Puget Sound non-Indian fishers and processors would not go below thirty-five percent. In an attempt to resolve the outstanding differences, the negotiations continued for almost twenty hours a day until exhaustion took over, and we finally agreed to meet in British Columbia within the next several months.

The failure to find a solution to the Fraser River division at the Lynnwood meeting was, in my opinion, a tragedy. This was because the U.S. leverage for an agreement that was reasonably equitable to the majority of the delegation would erode sharply over the next several years. Changes in the ocean environment would lead to an increased portion of the runs into the Fraser migrating down the east side of Vancouver Island. To make matters worse, the runs of silver and Chinook salmon originating in waters south of British Columbia would decline as a result of overfishing, environmental degradation of their fresh water habitat, and a lower survival rate in the ocean waters. Thus, the U.S. argument that there was a balance of interceptions would steadily lose credibility.

At the subsequent meeting in Vancouver, the two sides finally came closer. The declining state of important salmon runs in both countries had made the agreement of paramount importance. Reaching an accord required the United States to give up rights to several hundred thousand fish in the historical Fraser convention area and did not establish a new formula that took into account fish migrating to the river via the Inside Passage. The agreement also required the United States to set limits on interception in other areas and take steps to enhance salmon runs in the Puget Sound region and southeast Alaska.

Mike and I polled our delegations to see if we had the political support to move the agreement through the U.S. Congress and the Canadian parliament. All of the Alaskans, the involved states, the sports fishermen, and most of the tribal groups supported the intended settlement. However, Billy Frank, who had been a major force in promoting Indian fishing rights in the Pacific Northwest, was unwilling to commit support from the tribes he represented. I believe that he supported the proposed agreement, but Indian politics made it difficult for him to pledge the tribes' support prior to more in-depth talks with them. The only major advisors who would not support the proposed settlement were the northern Puget Sound non-Indian fishers and processors. Under the agreement, they took the major hit in terms of fish allocations, and their refusal to sign was understandable. Given the strong support from the states, the Alaskans, and many of the fishing groups, I felt it essential to move ahead with the tentative agreement before the U.S. equity issue worsened. In the fall of 1982, the two sides reached an agreement that had been sought for the better part of four decades. Mike and I signed a memorandum of agreement and went home to convince our delegations to get on board.

The joint memorandum of agreement was signed in late October, but the results of the November election scuttled any chance of getting the necessary political support to push it through Congress. Governor Eagan had been defeated in Alaska and the new governor, Sheffield, had strong support from the Alaskan trollers and other fishing groups. His constituents convinced him that the Alaskans did not fare well in the proposed agreement. Within a matter of days, all but one Alaskan advisor had withdrawn their support of the agreement, arguing that some of the wording changes had altered its character. In reality, the change in political leadership held out an opportunity for a better outcome for the fisheries groups in Alaska, particularly the offshore trollers, and their support quickly waned. The only Alaskan who signed the agreement stood by his commitment was Clem Tillion, a well-known Alaskan senator and fisherman who had dedicated his life to improving fishery management in Alaska.

In December 1982, I was asked to present the proposed treaty to the North Pacific Fishery Management Council, knowing that, without the Alaskans, there was no chance of its being supported and approved by the commission. Nonetheless, it was my responsibility to give the most convincing presentation I could muster. When I arrived, the meeting room was overflowing with participants and observers. I felt like a platoon leader who urged his men forward, only to look around and discover that he was facing the enemy alone.

I took a deep breath and began my presentation by noting that the agreement was based on principles our nations had endorsed at UNCLOS and that there was a critical need to address the conservation problems confronting both the United States and Canada. It was a somewhat emotional appeal that must have stirred many of the Alaskans. The director of the Alaska Department of Fish and Game complemented me and noted that he believed that, with some minor adjustments, the agreement could still be put together. Later I had a personal discussion with the newly elected governor, who suggested I stay on as head of the U.S. delegation. The commissioners, with the exception of Clem Tillion and my son Bob, who had recently joined the council, voted against accepting the agreement—including those from the southern states who had supported the proposal initially. I had to assume that the rest of the

Alaska delegation found a rational basis for switching their support. As for Clem, he remains on my list of all-time heroes.

I came to the conclusion that unless the U.S. Congress and administration became directly involved and decided which groups of U.S. fisheries would have to cut back on their interceptions, the chances of reaching an agreement acceptable to both sides were few or none. I resigned my position and strongly cautioned the Department of State that the matter could not be settled at the regional level because of the highly disparate and entrenched interests among the states and the fishery groups. Someone with no historically vested interest had to arbitrate an acceptable U.S. position.

Within weeks, a Washington-based lawyer was chosen to lead the U.S. delegation. Two years later, in 1984, the countries came to an accord that severely reduced the U.S. access to fish headed for the Fraser and moderately improved the Alaskan access to fish they had historically caught. I attempted to compare the character of the signed treaty signed with the solutions that Mike and I had proposed two years earlier. The annexes were almost identical, but the number game was different. As a result, some U.S. fishers were better off and some worse. The cuts imposed on Puget Sound fishermen in the final treaty never would have been accepted at the local level. The new negotiator had the necessary national political backing to reach an accord. The important issue was that a treaty was finally in place, and the two countries could move forward to conserve fishery resources of common interest.

The forging of the treaty did not end the bickering between the United States and Canada; the dance continues. Each year, we redraw the swords and replay the battle of numbers. Each side remains convinced that the other side is not living up to its commitments and that its ox is being gored. On the positive side, there is now an arena where the countries can continue the debate and give greater attention to rebuilding stocks of salmon that have declined over a number of years. From what I have been told, the character of the science has improved and the political influence on the outcome of studies has decreased on each side of the border.

On the Home Front

During the 1970s, both of our children graduated from college, Bob from the University of Washington and Sue from Central Washington College in Ellensburg. Bob had some difficulty in deciding what he wanted to major in but finally completed his degree in economics. Sue, on the other hand, had her heart set on being a special education teacher and never wavered from her target. Sue stayed glued to her books, sailed through her educational experience, and was blessed with good grades. Bob, on the other hand, enjoyed his play time and struggled to maintain good grades. During his senior year, his academic performance was further challenged by his falling in love with a young Irish girl who was a good friend of Sue's.

When Bob graduated in 1972, the war in Vietnam was still ongoing, and we weren't sure if he would end up overseas. Bob didn't consider it evil but was not eager to fight in a war caused so much conflict in our nation. He came home one day during his senior year and told us that he had joined the Washington State National Guard. Not long after that he was sent to Fort Ord, California, for one year of training in the army. He remained in the guard until 1976. Ruby and I were thankful that he did not have to go to Vietnam. After returning to Seattle, he went to work for a finance company and started dating Dolly. They tied the knot in 1972 and moved to Renton to start their married life.

At the finance company, Bob was assigned to loan collections. After about a year, he told me he hated the job and was going to look for other work. Around the same time, an old friend of mine, Harold Lokken, was looking forward to retiring from his job as head of the Fishermen's Vessel Owners Association. I told him about Bob, who had taken some courses in natural resource economics at the U.W. Bob met Harold for an interview and was hired as a trainee for the manager's job. Two years later, Harold retired, and Bob became the new manager of the association. He was the third Alverson to work in the field of marine fisheries.

The association was made up largely of halibut and sablefish (black cod) line vessels, which fished from Oregon to the Bering Sea. It was one of the oldest fishing associations along the Pacific coast and was dominated by Norwegian fishermen based in Washington and Alaska. Because of the successful management of the halibut stocks, the fishermen were highly supportive of the international commission and its scientists. The manager's job was to help sell the fish caught through auction, represent the fishermen before the International Pacific Halibut Commission and in the conservation and allocation debates of the two Pacific coast fishery management councils, and carry out a variety of other business tasks. The fishermen's vessels were a mixture of the older so-called "halibut schooners," with the house aft and the workspace midship, and seine vessels designed with the house forward and working space aft. These vessels, which ranged from about fifty to eighty feet in length, caught and sold most of the halibut from the Pacific coast. The catches varied, but in some years they were in the vicinity of forty million pounds. The Canadians also caught a considerable portion of the halibut allocated by the IPHC.

Sue completed her work at Central Washington College in 1977 and received her degree as a special education teacher. She was soon working for the Highline school district in the greater Seattle area and loved her profession. After visiting one of her classes, I thought to myself, "There is a job that requires a lot of love and commitment." Before the end of the decade, she would meet and marry Wayne Wilson, Dolly's cousin. At the time of their meeting, Wayne was in the army and did not get out until a year or two after their marriage. He then became an apprentice electrician and later a journeyman electrician. Because Wayne was still in the army and based at Fort Louis, they moved to an area just south of Tacoma, Washington. By the end of the seventies, Ruby and I were home alone.

Ruby and I continued to make occasional trips to California to visit our folks, my brother and his wife Meche, and numerous other relatives, but the visits became less frequent. We spent most of our spare time at our cabin in Grapeview, where both of us dug steamer clams and spent hours fishing salmon and sea-run cutthroat. It was not difficult to catch a salmon or several nice trout and then grill them down on the terrace next to the beach. Four of our neighbors in Normandy Park had built summer homes or cabins in the Grapeview area, so each year we would rotate our Fourth of July celebration from one beachfront home to another. At first the gatherings had about twenty people, but over time the parties would include three generations and forty or more people.

Somewhat more affluent during this period, we began making annual trips to Hawaii. We usually spent some time on Waikiki and then several days in Hilo, where we visited Buddy and Kay, who were still living in the house his parents had built when he was a boy. Every time we arrived in Hilo, Buddy was there to greet us and demanded the right to show us around any part of the island we wanted to visit. The visits became a ritual, and Buddy and Kay returned the favor by visiting us in Seattle several times.

One time in Hawaii, Ruby and I were attending a marine debris meeting in Kona and settled into a nice hotel in a first-floor room overlooking the ocean. We had just gotten into bed that night when Ruby complained that something was pulling her hair. I dismissed the complaint until I, too, felt my hair being pulled. I jumped up and turned on the light in time to see two mice racing for the door. We called the hotel management and told them that was no way to treat an "island boy." They responded by giving us the wedding suite for one night and then located us in a room well above the ground floor.

The following day Buddy wanted to bring a gift to us. He asked me how many wives were attending the conference, and I told him about sixteen. Sometime before our evening banquet and dance, Buddy arrived from Hilo with sixteen beautiful orchids. When the dancing began, I gave an orchid to each of the wives and kissed them on the cheek. I made one serious mistake, however, and mistakenly gave an orchid to a lovely young lady who, unfortunately, was not a member of our group. Her husband did not handle the accidental gift very well and started an immediate argument with his wife that seemed to last the entire evening. So much for my "aloha spirit."

CHAPTER XIV
NATURAL RESOURCES CONSULTANTS

By the time the reins of the U.S.-Canada negotiations in 1982 had passed to my successor, I had been retired from the National Marine Fisheries Service for two years. When my option for retirement came up in 1980, the decision to quit the service was no longer a question in my mind. By early 1979 I had considered remaining as director of the center, terminating my government service and joining an academic institution, or establishing a consulting service. Regarding the third option, I had in mind an organization that would provide commercial and recreational fishers, environmental groups, and state and federal governments alternative interpretations of the science produced by and utilized in natural resource management.

One thing was not subject to reconsideration and that was remaining with the NMFS. I had been the first director of the Seattle center and had spent almost ten years attempting to organize, structure, and attract quality scientists to the institution. This effort had taken me further away from the scientific arena and deeper into administration and government bureaucracy. Furthermore, the Washington office, which was not supposed to design and implement fieldwork at the center levels, was busy doing just that, eroding the director's leadership. This had made me jaded, and I questioned a process that stifled many of the scientists working at the program level. I had greatly enjoyed my stint with the BCF and NMFS, but if I stayed any longer, I would leave bitter. It was not time to retire, but it was time to move on while I still had a good feeling about the government agencies that I had served.

By the end of 1979, I had put together plans to incorporate a consulting partnership and establish it under the name Natural Resource Consultants (NRC). In searching for partners I was lucky in that Roy Jackson, a friend who had been the director for the United Nations Fishers Division of the FAO, recently had returned to Seattle. I told Roy my plans and he was elated about the idea. Al Pruter, who had recently left the service, had already committed to join NRC. We had three individuals who were well steeped in fisheries administration, science, and management, but no one versed in fisheries economics. We needed someone who was well known in the area and who had contacts with many of the industry and government entities working in the natural resource arena.

This led me to call James Crutchfield, one of the leading fishery economists in the world, close to retiring from his formal teaching schedule at the U.W. Excited about the opportunities for NRC, Jim also agreed to join the partnership. My last and perhaps best punch for getting the company off to a good start was to call Gerry Grinstine, who was working for one of the large legal firms in Seattle. Gerry became an advisor who provided business guidance and assisted in the acquisition of contracts during the company's early years.

It was a small group but one, we felt, with enough stature to attract clients. Ruby and I looked for office space and finally decided on a complex on Salmon Bay, near the old building where I had worked with the state Department of Fisheries. We advertised for a secretary and, by the time I reached my NMFS retirement date, we were ready to launch Natural Resource Consultants.

I ended my service with the Department of Commerce in early December 1979. I had started my government career with the old Bureau of Commercial Fisheries in 1951 and, with the exception of a short stint with the Washington Department of Fisheries, spent a little over twenty-five years with the departments of Interior and Commerce. My retirement pay would be based on twenty-nine years of service, which included my years in the military. During my career, I had advanced to the top position in the BCF and had spent more than ten years as director of the Northwest and Alaska Fisheries Center under the National

Oceanographic and Atmospheric Administration. I had received the Gold Medal for exceptional science at both Interior and Commerce as well as the highest science award at NOAA.

From my perspective, I was leaving on good terms. Under the direction of Heater Heyamoto, the staff at the center threw me a great retirement party at the old Olympic Hotel, newly named the Four Seasons. As a going-away gift, they gave Ruby and me a trip to the Hawaiian Islands. Gerry Grinstine served as master of ceremonies, and Senator Magnuson's wife sat at the head table. The celebration included a number of friends, business and science associates, and members of the Seattle fishing industry.

When NRC opened for business in January 1980, I remained under contract with the Department of State as head of the U.S. team negotiating with Canada. However, this took up only about twenty percent of my time. To pay our overhead, we needed to develop a range of new job opportunities for the company. In the fall of 1979, prior to my departure from the NMFS, a number of the top Seattle crab fishermen who were active in the Bering Sea and the Gulf of Alaska paid me a visit. They were concerned that no one outside of government was tracking the science and management decisions that drove the crab quotas for the area. Even four years after the extended jurisdiction act, there had been little development in the major ground fisheries of the region.

Chief among the group was Arnie Madsen, a former classmate of mine at the U.W. Arnie had followed in his family's footsteps and entered the commercial fisheries business after graduation. He had done very well, as had his cronies, and owned one of the larger crab vessels. Aware that I planned to start a consulting firm, Arnie and the others asked me to become a science advisor to the group and help find ways to promote growth of the U.S. fisheries in the eastern Pacific region.

Unfortunately, Arnie was lost at sea later that year while he was crossing the Gulf of Alaska on the way to the Bering Sea. There was no distress call, no calls to anyone reporting trouble. The vessel just disappeared one night, and no traces of it or the crew were ever found. Soon after we had opened our doors, however, a group of Arnie's friends—including Konrade Uri, Sam Hjelle, Dennis Petersen, Kenneth Petersen, Stan Hovak, Magne Ness, Rudy Peterson, Einar Peterson, Reider Tyness, and Francis Miller—showed up at our Salmon Bay office.

All except Francis were Norwegians and were frequently referred to as part of the Ballard Mafia, probably because of their great importance to the region's fishing industry and their political influence. I had met a good number of them, including Einar, Magne, Konrade, and Francis, when they were trawl fishermen who made their living off the Washington and Oregon coasts. At that point, they were all barely scraping out a living, but their financial status changed after many of them helped pioneer what would become Alaska's second gold rush. The development of the Alaskan king crab fishery and its ultimate value in the market resulted in millions of dollars being poured into the construction of modern fishing vessels ranging from about 90 to 120 feet. During the 1970s, the crab fleet expanded rapidly, as did the landing of king crab. Those who were there at the right time and place did exceptionally well, and those who invested their income with care built small dynasties. The less frugal contributed to a booming bar and prostitute business that swelled the economy of Kodiak and other Alaskan fishing communities.

Our group of visitors was concerned about the future of the Alaskan crab fishery and about trends in the population of the highly sought-after king crab. They asked if I would look into the population models produced by the Alaska Department of Fish and Game and NMFS. They also believed that the 1977 extended jurisdiction act had not stimulated much, if any, growth in the mammoth ground fisheries of the region. With the exception of the crab fisheries, U.S. fishermen had been unable to displace the large foreign trawl and line fisheries that swarmed over the continental shelf areas off Alaska and all the way south to California.

The talent and experience of the group sitting around the table was unparalleled in the north Pacific region. They were what the fishing industry referred to as "highliners," fishermen who were productive, hard working, successful, and experienced. Almost all of them had participated in a wide range of fisheries, years of experience, and a reputation as knowledgeable seamen and were successful citizens. I did not want NRC to be an advocate for any fishing or conservation group, but I was interested in having a group of respected fishermen establish a nonprofit organization dedicated to: 1) improving the status of the north Pacific fisherman; 2) promoting conservation based on good science; and 3) supporting studies on the consequences of fishing on the exploited stocks and their environment.

After an afternoon of chatting, we agreed to establish the Highliners Association with the goals noted above. Each member agreed to put $4,000 biannually into the association. The money would be spent on projects jointly agreed upon by the members. This was our second contract, and it gave initial support to our biologically inclined staff. We still needed a project that would involve Jim. With Gerry Grinstein's help, we were fortunate to be introduced, to the staff of the Seattle Port Authority. The Port Authority wanted an evaluation of the potential infrastructure needs of the Washington-based fishing fleet in light of the passage of the FCMA.

Several months after establishing the NRC partnership, we had about $40,000 worth of work to keep us busy. By March, NRC was up to full speed, which made me rather pleased. Many of my friends in the service and industry had said that an independent consulting group could not make it in the fishery science world. But our critics didn't know that we had one great advantage over our competitors. Everyone on the staff was retired and had a good independent income. None of us would starve if we didn't make it and we did not have to take on work that would compromise our commitment to producing objective reports. We even established a policy that our clients could not use our reports in advocacy situations unless the full NRC report was made available. Clients were given, however, the right to bury NRC reports they had sponsored if they felt the results did not support their positions, something that happened more than once.

During its early stages of development, NRC had difficulty holding on to a good office secretary/manager, but by the end of the decade we hired Sharon Parks, who was wonderful. She became our bookkeeper, office manager, and all-around support for the staff and owners. After several years, we brought on board Scott Goodman, an excellent computer technician. We also hired Steve Hughes, Jeff June, and several other individuals, who worked with the group for a year or two before becoming NRC owners. This move broadened our scientific skills, as well as adding an individual (Steve) who was well versed in the technical and operational aspects of the north Pacific fishing industry.

NRC remained a viable enterprise well into the twenty-first century. It provided consultations to thousands of clients, including governments, law firms, the fishing industry, the U.S. military, environmental groups, fishery agencies, communication companies, sports fishing groups and the World Bank. Its history could fill several volumes, but I will confine my commentary to some of the more interesting events that involved NRC.

The Highliners

The Highliner group had a great number of attributes in common. Einar and Sam had immigrated to the United States from Norway, and the others were the sons of Norwegian families that had immigrated in the early part of the twentieth century. They were either brought up in fishing families or had begun working in the fishing industries of the Pacific Northwest as teens or young adults. All had started working on deck and worked their way up to owners and captain fishing vessels through hard work and involvement in a variety of fishing activities. Finally, most had made only modest financial gains until they had joined the Alaskan king crab fishery, which made them wealthy. One might assume that these similar backgrounds would have led them to a consensus on what was needed to improve their status. In fact, the aggressive and hard-working traits that had made them highly successful fishermen also spawned individuals with strong and independent views.

Developing a Highliners program acceptable to the group was not an easy task and led to lengthy and healthy internal arguments. It finally dawned on me that the only issue upon which there was complete agreement was that they wanted the fisheries of the north Pacific to be "Americanized." This meant that they wanted the U.S. industry to catch and process all of the fish taken within the 200-mile limit off the U.S. coast. It thus made sense that the first major program that NRC should undertake for the Highliners was the Americanization of the industry.

This led the NRC staff to examine and evaluate factors that had made the crab vessels highly successful. The reason was obvious and well known to the industry. The U.S. industry simply could not catch and process ground fish and other products that were traded within the global fish market. It had little or nothing to do with the lack of American know-how and everything to do with economics. Despite this rather discouraging factor, there was some evidence suggesting that the U.S. fishermen were very competitive when it came to harvesting bottom fish and their catch rates were impressive but not when it came to their overall costs.

In 1978 an enterprising fisherman, Barry Fisher, and a cold-storage operator, Jim Talbot, owner of Bellingham Cold Storage, put together a joint venture with the help of Wally Pereyra, who had worked with me at the center. This venture involved several U.S. catcher boats catching flounders and delivering their catch to Soviet processor vessels. This was not a simple operation as it involved working in heavy weather and then bringing in the nets, often filled with tons of fish, to the stern of the vessel. The catch in the cod end (the terminal end of the trawl net in which the catch was contained) then was transferred to the Soviet ship. Although the original venture struggled with certain operational factors, most of the difficulties had been overcome by the end of the season. At the end of the second year, it was apparent that the U.S. fishermen could harvest and sell their catch at about four cents a pound and still make money. Their catch rates were so high that they could afford to fish at the low prices offered by the Soviets. The success of the Soviet operation led us to examine similar arrangements with other foreign operators in the north Pacific.

In 1980, the Japanese were still the largest harvesters of ground fish along the North American coast, catching and landing in excess of a million tons a year. Any significant growth in the U.S. joint-venture fishery would have to engage and involve the Japanese industry. The recently modernized crab vessels were the only fleet that could undertake the joint venture, but they already were making lots of money and, at the time, were not overly interested in fishing ground fish at four cents per pound. However, one notable resource trend led to an aggressive Highliner program to initiate joint-venture operations in the north Pacific region.

In mid-1979, scientists at the NMFS and the Alaska Department of Food and Game (ADF&G) noted that, regardless of the high abundance of king crab in the Bering Sea, the levels of young crab entering the fishery (recruitment) were declining drastically. At the time, it did not appear that the decline was the result of fishing but, regardless of the cause, the result would lead to a brutal situation. The multimillion-dollar crab industry faced disaster. The extensive fishable king crab population in the Bering Sea would decline rapidly, and within a few years the vessels and processors would face economic ruin. When this sunk in, it galvanized the Highliners, and they were driven to find a profitable means to fish the Alaska ground stocks.

NRC offered the Highliners two approaches to accelerate and promote the development of the U.S. fishing industry in the Pacific Northwest. First, we could examine the FCMA to see if there were any amendments that would assist the industry's growth. Second, the Highliners could hold an industry-to-industry meeting with the Japanese. The Japanese had the biggest and most financially solid fishing industry among the nations without strong socialized governments. The meetings would be designed to convince the Japanese industries that U.S. catcher trawlers were so efficient that they would eliminate the need for Japanese catcher vessels. After a morning of debate, the group suggested we move forward on both fronts simultaneously.

We could not move the Japanese industry without a "stick and carrot" approach, so several of us began to kick around ideas about amendments to the FCMA that could stimulate joint-venture action in the Japanese industry. In the late 1970s, the industry had persuaded Congress to pass an amendment to the FCMA that would allocate a "total allowable catch" (established catch levels that took conservation principles into account) by three categories. The highest priority was given to vessels that caught and delivered their catch to U.S. processors. The second priority went to U.S. vessels that delivered their catch to foreign processors, and what was left over went to foreign fishing and processing vessels. Since there were almost no U.S. ground-fish processors in Alaska, with the exception of those receiving halibut and some black cod, it seemed reasonable to begin by promoting joint-venture operations in ground fish with the Japanese and Koreans.

The U.S.-Soviet success in joint ventures seemed to provide a partial answer. It was apparent that with the passage of the 200-mile fishery zone all the foreign participants wanted to retain or even increase their access to the extensive ground-fish resources off the coasts of Alaska, Washington, Oregon, and northern California. To take advantage of the three-tiered priority amendment, we needed to offer those foreign fishing industries that initiated joint ventures (e.g., the Soviets), or helped give U.S. shore-side processors priority, access to the total allowable level of foreign fishing (TALFF). With this in mind, we approached Senator Magnuson's office with a proposed amendment incorporating this feature.

Most of the Highliner group had little experience in lobbying on Capitol Hill, but it was a good time to learn. We set off to Washington with the majority of our Highliners and made calls on Representatives Dicks, Foley, and Pelly from Washington and Senators Magnuson, Stevens, Brough, and Kane. Although some of the Highliner members had difficulty in expressing themselves, I suspect that their strong Norwegian accents made their comments even more credible. Einar Pederson, who was the oldest member and often served as a spokesman for the group, didn't say a lot, but when he did the representatives and senators listened carefully. I recall one meeting in Senator Magnuson's office when we were pushing for the passage of the amendment. After listening to the group's presentation, the senator began to meander, talking about fishing issues he had dealt with during his years in Washington. Einar waited patiently for the senator to conclude his extended remarks, and said with a strong Norwegian accent, "Ya, Maggie, that's all goud, but ve want to know if you're going to take care of our amendment?" Maggie just smiled and said, "Einar, don't worry."

We had no idea how the proposed amendment would be received. The state department was still generating a lot of pressure to slowly phase out foreign fishing. We were thus elated when the amendment slid through both houses of Congress with little opposition. Somewhere along the line, probably in Magnuson's office, someone tagged the change the "Fish and Chips Amendment." If a nation had historically fished in the region and helped the growth of the U.S. industry, it would continue to receive fish from whatever TALFF was available.

This was just the ticket to gain entrance into the large Japanese market. With the amendment in hand, the Highliners began to organize the first industry-to-industry meeting with the Japanese. We were aware that our efforts were not appreciated by some elements in NMFS or the state department, but we were not sure why diplomats needed to be involved in the international fish business. The Highliner members contacted members of the crab and trawl fleets with a special interest in promoting joint ventures, which turned out to be between thirty and forty vessel owners. The next job was to bring on board a high-profile industry member to lead our planned negotiations. We settled on Ron Jensen, who had headed several major fish companies and been involved in the crab, salmon, and bottom-fish industries. Ron was about six feet two, handsome, and an excellent speaker who had been engaged in industry price negotiations for almost two decades. Well aware of the Soviet joint-venture operation and the Fish and Chips Amendment, the Japanese industry responded quickly and positively to the idea of a meeting. By late 1981, we had picked a date and settled on Seattle for the first formal session.

Other elements of the fishing industry had a small interest in the joint ventures, but it was not a high priority for them because the rapidly declining king crab industry had not yet had an impact on a significant number of the larger vessels. We started off our industry meetings in a conference room at a Seattle hotel. Each side began with short presentations about its concerns regarding the establishment of a U.S.-Japanese joint venture involving the massive Japanese pollack fishery in the Bering Sea. Despite the success of the project with the Soviets, the Japanese made it clear that they did not believe the U.S. catcher fleet had the experience and operational know-how to effectively carry out a joint venture. Therefore, the first two days of our meetings were enmeshed in debates about whether the U.S. industry could perform and if there was an adequate fleet of the larger (100–128 feet) catcher boats. Both sides continued to recycle their arguments with little progress, and many in our delegation thought the negotiations were headed for the scrap heap.

Both Ron and I had previously been in a number of negotiations with the Japanese, however, and we felt that concessions were unlikely until late in the negotiations. On the last day, the Japanese proposed a joint-venture quota amounting to almost 200,000 metric tons. This meant that over a half-dozen catcher boats, largely out of the crab fleet, could begin catching bottom fish and delivering it to Japanese at-sea processors. From the perspective of our industry, this was a major step in the Americanization of the Bering Sea trawl fisheries. Some Congressional elements, however, were not all that happy over the outcome. I considered this attitude largely sour grapes because they were not involved.

The 1982 meeting was the first in a series that extended into the middle of the decade. Although the meetings put the two sides in a highly controversial and adversarial situation, friendships and business relationships emerged over time that helped to foster the American industry. At the onset of the U.S.-Japanese meetings, the U.S. delegation was composed of only catcher owners, but the group later included both at-sea and shore processors. They greatly strengthened our negotiating hand. By the middle of the decade, these negotiations, and similar developments with the South Koreans, resulted in the phasing out of all the foreign catcher fisheries in the Bering Sea. Within five years, all the bottom fish of the region were harvested by U.S. catcher vessels, and a U.S. at-sea and on-shore processing industry was beginning to emerge. The coalition of fishermen and processors, although fragile at times, held together until it was no longer needed.

The lagging development of the shore processing industry and the delayed growth of U.S. at-sea processors were not due, as many suggested, to a lack of interest on the part of the historical salmon and crab operators, but to the economics of fishing and processing. Domestic labor costs and on-hand technology did not allow our industry to produce products competitive with foreign at-sea processing at that time. If American at-sea and on-shore processors were to access the large pollack market in Japan, they would have to import processing equipment and build a capacity to produce surimi (fish paste). Surimi served as the building material for a great variety of Japanese fish products and utilized over a million tons of pollack a year. The scope and character of the large Japanese surimi business was not well know to fishermen, so the Highliners asked me to go to Japan and glean all the information I could about surimi production, products, prices, and sales. After visiting all the major producers, I returned to Seattle and compiled my findings in a 120-page book.

Sara Hemphill, once a student in a class I taught at the U.W., had become a member of the North Pacific Fisheries Management Council. She read the report and suggested organizing a meeting in Anchorage to present the findings, along with other technical information collected by various industry members and academics. I agreed and the meeting was organized, largely by Sara.

When I arrived at the Sheraton Hotel in Anchorage, Sara said that they had arranged for a special suite for me. The lavish suite had a reception room and a large bedroom and bathroom. I was sitting on the bed when a tall, young blonde in a short skirt came in, looked at me for a few seconds, and said, "I am Bonnie," and asked if I wanted any company for the evening. I wasn't sure how she entered the room, because I had closed the door. I thanked her and told her no, and then returned downstairs to join Jim Crutchfield, who had come with me to Anchorage, and several other friends. I told them the story and said that I was sure that Sara had arranged for the call girl. They all laughed, but didn't believe my tale. A little later Jim went to his room for few minutes, and when he returned he said, "You will never believe who I met in the elevator—Bonnie, and she asked me the same question." It turned out that Sara was not the instigator of these propositions, and I never found out how Bonnie gained entrance to my room. The meeting itself was well attended and educational to its participants.

During the mid-1980s, the gains made in the industry-to-industry meetings, including a Japanese concession to buy from U.S. processors, were coupled with a significant rise in world prices for cods, hakes, and flounders. The U.S. fish processors came out of the blocks running. By the end of the decade, they had succeeded in phasing out all foreign processing and, with the three-tiered priority amendment in place, also displaced the over-the-side, joint-venture fisheries that had set the stage for Americanization of the region's fisheries. A multibillion-dollar industry had come on line in just about a decade. Its ability to efficiently catch and process the extensive bottom fish of the region surprised other sectors of the industry, as well as some government officials, but not the involved fishermen, who always were convinced that they were among the most capable and competent harvesters in the world.

As for the Highliners, all of the original members took part in either the joint-venture operations or in the at-sea processor industry. All played a significant role in the almost unbelievable growth of U.S. grounds fishery off Alaska and later off the contiguous Pacific states during the 1980s and 1990s. Konrade Uri, Maggie Ness, Einar Pederson, Sam Hjelle, Stan Hovak, and Reider Tyness joined in the early joint-venture fisheries, while Rudy Peterson, Francis Miller, and Dennis and Kenneth Petersen entered the growing at-sea processing sector.

The first and most important goal of the Highliners had been secured. After its formation in 1980, several other prominent fishermen joined the group, including Barry Fisher, a Harvard graduate, and Michael Jones, both fishing out of Newport, Oregon, and Wilhelm (Hardtack) Jensen and Robert Watson from the Seattle area. Several successful businessmen—not fishermen—from the industry including Ron Jensen, Hugh Riley, and Frank Stewart also became members. Each was selected because of his contribution to the Americanization process and stature in the industry.

Throughout the 1980s, the group worked as a cohesive organization, holding most of its annual meetings in Hawaii. But by the end of the 1980s, following the Americanization of the Pacific coast and Alaskan fisheries, the glue that held the group together began to weaken. After the American fishermen eliminated the foreign fleets from U.S. coasts, the focus switched from the "foreign devils" to the different U.S. gear types fishing the resources of the region. The new battle lines were between offshore processors and on-shore processors, catcher trawlers and at-sea processors, trawlers and line fishermen, crab fishermen and trawlers. Each group worked feverishly to gain greater access to the resources that over the past ten years had bumped up against conservation quotas. The quotas limited each group's access to resources that they had only begun to fish a decade earlier. Suddenly the playing field for the Highliners became much narrower as members found themselves in competition with each other. This set off a certain internal discord, which forced a change in direction for the group. I urged the group to focus on growing conservation issues instead of the Americanization of the fisheries.

One of my early expectations was that the NRC staff would influence the Highliners to take an active role in promoting conservation issues. The group agreed not to play politics with the quotas established by the scientists or the conservation regulations proposed by the council's scientists. A new goal focused on the growing marine debris problem. The Highliners became instrumental in the formation of the first international fishermen's conference dedicated to dealing with the marine debris associated with fisheries. Held in Kona, Hawaii, the conference brought together heads of fishing organizations, scientists, and fishery managers, from Japan, Taiwan, South Korea, the United States, and Canada.

Marine debris in the oceans was a growing problem resulting from materials lost from merchant ships, lost plastic packaging materials, plastic items discarded by beach picnickers, a variety of synthetic gill and trawl nets, glass and plastic floats, lost fishing pots, etc. The fishing industry was without question one of the major contributors to the increased levels of marine debris, which was carried into all of the world's major oceans, including the Arctic and Antarctic regions. Much of the marine debris resulting from fishing was the result of lost nets or portions of nets, fishing lines, damaged fishing gear, and packaging materials used by at-sea processors. The accumulation of plastic nets and other materials in the oceans was taking a toll on marine birds, mammals, and turtles, as well as various fish and invertebrate species.

The meeting focused on the quantities of debris from fishing vessels, the distribution of debris in the world's oceans, its resulting damage to sea life, and potential solutions. In a little over three days, conference members agreed that they were a significant part of the problem and should work together to find solutions. This required the industry to stop discarding non-biodegradable materials, support the international effort to ban the discarding of such materials at sea, and finally to agree on a code of ethics that would significantly reduce losses of gear and other debris from fishing vessels.

With the passage of the international laws prohibiting dumping of non-degradable materials at sea and the fishing industry's efforts to establish unloading areas for marine debris, the problem declined over time. Nonetheless, significant amounts of debris continue to be discarded into the world's oceans. In addition to the fishing industry, offenders include broad sectors of the shipping industry, recreational boaters, beachgoers, and waste draining into the oceans from rivers. One might say, all of us.

During the early 1980s, the capture and discarding of species having little economic value and species prohibited from landing by law became a significant social, economic, and conservation issue. There was the matter of waste. In some fisheries, the weight of discarded fish exceeded the weight of the fish landed for sale. In some shrimp trawl fisheries, the discards were over ten times the weight of the landed catch. Furthermore, some discarded species were of value to other fisheries and others were marine mammals, birds and turtles, some of which were threatened or endangered. The extent of the problem, however, was just emerging.

Because of the discarding problems they had confronted in the Bering Sea and elsewhere, the Highliners were very conversant with the issue. As a result, NRC recommended that they take a leadership role in educating other sectors of the industry on the nature and extent of the problem and potential solutions. The suggestion led to the first National Fishing Industry Conference on discards in marine fisheries.

Working with the Highliners group had been very gratifying for the NRC, but by mid-1990s the group had run its course. A number of the original members had resigned or died, and the factors that splintered the group had continued to grow, so the organization was disbanded. Still, NRC's volume and scope of work continued to increase. Our early successes led to requests for assistance from a broad sector of the fishing industry; banks; state, tribal and federal government agencies; a surprising array of international organizations; and ultimately various conservation groups. We even surprised ourselves and, although there were some slow periods, we kept busy for the most part. The partnership, which was later changed to a corporation, ended in the black almost every year.

International Efforts

During the twenty-three years that I worked with the NRC, much of my work originated overseas. A great deal of my time was spent on planes going to and from assignments, including trips to the South Pacific, Korea, Japan, the Middle East, Africa, much of Europe, Newfoundland, and my second home, the Hawaiian Islands. The work covered a great range of activities, including establishing national development and management programs, education, talks on the status of world fisheries, and conservation approaches to rational fish use.

Samoa

Sometime in the early 1980s, I got a call from the governor of American Samoa asking me to review their plans to promote fishing efforts by the Polynesian peoples. I was aware that there was a rather extensive high-seas tuna fishing and processing operation located in Pago Pago, and it was a mystery to me why they needed to stimulate local fishing activity. Despite this apparent paradox, I flew to Hawaii and then on to American Samoa. The flight to Pago Pago left Honolulu about 1:00 a.m. and arrived in Samoa early in the morning. When traveling on long flights I always asked for an aisle seat, but this time I had a center seat a few rows behind the first-class section. The seats on either side of me were soon occupied by two very large Samoans, who overflowed across the armrests into my space. It was going to be a long uncomfortable ride.

But I was wrong. It was a snug fit, certainly, but the company compensated for that. The man next to the window turned out to be a high talking chief of a village not too far from Pago Pago, and his wife occupied the aisle seat. They were delightful and very gregarious. I never knew the difference in stature between a high talking chief and the number one chief of a village or clan, and at the time it was of no real importance. They kept me in stitches during the entire trip and, when we weren't conversing, the wife was passing me something to eat from a large bag of food. When we arrived in Pago Pago, the high talking chief, whose name I have since forgotten, invited me to visit their village the next day.

I joined them the next evening in their village, about twenty miles from Pago Pago. We had a great time and, when the chief found out that I had lived in Hilo as a boy and loved raw fish, my stock skyrocketed. The chief had a number of friends in Hilo, and we enjoyed some tuna sashimi, which we dipped in lime juice, along with other Samoan dishes. He also gave me some advice regarding my mission in Samoa. He noted that there had been attempts to place local fishermen on the high-seas fishing vessels, but that this ran against their culture. It was a man's place to be at home with his family during the evenings, not off fishing, even if the pay was relatively good. Samoans needed to go back to fishing the local reefs and near-shore fishing grounds, he said, so that they could be with their families by nightfall. Working in the canneries was not a problem, but going to sea for a month or more ran against the grain of family life. After dinner, we went down to the beach to look at the village's collection of outrigger canoes and other small craft.

The next morning I met with the governor and his fisheries scientist, who had a good understanding of the local fisheries and the cultural preferences of Samoans. The government had a research vessel, about seventy feet in length, which he had been using to evaluate the availability of tunas and other pelagic species close to the island. He told me that it was often relatively easy to take good quantities of several tunas or tuna-like species in the waters relatively close to shore and asked if I would like to go out for the day and see what was available. Within an hour we were on our way out of Pago Pago harbor and trolling off the island's coast. We caught several mahi mahi, but it was soon apparent that the real goal of the fishers was to snag a marlin, which was relatively abundant off Samoa. We soon had a marlin on line, and the scientist spent the better part of an hour playing the fish, which made several spectacular leaps out of the water before it was landed. I thought the marlin would be taken ashore and photographed, but the crew was more interested in cutting it up so they could feast on raw fish. Instead of flavoring the fish with soy, as is done in Japan, they dipped the raw fish in lime juice. I joined the crew in the feast.

Several hours later we hooked another marlin; from its initial jumps, we thought it was somewhat larger than our first fish. I was invited to attempt to land the fish and sat in a chair on the stern of the vessel, which had a special holder for the rod. The hooked marlin made numerous jumps clearing the sea surface by four to six feet, spinning and twisting to free itself from the hook. It was a warm humid day and, after a half-hour, I realized that it was going to be a long contest. I was already sweating profusely and getting tired. The battle went on for at least another forty minutes before the magnificent fish was at the stern of the vessel and could be gaffed. I never got a picture of the beast as the crew almost immediately converted it into sashimi. I must admit a certain thrill in the contest, but at the end I was completely worn out. I came to the conclusion that I would never attempt to land another marlin. I felt absolutely done in by the contest, but more important was the dramatic and impressive attempt of the beautiful fish to free itself from the predator at the other end of the line.

On the fourth day of our visit, the scientist wanted me to go with him to Western Samoa (now Samoa) to see what the islanders were doing in the way of fisheries. Western Samoa is made up of two large islands, with the capital, Apia, on the one to the south. We jumped on a two-engine prop plane and made the short flight, landing on a runway surrounded by grassy fields edged by dense stands of coconut trees. While Western Samoa was a sovereign country and a member of the United Nations, it had operational ties to New Zealand, like a number of South Pacific nations. At that time, the culture of the island had not been significantly impacted by tourism; much of the population lived like their ancestors had for centuries.

The road from the airport cut across the island through large groves of coconut trees until it arrived at the ocean. Along the beach were a number of homes with grass roofs, completely open sides, and cooking facilities outside. People probably rolled down "tapa" mats to shelter the interiors during rainy weather. Outrigger canoes sat on the beach outside of most of the houses. I learned that the natives mainly fished and worked in their taro patches, although some worked in local businesses. There was a thriving subsistence fishery, but little in the way of commercial fisheries. We finally came to Apia and were put up in one of the tourist hotels, which, as I remember, was called Aggie Macs. There were no menus in the hotel restaurant, but it served delicious family-style dinners. The dining area was in a large building that included the check-in area. The rooms were grass-covered cabanas that were scattered in a grassy area with flowers and trees. At mealtimes, everyone walked in from their sleeping quarters to the main building and sat together at one long table.

The head of the fisheries department wanted to talk with me, and I assumed that he was interested in developing commercial fisheries around the islands. In fact, he was interested in talking about the UN law of the sea conference. We spent the better part of the morning going over the various fisheries proposals that were before the conference. Following our tête-à-tête, the scientist and I returned to Pago Pago. Before leaving American Samoa, I chatted with the governor about the potential of small-scale fishery development in the islands and the problem of waste dumped into the bay by the two large tuna canneries. Tuna waste in the harbor attracted sharks, which were a potential threat to divers, including tourists, who frequented a nearby reef.

The Cook Islands

By the late 1980s, I had traveled extensively to various parts of the world, but one of my greatest ventures was yet to come. Sometime after my trip to Samoa, I received a message from the Cook Islands inviting me to Rarotonga to review the island's fisheries and draft a plan on their development and management. The Cook Islands lie just east of Samoa and constitute a group of eight islands stretching from about ten to twenty degrees south of the equator. Two are volcanic islands, and the other six are coral atolls. Rarotonga has the largest population and is the seat of government. The islands are all inhabited by Polynesians. I was quick to respond to the request; having grown up in Hilo, I had many fond memories of the Polynesian people. Getting there was not an easy task as only two major airlines, New Zealand and Hawaiian Air, served the Cook Islands.

I was excited by the opportunity because the islands were not part of the tourist mainstream. Several friends told me that the atolls were gorgeous. I also knew that "Tap" Prior, the son of a top executive at Pan American Airlines, had moved to Rarotonga and was developing marine aquaculture for species that could be exported to the lucrative Japanese sushi market. Tap had operated the well-known Sea Life Park on the southeast end of Oahu. I got to know him at a meeting in Hana, Maui, with several members of NOAA, including Bob Able, who directed the NOAA Sea Grant Program. He had married the head trainer for dolphins and other marine mammals at Sea Life Park, but they had since gone their separate ways.

My invitation to the islands came from a former top-level director of the Canadian Department of Fisheries, who was now chief advisor on fisheries to the prime minister of the Cook Islands. The two of us had also met during my work with NOAA. I flew into the islands from Tahiti in the early morning, and I can remember peering out the window looking for the lights of the island, as the sun had not yet risen. We were late, and I thought to myself, "I hope all the navigation aids are working" because Rarotonga is a small mountain that breaches the surface of a very large ocean. Soon I saw lights lining the shore of a small island, and within fifteen minutes we were on the ground.

My hotel, the Edgewater, was on the beach. It had a nice view of a large area inside a reef several hundred yards offshore. Although somewhat beat from the long, overnight flight, I met the prime minister's advisor, and we went into the city, which consisted of a few restaurants, grocery stores, bars, and government buildings. The first thing that caught my eye was that almost all of the local girls wore flower leis around their heads, just like in the movies. The natives were shorter than the Polynesians in the Hawaiian Islands or Samoa, and they looked sturdy. After an hour's discussion, my host asked if I would like to take a ride around Rarotonga. To my surprise, it took only a little over an hour to circle the island, including a couple of short stops to view the sites. Unlike French Polynesia or the Hawaiian Islands, Rarotonga does not have spectacular mountains. Although a small ridge of hills and low mountains form the center of the island, it does not possess the beauty of some of the other islands in the South Pacific.

But it did have something that was very attractive to me. Chickens, and occasionally pigs, roamed freely, and the homes were scattered among coconut trees that fringed the island. There were only a few hotels, so the area was free of the tourist hordes that trample the Hawaiian Islands. The island and its fauna reminded me a great deal of Hilo, particularly the land-grant area next to my boyhood home.

Back at my hotel by early afternoon, I tried to get some sleep. The room was not air conditioned, however, and the temperature was about ninety degrees with a high degree of humidity, so I mainly rested and attempted to keep cool. Later in the afternoon, Tap Prior called, asking if I would go to church with him the next morning. I was not a steady churchgoer then, but observing some of the local religious activities seemed a lot more interesting than staying in the hotel. That evening, I ventured out to dinner at a popular bar and dining spot owned by a woman I had met on the plane. The place was filled with locals sitting at the bar, but unfortunately the owner was not feeling well and had not come in. I sat in the back room and ordered a fish dinner. Before long I heard a ruckus at the bar, and the management had to intervene in the brawl. I finished my dinner and left. That first night, the temperature didn't drop much, there was little air in the room, and I struggled to get even a few hours of sleep.

The next morning, I heard roosters crowing and looked out over the reef to a deep blue ocean. I had breakfast—fresh pineapple, papaya, tangerines, cereal, and spectacular omelets made to order—down at the water's edge near the swimming pool. At one end of the serving table, there was a continuous rolling toaster along with a choice of many juices. Most of the guests in the hotel appeared to be from New Zealand, although a few were from Germany. I didn't meet any fellow Americans, but there may have been several in the hotel.

Tap picked me up around 10:00 a.m. I did not know that it was "Missionaries Day," which the churches marked with a special celebration. Most of the churches, at least those of the same denomination, gathered in one location and celebrated together. On Missionary Day, too many people gathered to hold the service in the church, so it was held on an outdoor stage. There must have been a crowd of 300 to 400 seated in chairs, and all of the women wore flowered leis around their heads. We took a seat in the second row, below the middle of the stage and listened to the preacher for about twenty minutes. Then the scenario changed. Each church group in attendance gathered at the bottom of the hill near the church, and then proceeded to walk up the road toward the stage, singing a song. The procession included young and old, men, women, and children. The singing was spectacular. It was hard to believe that the entire flock could retain such harmony. It sent a shiver down my spine.

When each group reached the stage, they finished their song and then put on a play. The second group did the story of Adam and Eve. They had selected a handsome young Cook Islander as Adam and a beautiful young lady as Eve. During the play, the young lady made seductive and suggestive moves to the young man, and the crowd cheered and howled. After several Bible plays, which took up almost two hours, young women with trays came down the aisles passing out fruit, including bananas, pineapple, mangos, and guavas. A few minutes later, younger women came by with trays filled with a variety of meat, including chicken and pork. It was a different service than I had ever experienced, but I enjoyed every moment of the pageantry and the Polynesian festival.

After about three hours, Tap said, "Let's go to the rest area." This was a series of tents set up where one could get an ice cream cone, hot dogs, and other foods. The area was crowded with young children and mothers, as well as others taking a break from the service. It wasn't until we were having some tropical fruit juice that Tap let me know that the celebration would last most of the day. I thought to myself, these people know how to enjoy their spiritual ties. As I was scheduled to fly out early the next morning to Aitutaki, an island about an hour by air to the north of Rarotonga, we left long before the festival concluded. I had been treated to an event that will remain with me for the rest of my life. The second night was somewhat cooler than the first, as a gentle breeze was blowing in across the water.

My employer wanted me to go to Aitutaki, where local scientists were attempting to seed its lagoon with giant clams. At one time, giant clams had been plentiful in the Cook Islands, but they had been fished out over the last several decades. In addition, I was asked to discuss the nature of the region's fisheries with fishery council members. Our group left Rarotonga the next morning on a two-engine turbo prop plane. Aitutaki is probably no more than five miles across with a relatively small population. There are several tourist hotels. Most visitors were interested in the expansive lagoon, which was surrounded by a coral reef that extended from the south end of the island. The lagoon area was larger than the island itself. That afternoon, we took a tour of their fishing boats and landing facilities, which were quite primitive. I was surprised to find that the local Polynesian fishermen frequently took their fifteen- to twenty-foot open boats to a seamount area over a hundred miles from Aitutaki, using only a compass. The catch not used by the local islanders was shipped by air to Rarotonga. The north side of the island lacked a large lagoon, but did have a protective coral reef several hundred yards offshore.

Later in the afternoon, we met with members of the council that regulated fishing activities in the waters of the lagoon and adjacent to the island. There were about ten, mostly elderly, members in the group. They noted that fishing in the lagoon had taken its toll on tropical reef fishes and that there were few fish useful for subsistence or commerce. When I asked about the regulations in place to control the level of catch, the elders looked at each other, then acknowledged that there weren't any. Then they let me know that their real problem was small, mesh gill nets. I didn't want to push them on the issue of why the nets had remained a legal gear for harvest in Aitutaki because it was apparent that politics was played a similar role in the island's fisheries as it did in many commercial fisheries around the world. After all, most of the fishers were likely their own kin, children, or friends. Council members, however, did ask for my suggestions for the lagoon fishery. We planned to take a boat out into the lagoon the next day so I could gauge their concerns.

We returned to the hotel in time to watch the sun sink into the deep blue ocean. In the evening, some of the locals danced during our dinner of fish, pork, sweet potatoes, and a great variety of tropical fruits. My room was much like the one I had in Samoa—a hut with a thatched roof. Someone warned me not to be concerned about lizards on the walls or ceiling; they were a prominent part of the local bug control. The room did not have air conditioning, but even though we were closer to the equator, the temperature on that day was lower and the humidity more tolerable than on Rarotonga. Most likely, it was an aberration. Before retiring, I walked along the lagoon shoreline to see what I could of the sea life. There were butterfly fish and damselfish and a few puffers and angelfish in the channel between the big island and the small island that the hotel was on, but I could not see much in the shallow water that spread toward the large lagoon area. A number of relatively large crabs scampered back into the many holes along the sand and coral shoreline.

In my room, I checked out the ceiling and, sure enough, there were a half-dozen small lizards. They seemed harmless, so I settled in for the night; then it began to rain. Growing up in Hilo I became used to heavy downpours, but I had never seen or heard anything like the rain that that night. It sounded like a waterfall was crashing on top of the hut; if it had continued for more than half an hour, I would have needed an ark to get back to Rarotonga. Until that time I had wondered where the local water supply came from. I don't know how often it rained that hard, but if there were any reservoirs on the island, they would have been overflowing in a matter of minutes.

The warm tropical sun was well into the sky when we boarded a small boat of about eighteen feet the next morning. There were five or six of us on the tour. We first motored out to the area where they had planted a number of large giant clams. The lagoon was about six to ten miles across and about twenty feet deep, with an abundance of large coral heads that sprang up from the bottom like a forest of mushrooms. The clams were in an area where the bottom was sandy and free of coral heads. To get a good look at the plantings, I put on a facemask and dove down to the caged area on the seabed. The clams were spectacular even though they were far from full-grown. After checking out the clam bed, we proceeded to look over a significant portion of the lagoon. It was unbelievably beautiful and the large coral outcroppings were sensational. Only one thing was missing—reef fish.

One could see a number of small damselfish and an occasional parrotfish, but there were no clouds of butterfly, trigger, angel, puffer, or other reef fish to be seen. It was almost like a luxury swimming pool with no bathers. There were few areas where the snorkeling was more attractive but, in many respects, the fauna of the lagoon was disappointing. Yet, even though there was a scarcity of tropical fish, the cruise around the lagoon left me with an unforgettable feeling of being in paradise. In the mid-afternoon, we pulled up on a beach, perhaps a hundred feet wide, with brush plants that provided some shade. There we ate our lunch, swam in the warm lagoon waters, and rested, looking at the coconut palms that rimmed the tiny island.

That evening, I had an informal chat with the Canadian advisor to the government of the Cook Islands. We concluded that we should try to persuade the Aitutaki Council to terminate commercial and subsistence fishing within the lagoon and concentrate their local fishery in the offshore waters closely adjacent to the island and the historically fished distant-water grounds. That would mean outlawing gill nets and other fishing methods within the lagoon. We also talked about Rarotonga providing infrastructure support to the distant-water albacore fishery several hundred miles south of the islands.

Despite its relatively good location, Rarotonga had one major drawback. There wasn't a good natural harbor and almost nothing in the way of a marine facility that could meet the needs of the distant-water albacore fleet. The cost of building a safe harbor facility, considering the frequent hurricanes in the area, didn't make economic sense. During storm periods, the local boats fishing out of Rarotonga were hauled out well away from the shoreline. Larger vessels, such as the Japanese, Taiwanese, and Korean long-line fleets, did not use Rarotonga's small harbor. Furthermore, the general ocean productivity around the southern Cook Islands was low, and the area was not that attractive to distant-water operators, who felt the fees to fish there were excessive. Some of the Cook atolls farther to the north were in more productive waters, and there was some possibility of tuna and other pelagic fish operations within that region.

On the fourth day of the Cook tour, we flew back from Aitutaki to Rarotonga. I discussed my concerns with the prime minister and promised to have a report in his hands within twenty days. The thirty-page report included such topics as local shells exported for the button industry, making the Aitutaki lagoon a marine park, developing local fisheries for near-shore tuna-like species, and dealing with distant-water operators fishing in the islands' 200-mile limit. Like many consultant reports, its ultimate impact on the local government decisions was never officially reported to NRC. Regardless, the beautiful lagoon repopulated with tropical fish ultimately would be a haven for scuba divers seeking to see the splendor of coral-reef fish communities.

The NRC work extended over two decades. Although the South Pacific assignments were most rewarding in terms of my love of the region, two other endeavors are worth noting.

Djibouti, East Africa

During the late 1980s, Roy Jackson was contacted by a friend with a contract to undertake a resource survey off northern Somalia; he wanted NRC to do the field-survey work. The World Bank was funding the project and, because of the political instability in Somalia, it had decided that the work would be conducted out of Djibouti, a "city-state" located at the south end of the Red Sea and east of the Great Horn of Africa. At that point in history, it was situated between Ethiopia to the north and east and Somalia to the south.

The Djibouti project was administered by a retired FAO employee who had worked with its Division of Fisheries. Our job, as subcontractors, was to provide the scientific expertise to explore the continental-shelf waters off northern Somalia as well as the offshore ocean region. The goal was to identify the species of commercial ground fish and pelagic fishes of the region and to evaluate their abundance and suitability for commercial exploitation. Two seiners would be operated by Norwegian crews supported by Somali fishermen. The NRC staff was elated about the possible contract. Almost everyone on staff had some experience in resource survey work, but we had to overcome two major problems. First, Djibouti was one of the hottest places in the world, with summer temperatures soaring well over 120 degrees. In fact, during the warm months, much of the government and wealthier elements of the population abandoned the port city and headed to the surrounding mountain regions. Thus, the field project was scheduled to start in late fall and extend into May over a three-year period. Our second problem was one of staff availability, since none of the senior scientists wanted to leave Seattle for an extended project in Djibouti. Al Pruter, however, was interested in getting the project started with several young scientists we had planned to hire as full-time local staff. We discussed the matter with the project manager and agreed that Al would initiate the project along with the two scientists, Jeff June and Greg Small, we had hired away from NMFS. The others could be hired later.

The project was difficult from a logistic point of view. The two seiners had to be outfitted in Norway and installed with air conditioning. They then had to be supplied and sailed down the coast of western Europe, east through the Mediterranean, and down through the Suez Canal and the Red Sea to Djibouti. To get to Djibouti from Seattle, the team had to fly to Paris and then take Air France, which was the only major airline serving the East African port city. The ticket costs were scandalous. In fact, it was cheaper for us to fly around the world. But Djibouti had been selected as a project operational base partly because it was almost impossible to get vessel fuel from any of the other ports along the northern Somalia coast. In addition, Djibouti was just a much safer place for our staff than northern Somalia. Djibouti once had been a part of French Somalia and still maintained close ties with France. It had a certain stability, and elements of the French Foreign Legion were based there. Even so, to ensure greater safety for the project, we were allowed to rent a home within the secure port area.

The Norwegian seiners managed to get to Djibouti on time, and the NRC staff, including Al, got the work underway on schedule. When Pruter returned from the first leg of the survey after four months, he was twenty pounds lighter and looked like a released war prisoner. He said that the work was interesting and progressing very satisfactorily, but that dealing with the heat and working on deck under the sun fourteen or more hours a day took its toll on any excess adipose tissue. On a hot day, the crew could not work without sandals or shoes because the deck would burn their feet. Al reported that they took skipjack and yellow fin tuna in the Gulf of Aden and in the Indian Ocean, south of the Horn of Africa. Some of the long-line sets were particularly productive, yielding significant catches of large emperor fish, various tropical sea bass, and other bottom fish, most of which had a high value in the world markets. Handling and identifying the catch, which included a great diversity of species, required considerable time and care because the bottom-fish catches frequently included a variety of sea snakes, some of which were poisonous. The snakes would end up in the net, and it was sometimes necessary to shake them, with care, onto the ship's deck.

After Al returned to Seattle, Jeff June became our senior scientist on site. Both he and Greg lived in the house in the port area when ashore and in small staterooms aboard ship. After they had been at work for about five months, I went to Djibouti to look at the operation first hand, review the project's progress, and determine what the staff needed to complete their work successfully. I arranged for a February flight from Seattle to Paris and then south to Djibouti. On the way from Paris, we crossed the eastern Mediterranean, flew south over Egypt, and then the Sudan. The pilot announced on the audio system that we were over Khartoum, a name that had stuck in my mind since high school, perhaps from reading some of Kipling's works. We landed in Djibouti before sunup, and it felt like I had walked into a hot oven. The temperature was well over 100 degrees. Both Jeff and Greg came to meet me and drove me to the Sheraton Hotel, one of the few Western-style hotels in the city.

At the hotel, we had a cup of coffee and made plans for my visit. Jeff quickly let me know that it wasn't the greatest time for me to be in Djibouti as the United States had raided Kadafi's home in Libya. The local Muslim population was not very supportive of the U.S. action, and there was some concern about terrorist activity. The U.S. embassy had suggested that the Americans stay in the embassy compound until things cooled off. However, Jeff thought it would be better if we got out of town. I took a short nap, and Jeff and Greg picked me up in the company pickup truck around noon. We headed west toward a village on the border between Djibouti and Ethiopia. Jeff had been to the area earlier and knew a good restaurant there. The village was near to a large refugee camp that housed thousands of Ethiopians who had escaped across the border to avoid the ongoing war in their homeland. We soon began to climb steeply into the mountains, and the temperature became tolerable. We passed by the king's summer home and then dropped down onto an extended plateau. Every few miles, we saw large fifty-gallon drums on the side of the road. Jeff explained that a water truck would come by on a regular schedule and fill them with water for the local tribes. From time to time, we saw one or two women walking along the road with heavy loads of goods on their heads.

When we arrived at the restaurant, I asked Jeff why he thought it was safer to be away from the U.S. embassy and he answered, "If they are going to bomb or shoot someone, the first place they will think of attacking will be the embassy." The eating area was outside under a lattice roof, and the seats were benches pulled up to picnic tables. A few groups of people were sitting at several of the tables, including some French Foreign Legion soldiers and their wives or girlfriends. We spent most of the afternoon relaxing, eating a good meal of lamb, and having a little wine under the lattice roof. Around 4:30 p.m. we jumped in our pickup and headed back to Djibouti.

As we passed through the security control near the border, we came across two young Ethiopian lads who were looking for a ride to Somalia. The older one was about twelve years old and spoke relatively good English. He told us that his parents had been killed in the war and that he and his brother were trying to get to a village in northern Somalia where their uncle lived. The younger lad was about nine. They had walked and hitchhiked some 300 miles with little more than a couple of bottles of water and whatever food they could scrounge. We invited them to ride in the back of the pickup truck and hauled them about 40 miles to where the road split—one branch went to Somalia and the other continued toward the city of Djibouti. We dropped the boys off, gave them each a quart of water and a small amount of money, and wished them well. They seemed very mature for their ages and filled with confidence as they headed down the road, expecting to be reunited with their uncle.

We spent the next day reviewing the progress of the project. The survey work had been going well, although we were not getting the cooperation we had hoped for from the northern Somalis. The local government administrations were frequently uninformed regarding the project and, during one of our visits to Berbera, on the north coast of Somalia, the local port authority asked for our vaccination cards. When they discovered that the scientists and crew did not have up-to-date smallpox records, they demanded that they go to a local clinic and be revaccinated, even though smallpox vaccinations were no longer required in most areas of the world. The health officials wanted $100 for each individual.

When Jeff, Greg and the crew went to the clinic they found it unclean and were concerned about being scratched by dirty needles. This led to a discussion that, in the end, allowed us to enter the port without the vaccinations, but every individual who went ashore had to pay the $100 vaccination fee. The Somali crewmembers generally worked out well, and we were acquiring a wealth of new information on the diversity and abundance of ocean life off the northern Somalia coast. The next morning, I headed back to Seattle, convinced that our component of the project was moving ahead quite nicely.

Several months later I got a call asking if we were interested in taking over management of the project, which still had two more years before completion. Both Roy Jackson and I were surprised because we had known the project manager for some time, and he was well known and had a good reputation. Furthermore, he was a personal friend of Roy's, and NRC had not encountered any operational problems working with him. This made the question rather sticky, and put NRC in a "damned if we do and damned if we don't" position. We had taken on Jeff June and Greg Small for the duration of the project. If the administration of the project were given to another firm, we did not know whether they would retain us as subcontractors.

If we accepted the project management, however, it would be a slap in the face of a friend and respected peer. We got in touch with the World Bank officials to find out why they were sacking their project manager, and it sounded like a personality conflict. But in the end there would be too much uncertainty about NRC's future role if a new company took over the project. The only way to protect our on-site personnel was to accept the expanded administrative role. Thus we became the prime contractor for the Somalia work.

Our new role required us to handle all logistical requirements, including purchasing and acquiring sampling gear, fuel, and oil for the vessels; arranging for the sale or disposal of catches; replacing crew members; paying the vessel crew and the on-site manager; and dealing with political and legal matters associated with operating in Djibouti and Somalia. Jeff and Greg were familiar with the on-site manager, Ken Larsen, and the administrative officer, F. Hayek, a middle-aged Christian who had been forced to leave Jordan. They said these men had been doing a good job, so we retained their services, which simplified the transition and ensured continuity in the field program. Before taking on the expanded role, however, we met with the World Bank officials in Washington, D.C., and then Steve Hughes and I traveled to Djibouti to meet with the World Bank personnel there and work out the final contractual arrangements. Hughes had an excellent knowledge of vessel operations and resource surveys.

This trip took place in April or May, when Djibouti was shifting from very warm to hot. After we signed the new contract, Steve Hughes would remain and take charge of the surveys for several months. When we arrived, again at around 5 a.m., the temperature was over 100 degrees. Jeff told me it was a dry heat with low humidity, but it still felt like we had entered a furnace. We went to the Sheraton Hotel for a short nap, and in the early afternoon Steve, Jeff, Greg, and I began talks with Mr.

Chamberland, a member of the local World Bank staff. He had flown in the day before from Mogadishu, the capital of Somalia. Jeff, Greg, and Ken, who had been on the job in Djibouti for about a year, were of great help to Steve and me as we developed a workable plan. If everything went as planned, NRC would make a reasonable profit on the project, but unexpected costs are always possible when one is operating research vessels, particularly in such a remote area.

We made considerable progress the first day, and our group agreed to meet at a well-known downtown fish restaurant for dinner. After a short rest at our office/home in the port area, Steve and I set out for town about sundown, visiting an outdoor shopping bazaar that ran for several blocks. In one shop, which was operated by beautiful Ethiopian women, I bought a wedding scene made up of small cloth dolls and horses mounted on a narrow three-foot board. It remained a major conversation piece in our home until it was consumed by beetles, which had incubated (I think) and hatched some twenty years after its purchase. The bazaar was amazing in that the shops contained almost everything anyone might want to buy. The electronic hardware included radios, TVs, headsets, etc. I have no idea where the merchandise came from or who purchased it.

The restaurant was nothing fancy. In fact, its seating arrangement involved a number of picnic tables and benches. Several tables were occupied by French Foreign Legion soldiers, several Caucasian couples, and a number of locals, who came largely from two dominant tribal groups. The fish was cooked in an oven that was a circular cylinder clay pit about four feet deep and three feet across. Both the fish and flat bread, which was a specialty of the house, were slapped up against the vertical side of the oven wall by the cook, who first dampened his arm. The process looked exotic, the fish and bread were excellent, and we enjoyed a great meal. The next day we finished our work on a few sticky sections of the contract, and I returned to Seattle the following morning.

The project was probably one of the most interesting overseas assignments undertaken by NRC and yielded an excellent report on the fish resources off the northern Somalia coast. Their use by local natives was made difficult by the limited seasonal availability of the tuna resources and by the lack of onshore buyers and fishing infrastructure. Furthermore, the 200-mile fishing zone off Somalia was not patrolled and was raided frequently by distant-water operators.

Oman

Oman, like Djibouti, is not on everyone's radar screen. It sits at the far southeast end of the Arabian peninsula and stretches from the Persian Gulf southwest into the Indian Ocean and then swings northwest toward the Gulf of Aden. Actually, the far western border of Oman isn't that far from Somalia, and Arab dhow trading between Oman, Yemen, and Somalia ports had taken place for centuries. My introduction to this part of the world came as a surprise. I received a letter from the sultan of Oman asking me to chair a group of marine scientists he had chosen to evaluate the possible location for a college of marine sciences in his country. I would have expected the Oman's government staff to choose someone from Egypt or the United Kingdom because of their historical ties with these countries. Later, I was told that a Seattle consultant working in Oman had recommended me to the minister of fisheries.

The letter was somewhat vague regarding rate of pay, work details, and expected schedule. My interests in visiting that part of the world overrode my concern about having a signed and agreeable contract in hand. That might have been a mistake. Several weeks after accepting the job, and with no certainty about the payment or schedule, I headed off for Muscat, the capital of Oman. Before departing I wired my arrival time to the minister of fisheries, and I expected that the remainder of the team selected by the sultan would be there to commence work on the project.

I arrived in Muscat on a Saturday morning and was met by an officer from the ministry of fisheries. He took me to the local university, where I was put up in a very nice room in an on-campus visitors' hotel. The officer gave no details on the assignment, but said that I could take dinner downstairs and that there would be a meeting the next morning. I saw no other westerners during a walk around the campus or in the housing facility, and I had no idea if the other members of the task group had arrived or even who they were.

The next morning, I met Joe Baker, director of the Australian Institute of Marine Science, and Moustafa Fouda, who was from Egypt and a faculty member of the local Sultan Qaboos University. He told us that George Fulton, a fishery consultant from Seattle, and Professor Aboul-Fotouh Abel from the university would join us for dinner and lay out our work plan. For the remainder of the morning, we could take it easy and get to know each other. Baker and I exchanged information on the communications we had received from the government of Oman. Neither of us knew what our daily pay rate would be, when they intended to pay us, or how they were going to take care of our food and lodging. Furthermore, neither of us had received any substantive details on what they expected the committee to accomplish. "Well," I noted to my two colleagues at the table, "they said they had an important task for us and maybe we will just have to be patient."

On Sunday evening, we met for dinner at the faculty club, which had a good selection of both Western and Middle Eastern foods. George Fulton gave us an excellent rundown on what was expected of our team. Apparently the sultan had a deep interest in expanding Oman's technical and scientific skills related to the management of the fisheries and other marine resources off its coast. For some time, the government had been advancing the state of knowledge of its marine resources and the associated environment by supporting and promoting a number of aid programs generally carried out by ex-patriots. But the sultan was interested in building a greater national competence in the field of oceanography and fisheries. He wanted our team to comment on the feasibility of establishing a marine science center at a college in Oman and suggest a location for it. He was also interested in the nature and scope of marine sciences on which the new institution might focus. We learned that the sultan had a summer castle in Salalah and would prefer the new institution to be built there.

Early the following day, we got underway. As chairman of the technical group, I reviewed the tasks, scheduled the work, and then led the group in a discussion of the tasks we were expected to complete that week. A logical question from a Westerner's point of view might be, "Why build such an institution in Oman?" It was not known as a nation with significant fisheries, and it was rather isolated from important marine institutions around the world. On the other hand, Oman sweeps across almost 1,000 miles of the upper Indian Ocean. Considerable upwelling was known to occur, particularly along its western coast. There were large quantities of oil sardines along the coasts, and tuna and other pelagic fishes frequented the area on a seasonal basis. The fisheries of the region had been expanding, and there was considerable interest among foreign distant-water vessels in securing fishing rights in Oman's 200-mile zone. Oman's "remote" nature of the location was deemed a problem at the time, but with the rapid advance of information technology, today, no location is isolated from the scientific information produced around the world.

The two scientists from the university reviewed the general character of the local fisheries for the group. To give everyone a better perspective on the diversity of fish being landed in the region, we were scheduled to visit the fish auction and market that afternoon. By noon we had organized our plan of work and were ready to start.

The government had scheduled trips to Sur and Salalah, so our working time was somewhat limited. Thanks to George and the university scientists, we had a good overview of the sultan's interests and the extant marine science capacity within the country. To augment this briefing, we met with government scientists in Muscat. After a day's work at the university, we took Tuesday morning off to visit one of the larger fish companies in Muscat. It was my first look at the port facilities and the harbor area. The city of Muscat was built around a large crescent-shaped bay in which a number of Arab dhows were anchored, a very attractive setting. The fish house, located on the bay's edge, was filled with a great diversity of tropical fish, laid out by species on the floor, including tunas, various species of sea bass, groupers, flounders, and mackerel.

Joe and I were both caught off guard by a dozen or so dolphins that were about to be auctioned. When we asked how they had been captured, we were told that they were taken incidentally in the seine and gill-net tuna fisheries. Owners of U.S. tuna canneries had committed themselves not to purchase any tuna caught in association with dolphins. Moreover, the conservation community had made a major issue of bycatch of dolphins or porpoise taken in world fisheries. It was simply unacceptable. But Muscat was a long way from the canneries in California, and dolphins were just as edible as fish. We knew the tuna taken were exported, but we were not sure which nations or buyers were involved in their purchase. Despite the large catches of tuna taken in the Indian Ocean, little was reported or known regarding the incidental catch of dolphins by these fisheries.

The trip to Salalah was one of the high points of the trip. We took off from the airport in Muscat and flew almost two hours to the west. Most of the area, including a mountain range that bordered the coastal area, looked dry and desolate, but as we approached Salalah we could see green fields and rivers draining the mountains. The area apparently was blessed with somewhat greater rainfall than the area to the east. We were put up in a small but comfortable hotel for the night, and the next day we examined the large bay on the outskirts of the city. As we walked along the bay edge, we could see thousands of oil herring being dried on the beach and flocks of birds diving on baitfish close to shore.

We were told that tuna could be caught within a mile or so offshore, so Salalah looked like it could be a Mecca for sport fishermen. As we walked along the beach, twenty or so camels walked by us, seemingly indifferent to our presence. They just roamed freely through the area. We also were surprised to learn that the dried oil sardines were used to feed the camels and that, in the past, they had been carried hundreds of miles inland to feed camel herds in cities north of the coastal mountain chain. We all returned with a rather good feeling about the possibilities of establishing a marine science center in Salalah.

We continued to work on our report and then left for Sur, which was west of Muscat. Unfortunately, Joe Baker, Moustafa Fouda, and I were the only ones who could make the trip. We headed north through rugged mountain country in two cars with local drivers. The road was relatively good and apparently there was no speed limit as the drivers seldom went below eighty-five miles per hour. I admit to being nervous for the entire trip. We stopped halfway along for refreshments and then resumed our race to Sur. Much of the countryside was arid desert, but from time to time we could see a group of palm trees surrounding water pools in the otherwise dry riverbeds. Homes and other buildings frequently were situated near these oases. Sur was not nearly as large as Salalah and seemed to be more spread out over the coastal area. After several hours in the city, we motored back to Muscat at speeds that must have raised my blood pressure fifty points.

On Thursday, we put the finishing touches on our report. We had still heard nothing regarding payments for lodging and our hourly consulting fees. Both Joe and I were a little concerned because, once the report was submitted and we left the country, it might be difficult to get reimbursed. The next morning we headed for the minister of fisheries' office, where we were served tea before presenting our report. Our findings, in summary, were that any new marine science institution in Oman should be located in Salalah. Our choice was based on a number of factors, including access by air, location in relation to productive adjacent marine resources, and the local infrastructure.

It took about fifteen minutes to complete our oral presentations to the minister. We were asked several questions and then asked to wait a few minutes. The minister and his staff left the room, and several minutes later a man dressed in a dark suit entered the room with envelopes in his hand. He proceeded to call out the names on the envelopes and handed them to us. Upon opening mine, I found it filled with $100 bills. I counted the booty, and it was to the penny what I had anticipated from my earlier communications with the Omani government officials. Apparently, we had done our job well, and the evidence was in the envelopes. I have often wondered what our fate would have been if they had found our report inadequate or inconsistent with their expectations. No matter, Joe and I were on a plane the next day. I headed to Rome to meet one of our NRC partners, Mark Freeburg, and Joe was taking the long flight back to Australia.

During the twenty-three years I spent with NRC—until 2003—the organization generally thrived. We worked with the fishing industry throughout the United States; the U.S. Navy and Coast Guard; a wide range of maritime users, including cable companies, recreational groups, and port authorities; and governments and conservation and management groups around the world. In all, we served over a thousand clients. As a final note on NRC, we were selected by a law firm to serve as expert witnesses for the fishing industry in the great Exxon Valdez oil spill. I left the firm in 2004 at the age of eighty, but the company continues to serve as one of the best, if not the best, natural resource consulting firms in the Pacific Northwest.

Before ending the NRC story, I must note that one of the most challenging and interesting assignments was a request by the Canadian government to chair a team of scientists in an investigation of the decline in the great cod stocks that had been the backbone of Newfoundland's economy for centuries. For almost six months, I commuted between Seattle and Newfoundland attending a series of meetings, whose participants included John Gulland and John Pope from England. Both were well known in the marine science field and considered two of the best scientists in the world dealing with the dynamics of fish stocks. The work took the committee to almost all of the important fishing communities along the island's rugged and beautiful coast. In the course of our studies, we grew to have a great appreciation for the Newfoundlanders. They were hospitable and kind, and most of them had made their living from the ocean resources since the island was first occupied. For hundreds of years, Newfoundland's waters had been among the most productive fishing areas in the world. As a result, it was not easy for me, as chairman of the committee, to have to report to the Canadian government that the stocks of cod were declining and could not be subjected to the level of fishing that had been permitted in the past.

Several years later the cod stocks collapsed, and the fishery that had started in the sixteenth century had to shut down. It was a terrible blow to the coastal villages that had relied on fishing as their major source of income. The story of the Newfoundland cod-stock collapse found its way into fishing literature throughout the world, and it is frequently mentioned as a horrible case of mismanagement. There can be little doubt that fishing was a major contributor to the doomed cod stocks, which by 2005 had not recovered. Who is to blame will be debated for years to come, similar to the assassination of President Kennedy. Early in 2005, I had the opportunity to review a paper by one of the scientists from the St. Johns Government Fisheries Laboratory. He assessed the cod stocks and attempted to trace the history of their decline. In the end, it seemed that the fishermen, government scientists, and government administrators all shared in the "crime" that led to the fisheries demise.

CHAPTER XV
HOME ALONE

When the 1980s came along, Ruby and I were back where we started, that is, there were just the two of us at home. There was a certain emptiness and loneliness in the home, but there was an added bonus. Now Ruby and I could travel together. We had managed to slip in a couple of trips before the children left, but travel for me had most often been alone.

By the mid-1980s, we had four grandchildren: first Aimie, Bob and Dolly's daughter; second and third Diane and Steven, Susan and Wayne's children; and last, but certainly not least, Tara, also born to Bob and Dolly. Ruby and I decided to take the grandchildren to Hawaii for a vacation. Tara was only about five while the others ranged up to eleven years of age. We knew that if we waited much longer some of our grandchildren would be in their teens, and it would be more difficult to plan a trip that they would all enjoy. I wanted my grandchildren to see where I grew up in the islands. It was no longer the paradise in which my brother and I had lived, but there was considerable charm left and I could still teach my grandchildren about my past.

We decided to spend a week on Waikiki and tour Oahu before going on to Hilo and visiting with Buddy (a member of the Breakwater Gang) and his wife. We wanted waterfront quarters on Waikiki, and Ruby found a great suite at the Waikiki Sheraton that looked down the beach toward Diamond Head. It was an extra-large room with a big bath and the required sleeping arrangements. In Hilo, Buddy got us rooms that looked out on the bay and had a great view of Coconut Island and the breakwater. I am not sure of the hotel's name but think it was the Hilo Bay. We were scheduled to go in June when the older grandchildren were out of school. It was not the first trip we had taken with them, as we had made several prior visits to Disneyland and my parents' home in Escondido. But this was the longest trip and required a lot more planning.

The trip from Seattle was about five and a half hours long, and the grandchildren's engines were running full force during the whole flight. The meals were somewhat better back then—they have been in a free fall ever since I started flying commercially in the 1950s—and the kids thought they were being fed royally. When we arrived in Honolulu, we rented a van, and drove to the hotel. When the kids saw the room and the view of the beach and Diamond Head, they were overwhelmed and thought they were in fantasyland.

Aimie and Diane, being less than a year apart, paired off well and were good swimmers, so we didn't have to keep as close an eye on them as we did on the two younger ones. Steve also was an excellent swimmer, and even Tara could hold her own, although her swimming skill was somewhat limited. But Steve assumed that he had all the talent needed to go out to the far surf and ride his rented board to the beach. He escaped our bondage once, and we suddenly realized that he had not kept his promise to ride the smaller waves near the beach. We watched as he paddled his board several hundred yards offshore and lost sight of him. We set off in search of the girls, so that one could rent a board and go after him. Fortunately, by the time we had found them, Steve had returned with great stories of how he had ridden in on several large waves wearing his life jacket, just as Ruby demanded.

That same evening we crossed over the main street to the International Market, which from our perspective had gone downhill over the years, but we knew our grandchildren would enjoy it. They loved the large banyan tree and wanted to see if anyone pulled a pearl out of the oysters being sold to the tourists. Most of all, they were smitten with the parrots on display and begged to have their pictures taken with a bird on their shoulders. Unfortunately, the peddler only had three parrots, so only the girls got their picture taken. Steve, who had suggested the photo, didn't want to be in the picture unless he also had a parrot. He felt put out and remained out of sorts until the next day.

On our second day, we drove to Hanauma Bay, rented some swim fins, face masks, and snorkel gear for the kids and climbed down into the ancient volcano. We spread our towels out on the sandy beach. All of the kids were thrilled when they were greeted at the water's edge by a swarm of diverse tropical fish looking for a handout. They fed peas to the fish, which was allowed at that time, and dove among them before swimming out into the "keyhole"—a sandy area in the middle of the reef. Ruby and I were amazed at how quickly they adapted to their snorkels. Tara and Steve, who were required to wear life jackets, bobbed around like turtles crossing the shallow coral reefs. They stayed in the water for several hours. Steve, as usual, gave us a bad time because he saw no reason to stay inside the reef and I had to swim out to bring him back to the safe zone.

After almost four hours at the bay, we headed off around the south end of the island to Sea Life Park, where we treated our grandkids to the various marine shows, including the dolphin and whale performances. We didn't spend a great deal of time there because we wanted to get on to the Polynesian Village. On the way we stopped at a small fruit stand where I was able to purchase five-inch portions of raw sugarcane for the kids to try. They thought it was super. During my Hilo childhood, we often cut and savored cane from a field near our house. Even Ruby chewed away on her cane and thought it a special treat.

At the village, the kids rode in the outrigger canoes and saw the volcano display. To Steve's delight, there was another parrot picture man. He got his photo taken but, to his dismay, the parrot decided to relieve itself on his shoulder, leaving a large white mess on his shirt. Before the day ended, we stopped at the Dole stand to eat our share of pineapple wedges, drove over the pass to the Pearl Harbor area, then back to the Sheraton. During our short stay in Waikiki, the grandchildren managed to get a good sunburn each day, but Ruby rubbed them down with vitamin E and they seemed fine the next morning.

Late in the first week, we packed up and headed for Hilo, where Buddy met us and plans were made for the next day. We were to leave the hotel early and head up to the volcano at Kilauea. The volcano had been active for several years, and lava was flowing to the sea down toward the

Kalapana black sand beach. It had actually overrun most of the villages and was pouring into the sea somewhat south of the old town. We left the hotel about 8:00 a.m. and headed up the grade to the volcano. On the way we stopped to take the kids through the lava tubes—a big tourist attraction—then went on up to the Volcano House, where Ruby and I had coffee and the kids had drinks. We then drove around the large crater to the viewing area and looked down into the fire pit or cauldron. Although there was a lot of steam, the main eruption and lava flow were down the mountainside. We continued until we reached the black sand beach and drove to the end of the road, where the beach was closed off to traffic.

The grandchildren wanted to get closer to the lava flow, so Buddy took them on a trail that led across some of the recent, but cooled lava, to a point where the police would not let anyone go beyond. From there they could see the red lava plunging into the sea and great plumes of steam rising up from the ocean. It was the highlight of their trip. On the walk back from the viewing site, all of the grandchildren wanted to collect some of the light "foamed" lava rock, but Buddy noted that it was inappropriate to take anything away that the volcano god "Pele had produced. Besides, Buddy noted, "It is bad luck to remove the rocks." The kids threw back the lava rocks they had picked up; at least, we thought they had. We returned to Hilo and went to Buddy's home for dinner. Buddy lived in the same home he had when he was a child, although it had been moved from its original site to a lot in the land-grant area.

The house had two stories, but Buddy and Katherine (Kay) lived downstairs where one side was open to the tropical foliage surrounding the home. There were chickens running around the yard, and some would occasionally walk into the house. There was a large table, which easily sat twelve. Kay asked four of their young grandchildren to join the feast, which made an even dozen. While Kay prepared the meal, the two sets of grandchildren went outside to play. They soon collaborated in an effort to take the husk off a ripe coconut. Buddy and the rest of us watched through the window as the youngsters worked feverishly, but with little success. After they had struggled for almost thirty minutes, Buddy went outside, took the coconut over to a spike that protruded from a heavy plank and with two quick movements removed the husk and handed our grandchildren the coconut. They all joined in drinking the milk and eating the white coconut meat.

After we had chatted for a couple of hours, Kay filled the table with a fruit salad, pork, chicken, sweet potatoes, fresh-baked biscuits, and more. There was enough food to feed a proverbial army. We sat down and gorged ourselves on a great meal. As I watched the two families partake in the festive affair, my mind raced back to the days when Buddy, my brother, and I played, fished, and ate together. The grandchildren had no concerns about race or culture and were enjoying themselves. It seems that children are always much more adept at crossing these boundaries than are adults.

We finished dinner well after dark and returned to the hotel. The next day we took a plane to Honolulu, where we would catch our flight back to Seattle. When we arrived at the Northwest counter, we found that the flight had been cancelled due to mechanical problems. Northwest had booked us on a Southwest Airlines flight to Phoenix. From there we were booked back to Seattle via Reno. Ruby and I were devastated because it meant that we would not arrive home until late the following day. However, the grandchildren were excited, pointing out that they would get to visit two more states; for them, the extra time was of no concern.

The Southwest flight left about 9:00 p.m., so we sat in the airport for over eight hours. We had no sooner left the ramp than the pilot announced that there was a problem with the hydraulic system. He was going to taxi to a hangar area where the problem would be fixed in minutes. As you might expect, the minutes turned into a couple of hours. We were really off to a bad start. On top of the delays, our seating had been fouled up and the grandchildren were spread all over the plane. Hence, Ruby and I were not in the greatest mood.

We arrived in Phoenix around 7:00 a.m. the next morning and went to the waiting room, which was mobbed; it had gotten so hot that a number of flights were delayed until the air cooled. When the air was too hot, the planes didn't get the lift necessary for a safe takeoff. Even with air conditioning, the airport was very warm. The grandchildren demanded to be taken to the entrance so that they could go outside and say they had been in Phoenix. They ran out about forty feet from the main entrance and then came back inside. The temperature outside was over 110 degrees, and it was only a little after 8:30 a.m. It didn't take them long to recognize that they didn't want a lengthy visit outside the airport. Our plane to Seattle didn't get off the ground until sometime late in the afternoon. We finally arrived around 9:00 p.m. after a short stop in Reno, almost thirty hours after we entered the airport in Honolulu.

Our grandchildren's parents were waiting at the airport to pick them up. When Tara saw her mother and father, she ran for about thirty yards toward them, and then suddenly reversed her course. She ran back to Ruby, put her hand in her pocket, pulled out a small lava rock, and said, "Here, Grandma, I don't want any bad luck." Ruby took the stone, looked at me, and said, "What shall I do with it?" I replied, "Just put it on the floor over next to the wall; we don't want any bad luck either." Ruby did just that. As I recall, the next day they had to close down the airport because of a threatened or actual hijacking incident. I do not know if the rock was involved, but we never ruled it out.

Throughout the 1980s and 1990s, Ruby traveled with me on business trips to Athens, Bergen, Rome, Tokyo, Fiji, Amsterdam, and Sydney. From my perspective, it made traveling a lot more fun, and Ruby seemed to adapt easily to wherever we went. These were all fun trips, but none held a candle to our trip to Japan's northernmost island. Shortly after I established NRC, I was asked by a Japanese long-line association to serve as a technical consultant to their group. It had about twenty-two vessels that had been fishing in Alaskan waters at the time the U.S. extended jurisdiction act was passed. Most of these boats fished sablefish as well as some rockfish, large flounders, and cod. Almost all of these species had relatively high market values in Japan, and U.S. fishers were bound to move rapidly into these fisheries. In several years, the expanding U.S. fleet would result in the exclusion of the Japanese line vessels fishing these species.

A crafty, shrewd, old gentleman called Nakamura headed the group. He was one of the Japanese industry leaders whom Dr. Fukuhara had met during our tour to educate the Japanese industry on the details of the Fishery Conservation and Management Act. Nakamura wanted us to come to Kushiro, Hokkaido, to discuss the consulting arrangement. Frank and I decided we would make the trip contingent on our being able to bring our wives. Nakamura approved the request, and in the late spring, sometime in the early 1980s, we made arrangements to travel to Kushiro.

The flight to Tokyo was now nonstop and we could fly directly from Seattle into the new airport, which was about forty miles out of town.

Ruby and I met Frank and his wife, Marcy, at the Seattle airport, climbed into a 707, and took off for Tokyo. The flight took a little over nine hours, which gave Frank and me the opportunity to discuss the NRC contract and tactics we might suggest to the group. We landed in Tokyo the next afternoon, losing a day as we crossed the international date line. One might have thought the association would send a car for us, but getting to and from the airport is a mess, so we took the bus that stopped at the major hotels. We were dropped off at the Okura Hotel in the late afternoon, took a nap, and then had an early dinner before retiring.

The next morning, a car picked us up and we were driven to the association's Tokyo office for an all-day meeting. They arranged for a chauffeur and limousine to take our wives on a tour of the city, including a luncheon at a very swank restaurant. Meanwhile, Frank and I joined a group of about fifteen long-line vessel owners. The long-line group was one of the only Japanese fishing associations that supported the U.S. unilateral jurisdiction. Their entire fishing operation was conducted off Alaska, British Columbia, Washington, and Oregon.

They first asked us to review the priority-access provisions of the act; then, after we returned to Seattle, we would develop a chart or graph estimating their remaining tenure in the U.S. fishing zone. This was to be coupled with a table showing the rate of reduction in fishing for each species they targeted. They also noted that, under an amendment provision of the act, some nations would be given special consideration for access to the U.S. fishing zone if they made efforts to aid the growth of the U.S. fisheries. They wanted us to suggest some options they might wish to consider. When the day was over, we had a list of their requests and the time-lag factor was beginning to have an impact on both of us. When we returned to the hotel, our wives had just gotten back from their all-day tour of the city and were ecstatic about what they had seen. Frank and I were ready to retire, but our hosts had already made arrangements for dinner at a Kabuki restaurant.

The next morning we were off to Kushiro, situated on the southeast coast of Hokkaido, with about a dozen members of the association. Fortunately, as throughout much of the world, air travel was improving and the flight was nonstop. In earlier years, we had to stop at Sapporo and then take another flight to other cities around Hokkaido. After a few minutes at our hotel, we were whisked off to the association office for another meeting, leaving our wives to rest or go shopping.

Frank and I spent the afternoon reviewing the new extended jurisdiction with Nakamura and his associates. Following this review, the members requested that we give our best guess about their future in the newly created U.S. EEZ. We decided to give them best- and worst-possible scenarios. Frank noted that the U.S. line and trawl fishermen had set their sights on Americanizing the sablefish or black-cod stocks largely because they had such a high market value in Japan. In addition, the U.S. fishers also were becoming increasingly interested in the rockfish market, also highly valued in Japan. We anticipated that the fishing for cod and large flounders was likely to survive the longest. In the worst case, the Japanese could be phased out of the U.S. fishing zone within three years. Even if U.S. fishery development was somewhat slower than we anticipated, it might be no more than five years. As we watched the looks on the faces of the vessel owners, it was obvious that they were not very happy with the forecast. Some of them probably would have liked to shoot the messengers. We concluded our presentation by assuring them that when we returned to Seattle, we would complete our work, and send them an official report of our findings. When the meeting was over, they took us back to the hotel to pick up our wives and then head out to a superb dinner.

When the dinner party was over, Nakamura told us that they had made special arrangements for the following day. We were to meet again with the association members until about 3:00 p.m., and then we would visit a hot spring area in the nearby mountains. The following morning Frank and I went over the reasons why we felt that the U.S. industry would invest in the vessels and technology required to develop their capacity to fish sablefish and rockfish. I noted that the capital could be easily acquired for the fleet's development. At exactly 3:00 p.m., we were taken to our hotel to pick up our luggage. Then Frank and I left the city with two members of the association and began to wind up a mountain road to the hot-spring resort. We were on the road several hours before we arrived at a small village next to a large lake.

Inhabited by a number of the aboriginal native peoples of Japan, the village was a major tourist attraction as well as a popular visiting site for the Japanese population on Hokkaido. The long-line association took us to a wonderful Japanese inn, where we joined our wives, who had been taken there earlier in a caravan of limousines. They had stopped to look at a number of tourist sites on the way, and both Ruby and Marcy said that they were treated like queens. As soon as we had unpacked, a Japanese woman entered our room and worked hard to persuade Ruby that she needed to go to the inn's hot spring bath so that she would be clean for the party planned for that evening. But Ruby wasn't about to disrobe and bathe in the nude even if it was just with other women.

Around 7:00 p.m. the four of us were escorted to a dining room that must have been about thirty-five feet wide and eighty feet long. Down the center of the room was a long, low table, about two feet high, elegantly set for about forty people. About twenty members of the association were seated on each side. Across one end of the long table was a short table, making the shape of a T, which was set with four places. We were escorted to those seats, and when we sat down, Japanese style on pillows (I had to put my legs to one side or I would never have been able to get up), the vessel owners stood up and clapped.

Nakamura said a few words of welcome, gave us a little history of the village, toasted us, and then the ceremonial dinner began. It is hard to describe the various courses that were served. Each came on a separate platter, and every course was artistically arranged with flowers on the plates. One course followed another, and in total there must have been fifteen or more individual dishes. We were given traditional Western utensils as well as a set of chopsticks. Having a considerable experience in their use, Frank, his wife, and I used the chopsticks. Ruby also used the chopsticks, even though she had little previous experience, and tried valiantly to feed herself.

But Ruby had a much bigger problem to overcome. She had never eaten raw fish or other marine life, which was a common component of many of the dishes served, and she wasn't about to do so. She pushed the food around on many of her plates in an attempt to show her participation and not insult her host. The masquerade went on to the final course and I thought Ruby had done a marvelous job of pretending that she was partaking in the dinner. Our hosts, however, had been quietly observing her reluctance to consume the superb Japanese food. They also recognized that it was her first trip to Japan and that the food, although beautifully served, was foreign to her. Just as the last course was being presented, two young Japanese women in their kimonos came down the length of the table, bringing Ruby a plate with steak and a baked potato. Ruby blushed a little, and everyone laughed and applauded. She did not go home hungry and has always remembered the hospitality shown by her Japanese hosts.

Our hosts presented our wives with some gold and pearl earrings before the festive affair ended. The next day we headed back to Tokyo. Frank and I had grown to admire Nakamura, who was not only kind but had the intelligence to lead his organization in a manner that would extend participation in the U.S. fishery zone as long as possible while planning for other options when his vessels were phased out of the U.S. EEZ. Nakamura led the Japanese long-line association for about four more years, before he died in his home in Hokkaido. Prior to his death, he came to Seattle and had dinner with the Fukuharas and Ruby and me at our home. He was fond of baked potatoes, so we wrapped a small football in aluminum foil and presented it to him on his plate. He thought he had been given the largest potato in the United States. We all had a good chuckle before it was replaced with the real thing. The Japanese long liners were phased out of the U.S. fishing zone by mid-1985.

The 1980s were good to Ruby and me, especially in the financial arena, but the '90s were a period of prayer and of trusting God regarding my health and that of other family members. In the early part of the decade, Ruby and I decided to take our grandchildren to Disneyland and made plans to go south after school was out. In the meantime, I went to the University of Washington Hospital for a regular checkup since I had a tendency toward high-blood pressure. The checkup went well, but the doctor recommended that I have a colonoscopy, given my age. I had the exam the following week and left the hospital thinking the exam had gone well. They had found a polyp, but the doctor said that there was only one chance in a hundred that it was malignant.

Several days later the doctor told me that the polyp had metastasized and that I had colon cancer. He urged me to have an operation as soon as possible and said that it could be arranged for the next week. I gave it a lot of thought, but decided that it would have to wait until after we had taken the grandkids to Disneyland. The next week we loaded them on a plane and headed for Orange County, California. We had a great three days there with the kids and four of their second cousins, who lived just two miles away in Anaheim. Ruby and I joined the group of eight children and my cousin in all the rides. When the fireworks were over late in the evening, we took the children to a nearby restaurant where they laughed and made so much noise that I think the remaining customers abandoned their meals and went home. The blood relatives had bonded.

Upon our return I entered the hospital and underwent a number of tests including a CAT scan. The scan showed nothing, and I went into surgery the following day. The doctors remained less than optimistic and removed twelve inches of my colon. The surgery went well, but apparently a number of blood vessels normally attached to the upper colon were attached to the section that needed to be removed. These had to be relocated, which prolonged the surgery for some time, resulting in considerable blood loss. As a consequence, it was likely that I would need a transfusion. I was hoping that it could be avoided because there was great concern at the time about HIV since blood checks were not very accurate. My blood pressure was low, and I was kept in intensive care until my body began to manufacture and replace the blood that was lost.

When I woke up in the recovery room and realized the length of the operation, I immediately decided that I must have had extensive cancer. My wife, the doctor, and some attentive nurses took turns trying to convince me that everything had gone well. The doctors checked the texture and appearance of the removed section of the colon and could not detect any evidence of cancer, but I would have to wait a couple of days until the lab completed a biopsy. Thank God, they found no sign of cancer in the removed section. It had been limited to the polyp. The doctors were so sure of their findings that they didn't even request any postoperative radiation or chemical treatments. I had to go back for checkups for several years, but no further evidence of cancer was found.

I had beaten cancer, but the 1990s still were not good years. In 1993, my brother, who was suffering from diabetes, called me one evening and said, "Lee, I just took some medicine and have become deaf in one of my ears." The next morning his wife Meche called to tell us he had been taken to the hospital and diagnosed as having had a stroke. Ruby and I jumped on a plane to San Diego, rented a car, and rushed to the hospital in Escondido. By the time we arrived, Frank was lying in a bed with a hole in his chest to facilitate breathing and his monitored diabetes values had surged far above an acceptable range. For a day or two, we thought this could be controlled, but there was little progress. Frank was a paratrooper to the last moment of his life. He looked at me and said, "I have already told Meche that if I can't breathe on my own I am going to have them pull the life-support systems, including oxygen. I don't want to live like this." Ruby and I left that afternoon, and several hours later Frank's life support systems were disconnected. He lasted only a few hours after that. Although we had not lived close to each other, we often visited and, as we grew older, we had drawn closer together. It was a hard loss for me.

The next year, 1994, I had a minor heart attack, which led to a triple-bypass operation. This time, the care I received at the hospital was less than ideal. When I talked with my surgeon prior to the operation, I made it clear that I had been on a daily aspirin treatment for some time and would not need a blood thinner before the operation. I also noted that I was very allergic to morphine and all of its derivatives. The operation started around 8:30 a.m., and I did not regain consciousness until around 6:00 that evening. I had been given heparin, a blood thinner, and continued to bleed for several hours after the operation. Around 5:00 p.m. the doctor told Ruby that, if I didn't stop bleeding, they would have to reopen the heart chamber. They had already given me a substantial amount of blood. Ruby got in touch with her prayer chain and called our relatives and church and asked them to open a channel to the Almighty. By 8:30 p.m., the bleeding had stopped, and I was in the recovery room waiting for the tubes to be removed from my stomach.

After I became conscious, I could hear the nurses and their aides talking about the operation. I could not speak and my eyes were most likely shut, but their remarks came in loud and clear. One of the nurses said that the doctor was upset that I had been given heparin, and another nurse said that she had forgotten to take some test the doctor had requested. I was then taken to intensive care. I was having a hard time breathing as I had reacted to the anesthetic or something else that was given to me. This lasted for several days, making it difficult for me to sleep or be in a prone position. When I tried to get help, a nurse's aide came in, but she couldn't understand enough English to give me what I needed. Things weren't going very well, so Ruby decided to stay at the hospital and slept in a reclining chair next to me for a week. My breathing remained a problem until Ruby and one of the nurses, discussing my allergic problems as they went over my medical regime, saw that I was being given morphine for pain. This was stopped, and within a day I began to breathe normally. I concluded that the biggest threat to one's life is staying in a hospital. Others in the hospital must have had better care or the mortality rate would have been extreme, but, then again, I survived. Later in the decade, I had a hernia operation, but that went very well. What about Ruby? Well, she seldom ever got sick; she comes from tough stock and good genes.

The Normandy Park Clan

Our social lives and ties were splintered into two major groups: those associated with my work and our neighbors within Normandy Park. The work friends included scientists fishermen and processors I met at meetings or for whom I conducted research.

Those in Normandy Park were mostly emigrants to the Pacific Northwest with few relatives within reasonable commuting range of Seattle; thus, like us, they were looking for friends to enhance their social lives. Our closer friends in the area included five families who lived within several blocks of our home and had children in the same age group as Bob and Sue. Our activities would include dinners, dances, picnicking, and special celebrations on national holidays. You may recall that sometime in the 1970s we built a summer home on Case Inlet, the long fiord at the south end of Puget Sound that curls back to the north and comes within several miles of Hood Canal. Well, we were not the last of the Normandy Park clan to fall in love with the area. During the last three decades of the twentieth century most of our friends eventually bought lots and built summer homes on Case Inlet. This tradition has continued as some of the children of the original five families also purchased homes there.

Our Fourth of July celebration, which started in the Berrymans' backyard, first shifted to the Bricks' Case Inlet lot (the first of our group to buy in the area). Subsequently, it was rotated on a yearly basis to each of the Normandy Park homeowners with summer retreats on the inlet. At the beginning, the clan was comprised of about twenty people. As the years passed and the number of owners and their progeny increased, the parties continued to grow. The last July 4 celebration that we hosted had more than fifty participants and four generations. New recruits that followed the Bricks included, ourselves, the Millers, Osbuns, Soltvigs, Dittys, and the Berrymans' children. The July 4 parties were unique in that each family hoisting the event usually supplied some special food, such as freshly grilled oysters, Indian baked and smoked salmon, and special cuts of beef.

Why was the Case Inlet area so attractive? The inlet, which runs to the north, offers those who live on the west side a clear view of Mount Rainier. The east side has a great view of the Olympic mountains and the evening sunsets. It is not nearly as crowed as the more popular Hood Canal waterway. In addition, in the early years (1970s–80s), there was an excellent population of sea-run, cut-throat trout, and it was a short trip by boat to several excellent salmon fishing areas within Puget Sound. Most of the clan had great clam and oyster populations on their beaches. And because there were several good sized islands at the mouth of the inlet, the waters were frequently smooth and excellent for water skiing and other water sports. I believe all of the families had ski boats and/or fishing craft. With these attributes, it was hard not to fall in love with the area. Because Ruby and I enjoyed our summer home on the Inlet we never allowed a phone to be installed. Still, family and office emergencies always seemed to get solved without our input.

CHAPTER XVI
ONE LAST VISIT TO THE RIVER OF NINE DRAGONS

Following World War II, I had rekindled my friendship with Jim Murphy, who was living in California, and accidentally found Verne Benedict in Seattle. Later, by chance, I met Bill Miller, who had served in Kunming, China, with our SACO group, on the U.W. campus. Subsequently, I heard from Duncan, who had been my roommate at Camp Six. On the other hand, I had no formal ties with anyone in mainland China or Taiwan, but I followed with interest the events in the two Chinas after World War II. This included the takeover of the mainland by communist forces, Chiang's exit from what had been Amoy, and the communist/nationalist debates and philosophical differences. My interest was ignited and fueled by the memories of my experiences in mainland China. Details have dimmed over time, but my interest and concern for the Chinese people has remained steadfast. Throughout most of my career, I had hoped that my work and involvement in international fisheries sciences would lead me back to China. I am grateful that this did occur.

Several months before my retirement, I was asked to go to Taiwan and discuss fishery problems of concern to both the United States and the Republic of China (ROC). Their high-seas gillnet fishery in the north Pacific was known to take incidental catches of marine mammals, birds, and salmon. It was not China, but it was as close as I could get, and the trip was a real prize for me. After landing at the airport outside of Taipei, I took a taxi into town and saw rice fields, water buffalo, and old earthen homes that were typical of southern China. A myriad of memories came into my mind as the war years flashed by somewhere in my cerebral cortex. After several days of meetings with local fishery scientists, I called Shen-ya Ye, a student in one of my classes at the U.W. who was now a professor at the local university. He met me at a local restaurant with his entire family; there must have been twenty-five or more people and three generations at the table.

Since my departure from Shanghai, the face of Chinese politics had undergone a dramatic change. Mao had swept down from northwestern China and defeated Chiang Kai Shek's armies, which had been severely weakened from the years of war with the Japanese and received only meager support from the United States. Chiang and his forces had retreated south to the delta of the River of the Nine Dragons. This part of China was very familiar to me. Chiang had moved troops and government personnel from old Amoy (Xiamen) to old Formosa, later renamed Taiwan, across the strait. There, he held onto the ROC and dreamed of one day reuniting the two Chinas. In the interim, Mao superimposed a communist regime across the ancient mainland, now the People's Republic of China.

At the time of my trip to Taiwan, I was unaware that a SACO veterans group had been formed. Over a period of years, they had developed special contacts within the Military Intelligence Bureau (MIB) of the People's Republic of China. As a result, I missed out on meeting with some of the Chinese members of SACO. My trip was limited to visits with the university scientists and fishery managers in the government. Then in the late 1980s, long after I had left the government, Bill Miller told me about the SACO veterans group and suggested organizing a reunion in Seattle. He asked Verne (Benny) Benedict and me, as well as other SACO members living in the state, to give him a hand.

The reunion was held at the Airport Marriott Hotel, and Jim Murphy and his wife, along with several of my buddies from Camp Six, were in attendance. I took care of the program, getting one of our state representatives to speak at the dinner. Ruby, Benny, and his wife Jackie helped at the registration desk. It was a learning session for me because

I knew very little about the special arrangements and friendships that had been formed between SACO and ROC's Bureau of Intelligence. The ROC invited and paid the way for all of the SACO vets to go to Taiwan and treated them like kings. The Seattle reunion flushed out a great deal of history that had been played out some forty years earlier, as did my trip to Taiwan. The two events reinforced my memories of a very special time in my life.

The year after the Seattle reunion, the ROC invited the SACO group to Taiwan again. Fortunately, Jim, Verne, and I and our wives were chosen by the SACO vets as participants. The trip would have made a great Disney movie. We were swept past customs and immigration by military officials and quartered in the Grand Hotel in Taipei. Each of us was assigned a room boy to take care of any special needs. Our room was large and elegant. It had tile floors with throw rugs, a king-size bed with a beautiful oak frame, and a bath and dressing room as big as most hotel rooms. We were told to charge any food, drinks, and services that we wanted. The ROC arranged trips to different parts of the island by air and bus, and everywhere we went we were treated like royalty. The talks given by various schoolteachers, city leaders, Taiwan government officials, and military authorities were impressive and heart wrenching. All emphasized their indebtedness to the American naval forces that fought in China during World War II. Speakers and audience members had tears in their eyes.

Ruby and I were amused by the room boys, who watched whenever we left to see if we needed anything. When we returned, they seemed to be in the same place, eager to pick up our packages, run errands, or answer any questions. Once, when we returned with Jim, Benny, and their wives, we decided to go to Benny's room for a drink. We were about twenty yards down the hall when one of the houseboys ran to catch up with us and said, "You are going in the wrong direction." We told him that we were not going to our room, but to the Benedicts'. He said, "Oh yes, they are only four doors down from here, let me show you the way."

When we traveled throughout Taiwan, it was always on a bus, with a military escort, a local police escort, an ambulance, and a doctor to take care of anybody who might get sick or hurt. When we visited a military establishment, their servicemen stood smartly at attention and saluted as we passed. At the banquet dinner, our hosts lavished gifts of jewelry and watches on their SACO visitors. If our wives wanted to go shopping, escorts took them to shops where they were given special discounts. We were amazed at the Chinese hospitality and by their genuine feeling of indebtedness to the Americans who had fought to defend China four decades earlier. At one junior-high school, the teacher told the class about our history in mainland China, which brought tears to her eyes and ours. The ROC minister of defense was so impressed with the historical work of SACO and the American effort that he had two medals struck, one for general service in China and one for those who participated in the war effort. I received both.

Following my initial visit to the ROC, I was invited back two other times, once with my wife and once alone. Each time the government picked up all the costs, and we were showered with gifts and treated like war heroes. The Chinese showed a commitment to the American veterans that we never expected back home. At the end of each trip, they presented each individual or family with a photo album that captured all of the important events and sights. Somehow, despite the fast pace of the trips, I always found time to break away from the group and visit Shen-ya, his wife, and their two children.

During my several trips to Taiwan, it was easy to see the changes in the country's economy as well as the growing strength of democracy. During our first trip, there was little criticism of the ROC government, which obviously had strong ties to the military. Many of those who had escaped from the mainland believed in the One-China policy and their hopes of ultimate reunion were apparent. After all, many had been born in mainland China and still had relatives living there. Politically, however, Taiwan and China remained sharply divided. Interestingly, despite these ideological differences, illegal trade between Taiwan and Fujian, the closest coastal province on the mainland, continued. Both governments were aware of, but ignored, the trade that took place on fishing vessels and other small craft. There obviously were economic benefits that were valuable to both sides. In our later visits to the ROC, the newspapers became more and more critical of government policies, and it was clear that the freedom of the press was growing.

Taiwan Sailors Saluting SACO Members and Author Returning Salute.

Return to the River of Nine Dragons

The ties of the SACO group were probably somewhat of an embarrassment to our government, which was moving steadily toward stronger ties with the People's Republic of China (PRC). I greatly enjoyed the hospitality and friendships developed with the people of the ROC, but I wondered what was occurring on the China coast across the Strait of Taiwan. It was just a short hop from Kaohsiung to Amoy and Fujian, where I had fought in World War II. I understood that a great deal of development had taken place on the Pacific coast of eastern China. In early 1990, fifty years after my experiences in China, I was invited to mainland China, but the itinerary did not provide an opportunity to visit the areas where I had been stationed and there was no offer to pick up the travel costs. I turned down that opportunity, but I still had a deep-seated desire to return to the region where we had fought during the war. Thus I was elated in early 1993 when I received another invitation to visit mainland China and give several lectures at the University of Shanghai and the College of Fisheries in Xiamen (pronounced Shamen), the name the PRC gave to Amoy. The trip sounded wonderful, as it would give me a firsthand look at all the developments that had taken place.

In filling out the paperwork for the trip, I made sure China's officials were aware of my history in China during the war, my ties to SACO, and my trips to Taiwan. I left Seattle filled with nervous excitement and anticipation about returning to my war haunts. My mental computer chip, although somewhat tarnished, gave me a reasonable data set on Shanghai in the mid-1940s. I could remember sailing down the Huangpu into the Yangtze and back to San Francisco. At that time, the Huangpu River was crowded with small sampans, powered by sail or oar, making their way to larger vessels anchored in the river or tied along the Shanghai waterfront. Where the Huangpu joined the Yangtze, the masts of ships sunken during the war made the area look like a deciduous forest had grown up in the river.

I was in Shanghai from September 1945 through February 1946. The physical aspects of the city were not greatly scarred by the war, but its social-cultural activities, as well as the governing infrastructure, suffered greatly, after almost a decade of Japanese control. The shops in the fall of 1945 had little or no goods. Many had been closed, and poverty was common among the greater portion of the city's Chinese population. The downtown area was reasonably modern for the 1940s, but transportation was largely via bicycle rickshaws. Few cars were in the streets, but trucks and military vehicles were common. At that time, the Park and Palace Hotels were the "in" places. The Park had been taken over by the U.S. military, but the Palace remained open to the public as an elegant old waterfront hotel.

This time, I arrived in Shanghai on May 25, 1993, and was picked up by a Chinese government official, Cui Lifeng, who would stay with me throughout the trip. Lifeng was a young man with a very charming personality, and he attempted to be as helpful as possible. I had been somewhat concerned about my reception by the communist government because I had gone to the ROC a number of times and had a World War II involvement with General Tai Li, an archenemy of the communist regime. Lifeng was young and apparently unfettered by my historical footnotes. He listened with courtesy and interest and was amazed at my China stories, all which had taken place before his birth. He was a great companion.

It was after dark when we arrived in Shanghai, but en route to the Sheraton Hotel it quickly became apparent that the outskirts of the city had blossomed into high-rise apartments, new offices, and hotels. I was struck by the degree of modernization. The Sheraton had a gorgeous lobby and great rooms by any standard, but it didn't feel very much like the China I had known fifty years earlier. In hindsight, I should have expected that. Still, I was eager to get started rediscovering old Shanghai and rekindling memories.

The next morning, Lifeng arrived to show me around town and make the final arrangements for my lectures at the University of Shanghai. Since my hotel was a long way from the university, he suggested that I move closer to the city center. By then I had realized that the Sheraton—a beautiful hotel with superb services—was a long way from "my China" and five miles from the old town center. On an impulse, I suggested the Park Hotel, hoping to return to a familiar piece of real estate. We agreed, and arrangements were made for my move. As we drove toward the Park Hotel, I was amazed at the number of modern-looking new buildings. I knew that fifty years had slipped by, but I was still caught off guard. It was evident that the size, character, and number of people in Shanghai had grown exponentially. The streets were swarming with bicycles, trucks, and pedestrians, all pressing to get to their destinations.

The bicycle rickshaws of the 1940s had given way to taxis whose drivers appeared to be training for the Indy 500. Although the traffic had been largely unstructured in the '40s, now it was an experience in survival. Those in cars, taxis, bicycles, and trucks seemed to pay attention to the traffic signals, but the pedestrians seemed to act independently of them. I concluded that the sounds of horns were only as reference signals that meant "take no sudden unexpected moves." Needless to say, the vehicle movements through town were painstakingly slow. Pedestrians and vehicles struggled and interacted in amazing ways, getting to their destinations in a kind of ordered chaos.

After a hair-raising ride, we arrived at the Park Hotel. The front of the hotel looked about the same, but they had redecorated the rooms. Most of the guests were Chinese, and Western facilities were minimal but, from my standpoint, very comfortable. In the early afternoon, Lifeng, the driver, and I set off to see the center of the old Shanghai area. A number of new buildings had cropped up, but there were enough of the old to provide elements of recognition. We headed down Bubbling Well Avenue, now Nanjing Road, toward the Bund. What was really different were the crowds of well-dressed people, frequently smiling and laughing, walking along the boulevards. The shops and department stores were filled with goods of all sorts and hordes of mostly female buyers. The drab gray of Mao seemed to have given way to the business suit, sportswear, numerous short skirts, and all sorts of western garb. All of the store mannequins were Anglo-Saxon—such discrimination!

It was apparent that the Chinese, like Americans, have a great love for children. It seemed like every fourth store was filled with a myriad of colorful children's clothing and toys. Youngsters on the street were elegantly dressed. The one-child national policy, which may seem harsh from our perspective, encouraged parents to place a great deal of value on their offspring. Values and the quality of life, at least in parts of east China, had changed. During the war years, children, particularly girls, were often sold, offered as payment for debts, or even abandoned as their families escaped the onslaught of the Japanese armies. Still, the potential long-term consequences of the one-child policy haunted and remained unclear in my mind.

When we reached the Bund, I could not see the Huangpu River. A twelve-foot dike with a walking path along its top obscured the view.

From the end of Nanjing Road, we could see a modern bridge connecting the two sides of the river. Later, I was driven through a tunnel built under the river to expedite traffic between the east and west banks. The situation was radically different from a half century ago, when we crossed the river in a small motorized boat or sampan with oars. On the other hand, the buildings along the Bund, including the Palace Hotel, had not changed a great deal. I spotted the building that the U.S. Navy used as a communication center at the end of the war. Somehow the tradition had endured as the building now housed a major radio station. Part of the older section of town had been converted into a tourist shopping center filled with vases, carvings, jewelry, silks, and embroidery. The prices were not much of a bargain, but I was treated to a great lunch and had no complaints.

On my second day in Shanghai, I presented two lectures at the Shanghai Fisheries University. The school had four specialized departments—aquaculture, fisheries engineering, food science and technology, and fisheries management—and had opened before the communist revolution. There were about 1,800 students and about 1,000 faculty and staff members. I lectured on the status of world fisheries and problems resulting from selective-fishing processes. It was clear from their questions and comments that they were well versed on resource management and conservation principles.

During my visit at the university, I gained an appreciation of the tremendous advances made by the country in the fields of marine fisheries and aquaculture. By the early 1990s, China was the world leader in the culture of freshwater fishes and had greatly expanded its marine aquaculture and its fisheries. Its aquaculture activities reportedly supported more than seven million workers. These gains had not come without associated problems, including pollution around fish and shellfish farms and overfishing of stocks, but these were no worse than in many other areas of the world.

The next day I was shown the various fishing facilities along the Shanghai waterfront and then given the afternoon off. I walked once more down the Nanjing Road and visited the old Palace Hotel. At age sixty-nine, the walk seemed much longer than it did when I was in my early twenties. The hotel still had much of its old world charm and elegance, although some modernization had obviously occurred. It did not seem to be a favorite of American tourists, who apparently preferred the new westernized hotels scattered on the outskirts of the city. Shanghai presented a very pleasant interlude, but my real expectations lay to the south, in Xiamen, Chiangchow, and the region where I had spent the war years. It was there that I expected to find a China more consistent with my memories.

On Saturday, May 29, I was taken to the Shanghai airport, the same field on which I had landed in 1946. Of course, the surroundings and the types of planes landing and taking off were much different. I was happy to see that the in-country airline, Xiamen Air, used a Boeing 737. When Lifeng and I boarded, however, it quickly became apparent that the seating was not designed for large people. I pulled in my feet and legs, and was pleased that it was a relatively short ride to Xiamen. The flight cost, about $80, was reasonable, considering that it was several hundred miles to our destination.

Shortly after takeoff, I started trying to spot familiar landscapes. It wasn't long before I saw the coastal range of mountains that ran from south of the Yangtze plain along the entire central and southern China coasts. From the air, it was relatively easy to identify the large coastal cities that were often adjacent to major river systems. Something new had been added to the scenery: a number of large reservoirs in the mountains just west of the coastal flat lands and a network of roads that ran west from the cities into the mountain regions. The rivers that had served as China's transportation routes for centuries now were supplemented by roads and a rail system.

As we began to descend to the airport in Xiamen, I saw the extensive delta region at the mouth of the River of Nine Dragons and later the islands nestled in the estuary. A series of roads connecting villages and cities stretched across the coastal plains running to the foothills of the mountains. Orchards and small farms filled the areas between villages. Nothing kindled any recognition of the trails and villages that I remembered from the past. The modern city of Xiamen had largely obliterated the city the Japanese called Amoy.

Alverson and Host at Shanghai Fisheries University

Xiamen is located in a subtropical region on a rocky mountainous island. The landing strip was located on the same field that I had flown out of fifty years earlier, but there were some new additions. As we headed downtown along a winding road, I saw tropical plantings, new apartment buildings, and, much to my surprise as we entered the city center, a modern Holiday Inn pushing more than twenty-five stories into the city skyline. Nothing looked familiar until we got to the old center of town near the waterfront. There, a number of the old buildings remained, and the shops were alive with people and products for sale.

The picture, however, was disturbed by a modern McDonald's smack in the middle of the old city. The famous American icon apparently had the same attraction as it does at home. Young Chinese children and their mothers were standing in line to buy fries, burgers, and soft drinks. I asked Lifeng to stop, and we bought a cup of coffee. It tasted exactly like the coffee at the golden arches in Burien, a town several miles from our home. The confluence of two cultures had begun.

The next day we took the ferry over to the island of Gulangyu. At the end of the war, a rag-tag group of Americans from Camp Six and Changchow (now spelled Zhangzhou) entered Xiamen and Gulangyu with their Chinese counterparts to take over from the Japanese occupational forces. The U.S. Navy used the island's Sea View Hotel as its headquarters and living area. As I recall, the island was then referred to as Goo Long Soo—a name that many of the oldsters still remembered. I aspired to locate the hotel, but again I was disappointed because it had been torn down or become unrecognizable. Since World War II, the island had been developed into a modern park with lush green tropical foliage interspersed with homes along the old roads and trails. No cars or vehicles were allowed on the island. The shops were filled with colorful children's clothes, and souvenirs. Lifeng and I walked to the viewing area at the top of the island and looked back across the water to Xiamen. It had grown into a beautiful city served by air, rail, and roads that ran north along the coast to Shanghai and south toward Hong Kong.

I spent most of my third day in Xiamen reviewing the academic program at the College of Fisheries on Jimei Island (connected and adjacent to Xiamen). The school was slightly smaller than the Shanghai campus and had a student enrollment of about 1,200. Lifeng and I ate at the student restaurant, but the lunch was not typical student food. Students brought their own bowls to the cafeteria where they were served rice with some greens, fish, or meat. The next day I gave two lectures that appeared to go over quite well although the translation might have confused some students. When the faculty in attendance heard that I had been in Fujian Province fifty years ago, that led to prolonged discussions between the old timers and me. One of the oceanographers about my age told me how he had come down from Changting to Changchow during the war. He had followed the same route that I had. We continued to exchange war stories and discuss the developments that had occurred in the region. He remarked, "You knew China and its poverty of fifty years ago and you can see where we are today." Perhaps this statement had a slight political overtone, but the reality of the changes in both Shanghai and Xiamen could not be ignored. Before departing, several professors urged me to return to China.

On Monday, arrangements were made for me to go to Zhangzhou to look for the World War II navy camp. During the war, it was a city of around 20,000 people and mostly two-story, brick-like buildings nestled along the river. We didn't walk or take a river ferry because a two-lane highway served the city. Driving along the river, we passed through banana plantations, field crops, and traditional rice fields. The main road from Xiamen to Zhangzhou crosses a long causeway linking the island to the mainland north of the river. About fifteen miles upriver from Xiamen, a bridge had been constructed over the River of Nine Dragons just above the confluence of its northern and southern branches. I had the driver stop at the side of the bridge, walked out, and looked for a long time up the river branch that led to where our camp had been situated along its banks.

We continued up the valley following the south fork of the river. New construction cluttered the road all the way up the river. We passed a number of brick-manufacturing plants that were busy supplying the demand for new building construction. When we reached Zhangzhou, I was again caught off guard. The village was now a city of about 350,000. Modern clover leaves enclosed flowered areas at many intersections. As in Xiamen, there were a number of high-rise buildings including a Western-style hotel. I searched in vain for the old school where we had been quartered during the war. It, too, had been swallowed up by the modernization of the eastern China seaboard.

I finally asked if the old Japanese bridge that crossed the river to the south was still in existence. It had been repaired and was now a walking bridge only. We entered the old shopping area that was familiar to me, and when we reached the bridge, I got out of the car. Lifeng and I walked out on to the bridge and stopped at its center. From there, I scanned up and down the river while a thousand memories flashed through my mind. I was about to return to the car when an old man carrying goods at the end of his "yo-yo" stick walked by with a typical swing to his steps. When he got close, I smiled at him and he did likewise, showing a few beetle nut-stained teeth. I turned to Lifeng and said, "Now, that's the China I remember from fifty years ago." He laughed.

Lifeng was a great guide. He had gone to the college in Xiamen and was very familiar with the region. After sunset, he took me down to the old section of town where a thriving night market was located on one of the streets near the river. The street was crowded with men, women, and children buying both local and imported goods. Members of the local fishing industry took us to a second-story restaurant overlooking the market area. The dinner consisted of a variety of local fish and shellfish products along with fruit and salads. During dinner, we relived the history of the region and its recent developments. It was there that I heard that Huaan had also changed over time, with two main roads and rail traffic now serving the village. During the evening, I told the local officials and fishing industry members about our raid on the island south of Xiamen. They immediately arranged for a government fisheries patrol vessel to take me to the World War II battle site, Woo Soo Island.

Xiamen College of Fisheries Faculty

On Tuesday we drove to old Shima and a waterfront dock that looked like the same one that we had used during the war. As we walked to the dock, someone bumped me rather sharply, but I paid little attention, as the street was jammed with people moving in both directions. When we reached the river, we boarded a sixty-five foot vessel that was headed downriver. The trip revived a lot of memories because along the riverbanks I could sense a lot of the China I had known, and I thought about the young woman who had helped me during the war. We passed sampans with oars and a number of older buildings, but the river also was busy with water taxis, sampans with outboard motors, tugs, barges, dredges, fishing boats, and cargo vessels.

At Woo Su, there were a large number of fish trawlers anchored off the island and others pulled onshore for repair. The island, which housed a Japanese radio station during the war, had been the target of an ill-fated Chinese-American raid. Now it was a major fishing village. As we approached, I felt shivers run down my back and thought that I was perhaps the only member of Camp Six that ever landed safely on the island.

The island constitutes somewhat of an enigma in that a major part of the catch taken by the vessels working from the region never lands in the PRC. Instead, the catch is transferred at sea to vessels from Taiwan and sold in the ROC. This inter-province or inter-country trade—take your pick—had made a number of the fishermen from Woo Su prosperous. I was taken to the home of one of the prominent fishermen whose two-story home and patio was a bit of elegance in the island's village. I asked Lifeng why this high seas trade between the two Chinas was allowed. Although the answer was somewhat circular, it seemed to suggest that both sides were aware of what was going on and both sides chose to ignore it. There were obviously benefits to both Chinas.

Before Lifeng and I left Woo Su, the locals treated us to a wonderful, informal dinner attended by a number of elderly people. They remembered very well the night of the battle and their disappointment when their fellow Chinese and the Americans were pushed back. We chatted endlessly and enjoyed revisiting history. I also introduced into the conversation more contemporary issues, such as the global problem of overfishing, China's rapid expansion into marine fisheries, and the tremendous efforts being made in the development of marine aquaculture. They acknowledged the problems of excess fishing capacity, overfishing, and pollution inherent in industrial aquaculture, noting that it was a major concern of their government. Despite my backgrounds and affiliation with the ROC, the people of Woo Su treated me with friendship and great hospitality. They seemed pleased that I had been a member of an American group that had fought against the Japanese during the war.

When we arrived back at the hotel later that evening, I found that my passport was missing. I searched everywhere, but it was just gone. Then I remembered the heavy "bump" on the street in Shima. It seemed quite likely that my passport had been lifted at that time, but I could not be sure. I discussed the matter with Lifeng, and he asked the people who ran the fisheries patrol vessel and the authorities in Woo Su to see if the passport had been left on the vessel. Unfortunately, it could not be located. Lifeng got in touch with the president of the university, other key faculty members, the mayor, and the leaders of the Xiamen fishing industry. They went as a group to the local city police to arrange for a permit that would get me back to Shanghai, where the U.S. consulate and Chinese officials could arrange for a new passport.

I had heard horrible stories of long delays for tourists who lost their passports. Nevertheless, I was hoping that a photocopy of my passport made before I left Seattle would help resolve the matter. While I was waiting and depending on Lifeng and others, members of the local fishing industry invited me to a dinner. The food and hospitality was exceptional, but my concern over the lost passport prevented me from enjoying what should have been a great evening.

The following day Lifeng and university officials arrived at my hotel room with the papers that would allow me to proceed to Shanghai. I never had to leave my hotel room to talk to the police or any government official. My hosts simply gathered en masse, gone to the police as a group, and returned with the necessary papers. And it only took two days. This would never have happened at home and most likely not in Xiamen if my determined colleagues had not gone to bat for me. Back home, Ruby got upset when she could no longer contact me and thought I was being held by the Chinese government. Unfortunately, I was not at the hotel, but out trying to solve my problems. In her fear she called Senator Gordon and asked that he intervene, but it wasn't necessary because the Chinese had solved the problem. The next day I left for Shanghai with Lifeng. When we arrived, he took me to the local Chinese authorities, who presented me with the documents required for a passport. Lifeng had made my entire trip a great success, and I shall forever be in his debt.

During the trip, my memories of the past were in constant conflict with my contemporary observations. The China I knew as a young warrior was rapidly fading into history, and its character lived mostly in the memories of the older members of its society. Thus I had a difficult time recapturing many aspects of the past, but there were moments. The people are the same; they continue to have a great sense of humor, to want to communicate, and to work hard. This was as true for the Chinese I met in the ROC as it was for the people in mainland China. Their language, with its many dialects, continues to entail endless discussions for clarification. The ideological differences between the two governments remain large, but perhaps the people, particularly in Taiwan and the adjacent province of Fujian, are closer than their leaders. I never raised the issue of Tiananmen Square and human-rights violations, but I hope that time and the people of China will resolve these issues.

My observations on China were limited to the special economic zones that exist along the eastern coast. I was told by many that the developments in east China had not occurred in many other regions. My trip to mainland China renewed and strengthened my memories of this great land and its people, and I shall take them to my grave. After my return, I drafted a report of my findings and experiences. The report, including numerous photos, was sent to the universities where I spoke, to Lifeng, and to friends in Taiwan. I attempted to keep politics out of the story, which seemed to go over well in both Chinas.

The Inauguration of President Lee

In May 1998, Bill Miller asked if I would like to attend the inauguration ceremonies for President Lee Teng-hui and Vice President Lien Chan in Taipei. The invitation came as a surprise as there were many other SACO members who were more deserving. Others had been asked, Bill said, but were not able to travel. He also noted that the Military Intelligence Bureau in Taiwan was pleased that I might attend. I agreed to go, and Bill and I flew together to Taipei. We went to a number of elegant affairs honoring Lee's inauguration, including a spectacular gymnastic and dancing exhibition at a sports pavilion on the outskirts of town. However, the most memorable event involved going to a theater that had a 360-degree screen where we watched a movie about a trip down the Yangtze River. From our seats we could look back, to either side or forward, seeing both sides of the river, boats in back of us and in front on us, etc. It was so realistic that I felt a tinge of motion sickness.

When we went to hear the inaugural speech, we were given seats closer to the president than the official U.S. delegation, reflecting the close bond that had developed between U.S. SACO members and the ROC. I read in detail Lee's translated speech. It appeared to be crafted in a manner not to rile Beijing, but I expected, correctly, that the official PRC line would be less enthusiastic. Lee was the first elected president born in Taiwan and not a member of Chiang's Kuomintang. After the inaugural speech, Bill and I met with the new president and vice president, who had both been partly educated in the United States.

Walking into the 2000s

My China connections remain intact. From time to time, I communicate with Shen-ya in the ROC and Lifeng in the PRC. Both seem to be doing well in their careers. Shen-ya is a full professor and department head in Taipei, and Lifeng has moved up the career ladder in Beijing. Sometime in the mid-1990s I began to write a book outlining my experiences in China and Taiwan. It took me the better part of four years to pull it together and find some interesting pictures to go along with the story. When I finished, I thought of taking it out to the U.W. Press where I had co-edited a technical book with one of my colleagues, but then decided it didn't have enough of an academic tone. For this reason, I let it sit on the shelf for several years. In 2004, I sent it to a publisher in Taiwan, who liked the historical content and published it in a six-part series in a magazine called *Biographical Literature*. I still hope that it will also be published and distributed in mainland China.

I am now classified as "elderly," but I retain many stored memories of the two Chinas and follow with great interest the developments in China as my life continues into this new century. I watch the papers carefully and read anything that is printed on the topic. Most years, I attend the annual SACO reunion and chat with some of the guests from Taiwan. Occasionally, I hear from friends in both Taiwan and mainland China. My parents and my brother are deceased, and only one of my childhood friends still survives. Buddy died early in this century and all of my war-time friends have passed on, leaving me quite alone with my thoughts. I face the future with my wonderful wife without trepidation. My memory chips keep me aware of my past, and my dreams often include visits to the River of the Nine Dragons. The China episodes in my life are deeply entrenched in my soul, and from time to time I have told my friends that I have a smidgen of Chinese blood in my veins.

Chapter XVII
Reflections

In the past year, I turned eighty and am now well on my way to my eighty-first birthday. At this point in life, one starts to recognize that he is not immortal, at least not in the earthly domain. My future time on this planet will be a small fragment of what has already elapsed. Although frequently we are told that you cannot live in the past, it is such an enormous part of my history that I often find myself slipping back in time, sifting through my life experiences. My early recollections take me back to about 1928 when the ice for our cold boxes came down the street on a wagon drawn by two horses. Commercial air flight was just getting underway, and there were no T.V., videos, VCRs, computers, or satellite radios. We listened to Jack Benny, Charlie McCarthy, Fibber McGee and Mollie, and Amos and Andy on a small dome-shaped radio.

My strongest memories are rooted in the Hawaiian Islands, particularly Hilo and the Big Island, but I also frequently think of Idyllwild, surfing at Mission Beach, our six-man football team, and the 35th Street gang. The war years, as you may have guessed, have a place all their own in my recall. And, of course, memories of my marriage to Ruby at her home in Maywood, California and her aunt playing loudly on the piano and singing "I Love You Truly" will never fade. Although I enjoyed college, I have few specific memories of that period beyond doing homework in our small apartment in Renton, Washington. My first job as a marine biologist involved sampling and tagging fish aboard small otter trawlers working out of Astoria and other Oregon coastal ports. Working aboard those boats and the *John N. Cobb* are some of my greatest professional memories.

My career began at the Montlake Laboratory in Seattle working with Don Powell, a close college friend. Because of issues I had with my boss several years later, I left the federal job and began working with the Washington Department of Fisheries, sampling catches from the trawlers that fished off the state of Washington and north along the Vancouver coast. I have great memories of the Seattle waterfront when at least a half-dozen fish houses purchased salmon, crabs, and bottom fish from small coastal fishing vessels. These are all now defunct, and it is rather sad that this aspect of the waterfront no longer exists. The coastal fishing vessels that remain have moved to Everett, Bellingham, and other ports. Like San Francisco, Seattle has pushed part of its heritage and culture out of sight. It was always a pleasure to go down to the ship canal and watch the host of seiners, trollers, and gill-netters make their way through the locks to the northern Puget Sound and offshore areas to begin their fishing seasons. Old men, young men, and often teenagers were busy on deck, tending nets or doing other needed work.

Trolling for salmon off Washington is now almost exclusively a Native American operation and much of the inside purse-seine and gillnet fishery has passed away as a result of changing international arrangements with Canada. A decision of the courts to allocate fifty percent of the salmon originating in Washington streams and rivers to local Indian tribes contributed to the shrinking of the traditional Pacific Northwest salmon fleets. Of course, in addition to the national and international changes in the access to the salmon resources, there were also changes in the salmon's freshwater environment and the consequences of excessive fishing. The construction of dams, forest practices, increased population density along a myriad of streams and rivers, and the use of water for agriculture and a variety of industries has all led to a deterioration of salmon's freshwater habitat. Added to these problems was the inability of the fishery-management institutions to reduce harvest in a timely manner.

On balance, it must be said that many benefited from the construction of the dams, which gave us cheaper energy and allowed for the growth of a significant agriculture production within the Columbia Basin as well as a variety of water-dependent industries. The fish and the fishers (Indian and non-Indian), on the other hand, were the major losers. We now have a vestige of the runs entering most Oregon and Washington rivers, and while these may be maintained and rebuilt, it will be only at costs that far exceed their economic value. Regardless of the cost, the people of the Pacific Northwest will continue to battle over salmon for cultural and social reasons and also because the inordinate funds spent on them generate a lot of jobs.

It should be clear that I'm not looking for a scapegoat. Besides, the number of salmon runs placed on the endangered species list have clouded the public's perception of the overall status of salmon stocks. The press covered these listings carefully without acknowledging the productive status of salmon runs throughout the north Pacific Ocean. Our public has been led to believe that Pacific salmon stocks in general are in great peril but, in fact, the runs of salmon into most of the rivers and streams of North America and Asia were at an all-time high from the late 1970s to the end of the twentieth century. It is at the southern end of their distribution, off California, Oregon, and Washington, that the salmon runs have suffered, but elsewhere, salmon continues to thrive. Washington State's salmon runs may be better managed and the harvest more equitably divided, but the traditional fishing fleets are rapidly disappearing and their culture is fading into the past. Many of the fishing boats are now relics that are moored down at the port of Seattle's Salmon Bay docks.

The decline of the salmon runs in the Pacific Northwest, however, pales in comparison to the events that have had an impact on marine fisheries on a global scale. A great expansion of world fisheries and national and international policies concerned with their development, conservation, and eventually overfishing, has taken place in the twentieth century, mostly since 1946. Rapid development and intensification of offshore fisheries (other than whaling) had its origins prior to World War II, with the advent of steam engines and advancements in refrigeration. At the beginning of the century, fish production was around five million metric tons, but it grew rapidly. During the war, larger vessels from Japan, western Europe and, to some extent, the United States participated in the war effort, transporting goods and monitoring the coasts. Once the hostilities were terminated, ocean fisheries began to grow at a fast pace and the catches of marine fishers soared.

The war, which ravaged and decimated many areas in the Northern and to some extent the Southern Hemisphere, also destroyed much of the high-seas fishing fleets. At the same time, it fostered and left a heritage of technologies that spurred the development of a complex fleet of modern high-seas vessels, including gillnetters, trawlers, long liners, and purse seiners. The post-war, high-seas fishing fleets speedily converted from steam to diesel propulsion, adopted a host of electronic navigation and fish detection devices, and began using strong and durable synthetic nets. The fishers of northern Europe and Asia quickly returned to their historical fishing grounds. As new competitors from the Eastern bloc nations arrived on these grounds and the fish stocks of the region began to dwindle, the solution was simple: expand to alternative, more productive, distant-water fishing areas.

During the late 1950s and early 1960s, marine scientists began to speculate on the potential of the world's oceans to produce animal protein to feed an expanding population. Estimates ranging from 50 million to 2,000 million metric tons emerged in the marine science literature. Looking back, it is indisputable that many fishery managers and politicians ignored, or were unaware of, the caveats that placed limitation on achieving the higher yields. Instead, they seized on, and extolled, the more optimistic estimates. By the late 1960s, the famous Stratton Commission report prepared for the U.S. Congress, titled "Our Nation and the Seas," echoed this enthusiasm proclaiming, "It would be realistic to expect production of marine products to grow between 400 and 500 million metric tons annually." Food from the sea quickly became a selling point for an expanded marine-science program in the United States and many other areas of the world, including several members of the United Nations. It was an era of great enthusiasm in the fields of marine research. Resource exploration became a significant component of many national fishery programs. Funds for exploration and documentation of the untapped living marine resources flowed from national governments, the World Bank, the UN, and private sources.

Throughout the 1960s, 1970s, and even the early 1980s, global explorations flourished, adding momentum to the international race to the sea. Government promotion of and industry commitments to expansion were significant, and the increasing fishing capacity and mounting catches reflected their success. In the two decades following World War II, global marine fish catches increased about 300 percent. It was a period during which the major world fisheries moved from national to international markets, and large fishery corporations became internationalized. However, by the late 1960s and in particular the 1970s, distant-water fishing operations began to engender their own set of problems. From the beginning, the distant-water fishing operations had generated a number of conservation and allocation disputes. As the fleets of foreign fishing vessels grew, the number of overfished stocks increased, while confidence in many of the international arrangements for conservation of fish resources dwindled. Eventually, the concept of freedom of the seas began to erode, and by the 1970s a number of nations were unilaterally extending their jurisdiction over ocean space to 200 nautical miles.

Unfortunately, the problems of overfishing that had accompanied the "colonization" of distant-water fishing grounds continued even under the auspices of national fishery management. The international revolt (supported by most of the nations within the UN) that led to extended national jurisdiction carried with it an array of conservation commitments. Nevertheless, the new managers seemed to adopt all of the institutional and policy failures that had plagued fishery management for decades. Referring to the overfishing problem, Dr. Edward Miles at the University of Washington noted, "We are forced to conclude that there is no necessary connection between extended coastal state control of and improved fisheries management."

In retrospect, I regretfully came to the conclusion that the underlying social and economic pressures that galvanized the growth of excessive distant-water fishing capacity had continued to fuel fisheries development and management long after extended jurisdiction. This was not just a problem for developing nations, but also for many developed countries, including the United States. It was not until the late 1980s and early 1990s that a small number of scientists began to sound the alarm that overfishing was becoming a major problem. I added my own opinion to this matter by writing a paper with my good friend and colleague, Dr. Larkin of the University of British Columbia, pointing out that close to a third of the ocean's major fish stocks were overfished and that the problem could become worse if the global conservation ethic did not change. During the 1990s, a number of scientists pointed out a number of ecological problems resulting from fishing.

Between the late 1980s and the end of the century, fisheries and the environmental sciences became highly politicized. Influential and well-financed environmental and private foundations began to push for much needed conservation and environmental legislation. Unfortunately, there was no middle ground: you were either on one side or the other. The public soon became engulfed in a news media blitz that frequently overstated the character of the issues and was inconsistent with available science. In many cases, the scientists themselves joined the battle by using the press to promote their views and perhaps seek notoriety.

During my career, I worked as a marine biologist, fishery administrator, and finally, for more than twenty years, as a fishery consultant. Over my fifty-five years as a marine biologist, I published more than 150 articles and three books dealing with many aspects of fishery science and management. In the early stages of my involvement in the fisheries field, the oceans were portrayed as the earth's last frontier, an area less known than the surface of the moon, a reservoir that would provide the needed protein for the growing world population, and a new source of energy to augment dwindling fossil fuel supplies. It all seemed romantic and encouraging, but this optimism has long since faded. More recently, the oceans have been characterized as a global cesspool for a variety of human waste, a dying biological system whose living resources have been contaminated, overexploited, and depleted. More and more species are said to be threatened with extinction. There is a modicum of overkill in some of these assertions; nevertheless, there is enough truth in them that it is not a very encouraging story to pass on to my grandchildren.

When I attempt to sort out the good, bad, and ugly aspects of fisheries science and management over the past sixty years, I suspect that those in my age bracket can find most of the answers just by looking in a mirror. Over this time period, I have been enthusiastic about the potential of the world's oceans to feed the world's hungry, supported the development of more effective harvest systems, helped to promote funds for ocean explorations, lent my support to extended national jurisdiction, and encouraged the Americanization of fisheries within the U.S. EEZ. Later in my career, I joined with others in sounding the alarm that fisheries were being mismanaged and overfished, and pointed out that government subsidies in the United States and around the world were supporting excessive fishing. In the early 1990s, I coauthored a book that focused attention on the large quantity of discards that were occurring in many fisheries. Along with other scientists, I noted the deaths that were not accounted for when the impact of fishing on target and non-target species was calculated. My views and understanding of the ocean's ecosystems and their capacity to produce food and goods for humanity were continuously reshaped as new information challenged historical perspectives.

Over the past decade and a half, there has been a growing propensity for the news media, fishery managers, and scientists, both in government and academia, to point their fingers at the world's fishing industry and paint it negatively. Fishers are seen as greedy, irresponsible, and aggressive exploiters, morally bankrupt culprits responsible for the widespread overfishing and destruction of the bottom (benthic) ecosystems of the continental shelves and slopes throughout the world. My perspective on these condemnations is undoubtedly biased. When I was five years old and living on Tatoosh Island at the entrance to the Strait of Juan de Fuca, I would go out on a salmon troller and picnic on a small beach with groups of fishermen. When I was nine, I spearfished along the breakwater in Hilo, Hawaii, and as a teenager I dove for abalone off the southern California coast. My first post-war job was in a small tuna cannery in San Diego. As a marine biologist, I conducted research or visited small gillnet boats, sampans, trollers, small trawlers, factory-trawlers, crab boats, and long liners. During these adventures I drank a lot of very black coffee, swapped stories, and argued with the crews. As a result of these experiences, I have a very different view of the fishers.

Some segments of the fishing industry have been irresponsible, and there is little doubt that there are greedy elements. But greed is pervasive in many sectors of our society, and I suspect that this includes the scientific community. Do you really believe all of the conflicting stories floating around concerning global warming caused by the burning of fossil fuels, ozone depletion, super volcanoes about to erupt, giant tsunamis rushing across the Atlantic, rapid depletion of the world's oil supply, asteroids hitting the earth, and the loss of ninety percent of the world's large fish predators? There is no conclusive scientific evidence for most of these issues, and there are arguments supporting and questioning all of them.

But there is also the possibility that some academics and universities have found that private granting organizations, the National Science Foundation, and Congress are more likely to provide large sums to those who deal in the catastrophic aspects of the future of global resources and the environment. They would undoubtedly not agree that greed is involved; they are just competing for billions of dollars and, of course, publicity and reputation.

The image of the world's fisheries and fishers has been tarnished with a broad brush, but is it a fair and accurate assessment? The responsibility for management of the ocean's living resource is vested in national and international government entities, with a few exceptions. These government management agencies establish and enforce the rules of the game and, hence, bear the responsibility for the historical course those fisheries have taken. Without question, the fishing industries during the 1950s through the 1980s placed heavy political pressure on government managers. But if in the past the industry overly influenced those responsible for management and enforcement, including the politicians in Washington and other national capitals, then the fault, at least in part, lies with the management entities and their political masters. It seems incredible to me that some of the same congressmen who, two decades ago, promoted fishery development and legislated subsidies for the U.S. fishing industry, are now avid supporters of conservation advocacy groups. Some of these groups would like to eliminate almost all of the large-scale fisheries in the world. These congressmen have now climbed onto high moral ground and are contemptuous of those they urged to go to sea and compete.

For those of us who remain concerned about the future of the ocean's living resources, there is some hope. During the 1990s and into this century, ocean management entities began to make significant strides in controlling levels of fishing, waste, and habitat degradation. In addition, we have a powerful deity on our side. Several years ago, a young man from a religious group knocked on my door in Normandy Park and gave me a pamphlet called *Awake* with an article titled "The Oceans Reveal their Deepest Secrets." I took it to work and read it while relaxing with a cup of coffee. The material on the wonders and mysteries of the deep was well written, educational, and consistent with facts related to deep-ocean vents. In it were some interesting prophesies regarding those who continue to contaminate and overfish the ocean's fishery resources. Perhaps we can take refuge in its concluding observations:

"The more we learn about the earth and its living marvels, the more we ought to develop a respect for this truly dynamic planet. Of course, people demonstrate their love for earth, perhaps by getting deeply involved in environmental issues. Sadly, however, these people face insurmountable obstacles including human greed and ignorance, which the noblest intentions cannot eradicate. Nevertheless, sincerely concerned persons can take comfort in the Bible's promise that God himself will soon act by eliminating all greedy, godless individuals who are ruining the earth (Revelations 13:18:2 and Timothy 3). Therefore, the creator whose name is Jehovah will institute a programme of rehabilitation that will thoroughly cleanse the earth and turn it into a paradisiacal fountain of life."

An interesting prophecy, and I do not expect that God will submit an environmental impact statement—or perhaps he already has. What about my spiritual life and beliefs? Well, it may seem rather an enigma. My parents seldom attended church and never encouraged my brother and me to attend Sunday school, but they were believers. With eight years of indoctrination into the sciences and fifty-five years working within the scientific community, which has more than its share of agnostics, you might suspect that I would be a skeptic. But I have always been a confirmed believer in Jesus Christ and the Father. I recall as a boy walking by a church on Sunday, listening to the choir sing, and wishing I could be a participant. This feeling of wanting to be among the believers was so strong that when we lived in Hilo, I would occasionally walk several miles to a Sunday school to join the other children listening to the gospel. One day while playing football at my grammar school, the ball went out into the street. I ran full speed between two parked cars and was just about to enter the road in pursuit of the ball when I suddenly heard this scream inside my head: "Stop!" I put on the brakes like I was about to go over a cliff, and a car sped by right in front of me. No, it wasn't one of the other kids or someone else watching me. The instructions came like a horn blast from within me. Ever since, I have been a devout believer, although my behavior has not always followed the Christian ethic.

My passage through college and graduate school, in some instances, budged me off the trail, but before long I was back on track. My readings of Darwin's books, *The Origin of the Species* and *The Voyage of the Beagle*, and scientific articles on the mathematics of mutations and chemical evolution were interesting and stimulating. I also read with vigor all the papers that crossed my desk concerning the origins of the universe, the big bang, and string theories, but I remain at home with God. No doubt many of my colleagues would classify these beliefs as the thoughts of a feeble old man, but to each his or her own.

By this time my wife and I had entered the 1990s, our grandchildren had become adults and were busy with their own lives. As a result we saw less of them and at times we found ourselves a bit lonely, but this problem was soon resolved when a Korean family named Song, moved into a home just across the street from our house. They had two beautiful young girls, the older one, Jane, was five years old and the younger, Gina, was three. Jane could speak English, but her sister could not. The father only had a meager grasp of English, while the mother could communicate quite effectively.

During those years, I made a habit of walking around the park in front of our home. I would usually make fifteen trips around the park to get my daily exercise. For some reason, the girls took pleasure in following me around the park and when they got tired they would cut across the park and meet me on the other side.

We began to have short talks, and I learned that they had only lived in the United States for about a year. Most of their relatives, including their grandparents, lived in Korea. I asked them if they would like Ruby and me to be their adopted American grandparents and they were delighted.

We became very close to the two girls and made friends with their parents. I began to teach English to the father, Sang, and help him study for his citizenship test. Jane and Gina came to our house for help with their homework. Over time we began to celebrate their birthdays, and the family would join us for holiday dinners, The Songs always brought a turkey for our Thanksgiving dinner. The two girls showed up on our porch bringing us flowers on our birthdays as well as Mother's Day and Father's Day. Their arrival added a missing element to our lives, and we remained close to them as they grew into their teens and graduated from high school. One is now a student at the U.W., and the other is about to enter Central Washington College.

It was deja vu; they were moving on to fulfill their own dreams, and we saw them less often. But the story has a great ending. In 2005, the Songs brought two of their relative's young children to Seattle to be educated and learn English. Kelly, the older of the two, is a young teenager, and her cousin Daniel is a few years younger. Both now come to our home to study English, and we have the enjoyment of two new grandchildren. Kelly has turned out to be a real charmer. She joined us like we had always been a part of her family and surprised us by discussing all her school and other problems in an open way. As for Daniel, he is a boy struggling to find new friends in his newly adopted country. Thus a Korean emigrant family and their children added spice to the latter years of our lives.

It is now spring in Seattle, the rhododendrons, tulips, and daffodils are out as are the wild dogwood blossoms. It is fortunate that the stocks of salmon, ground fish, and crab off Alaska have not been overfished and remain an admirable example of what good science and precautionary management can accomplish. As a result, there is still a robust fishing fleet based in Seattle that fishes the rich waters in the Gulf of Alaska and the Bering Sea. It is time for Ruby and me to drive down to the Ballard Locks and watch the fishing boats move out into Puget Sound. Then we will drive toward Bothell and see our four-week old great-grandchild, Kollin. And as I completed the writing for this book, my granddaughter, Aimie, gave us a great-granddaughter named Madeline LeeAnn Undseth.

Dedicated to:

My wife Ruby

Fishermen of the world's oceans

Those scientists who investigate and provide information on the consequences of fishing the oceans' living resources and their ecosystems

Breinigsville, PA USA
27 August 2009
223090BV00001B/4/P